High Resolution
NMR Spectroscopy
of Synthetic Polymers
in Bulk

Methods in Stereochemical Analysis

Volume 7

Series Editor: **Alan P. Marchand**

Department of Chemistry

North Texas State University

Denton, Texas 76203

Advisory Board

© 1986 VCH Publishers, Inc. Deerfield Beach, Florida

Distribution: VCH Verlagsgellschaft mbH, P.O. Box 1260/1280, D-6940 Weinheim,
 Federal Republic of Germany

USA and Canada: VCH Publishers, Inc., 303 N.W. 12th Avenue, Deerfield Beach, FL 33442–1705, USA

High Resolution NMR Spectroscopy of Synthetic Polymers in Bulk

Edited by

Richard A. Komoroski

Richard A. Komoroski
University of Arkansas for Medical Sciences
Little Rock, Arkansas 72205

Library of Congress Cataloging-in-Publication Data

High resolution NMR spectroscopy of synthetic polymers
  in bulk.

  (Methods in stereochemical analysis; v. 7)
  Includes bibliographies and index.
  1. Polymers and polymerization—Analysis.
2. Nuclear magnetic resonance spectroscopy.
I. Komoroski, Richard A., 1947–   . II. Series.
QD381.H54 1986   547.7'046   86-5496
ISBN 0-89573-146-0

ISBN 0-89573-146-0   VCH Publishers
ISBN 3-527-26464-7   VCH Verlagsgesellschaft

Preface

The last 12 years have seen sustained growth in the use of high-resolution NMR techniques for studying bulk synthetic polymers. Initial efforts used standard ^{13}C NMR techniques to study polymers well above their glass transition temperatures. The development of methods for efficiently obtaining high-resolution spectra of rigid solids opened the door to glassy polymers, the crystalline regions of semicrystalline polymers, and insoluble, highly cross-linked polymers. Whereas solution-state ^{13}C NMR was becoming the technique of choice for the determination of polymer microstructure, solid-state ^{13}C NMR was developing into a powerful method for studying polymer structure, conformation, and morphology. Moreover, polymer chain and side-group motions could be analyzed in detail through spin-relaxation time measurements.

HIGH-RESOLUTION NMR SPECTROSCOPY OF SYNTHETIC POLYMERS IN BULK brings together the various approaches for high-resolution NMR studies of bulk polymers into one volume. Of necessity, heavy emphasis is given to ^{13}C NMR studies both above and below T_g. Standard high-power pulse and wide-line techniques are not covered because these have been treated adequately elsewhere.

The book is intended primarily for polymer scientists concerned with the structure and morphology of solid polymers. The emphasis is on the fundamental aspects for both totally amorphous and semicrystalline polymers. However, some analytical applications are treated at various points. The book should be valuable to NMR spectroscopists and others desiring an overview of this important field. The monograph is meant to serve as a guidepost for future research as well as a reasonably comprehensive treatment of the work to date. Techniques at the frontier of NMR research that hold promise for bulk polymer studies are mentioned.

Chapter 1, by R. A. Komoroski and L. Mandelkern, provides a brief introduction to the field. Some concepts basic to polymer structure in the solid state are reviewed. The strengths of high-resolution NMR are outlined, with frequent comparison to the study of polymers in solution. Chapter 1 details the types of information to be expected from high-resolution NMR studies of solid polymers, with emphasis on that which is unique to the method.

Chapter 2 covers the NMR interactions operative in the rigid, solid state and the methods (dipolar decoupling, magic angle spinning, and cross polarization) used to produce high-resolution ^{13}C spectra. The combined effects of molecular motion and coherent averaging are discussed briefly. Finally, a number of advanced techniques for extracting additional information from the high-resolution, solid-state spectrum are described.

The next three chapters cover the three major types of polymers, although some overlap is unavoidable. This division is convenient from the points of view of both the polymer chemist and the NMR spectroscopist since both the polymer type and the NMR technique used are determined by chain mobility. Chapter 3, by J. R. Lyerla, describes the many aspects of glassy polymers, an area of intense study by high-resolution NMR. In Chapter 4, R. A. Komoroski covers the application of high-resolution techniques for both liquids and solids to polymers well above their glass transition temperature. This includes totally elastomeric materials as well as the rubbery regions of heterogeneous systems. The important and complex topic of semicrystalline polymers is addressed by D. E. Axelson in Chapter 5. Here Axelson covers both the crystalline and the amorphous regions, with emphasis on the ability of high-resolution NMR techniques to probe each region independently.

Each of the last five chapters covers a topic of importance to NMR studies of solid polymers. In Chapter 6, which functions somewhat as a supplement to Chapter 5, D. E. Axelson describes the observation of specific conformations of polymer chains in the solid by high-resolution ^{13}C NMR. The influence and measurement of polymer motion are examined from various aspects in several places in the book. In Chapter 7, A. A. Jones coherently treats the measurement and modeling of polymer motion by concentrating on a particular polymeric system, the polycarbonates. The NMR results for the polycarbonates are ultimately related to dynamic mechanical and dielectric results. The particular strengths of NMR for studying oriented polymers are covered in detail in Chapter 8 by A. J. Brandolini and C. R. Dybowski.

The last two chapters concentrate on two NMR techniques that have much potential for probing polymers in the solid state. In Chapter 9, B. C. Gerstein describes the applications of multipulse NMR techniques for obtaining high-resolution spectra of abundant nuclei (^{1}H and ^{19}F) in solid polymers. L. W. Jelinski outlines the application of solid-state deuterium NMR to polymers in the last chapter. Although deuterium NMR is not a high-resolution technique, it is included here because it can be a powerful probe of polymer mobility and it often nicely complements results obtained by high-resolution methods.

I thank all the authors for their excellent contributions and for their efforts toward making this work an organized, coherent whole.

Richard A. Komoroski

Contributors

David E. Axelson
Department of Energy, Mines,
  and Resources Canada
Coal Research Laboratories
P.O. Bag 1280
Devon, Alberta, Canada

Anita J. Brandolini
Mobil Chemical Company
Research and Development
Edison, New Jersey 08818

Cecil Dybowski
Department of Chemistry
University of Delaware
Newark, Delaware 19716

Bernard C. Gerstein
Department of Chemistry
Iowa State University
  and Ames Laboratory
U.S. Department of Energy
Ames, Iowa 50019

Lynn W. Jelinski
AT&T Bell Laboratories
600 Mountain Ave.
Murray Hill, New Jersey 07974

Alan A. Jones
Department of Chemistry
Clark University
Worcester, Massachusetts 01610

*Richard A. Komoroski
BFGoodrich Research
  and Development Center
9921 Brecksville Rd.
Brecksville, Ohio 44141

James R. Lyerla
IBM Research Laboratory
K42/282
5600 Cottle Road
San Jose, California 95193

Leo Mandelkern
Department of Chemistry and
Institute of Molecular Biophysics
Florida State University
Tallahassee, Florida 32306

* Departments of Radiology and Pathology, University of Arkansas for Medical Sciences, 4301 West Markham, Little Rock, Arkansas 72205.

Contents

1

OVERVIEW OF HIGH-RESOLUTION NMR OF SOLID POLYMERS

Richard A. Komoroski

B F GOODRICH RESEARCH CENTER,
9921 BRECKSVILLE ROAD, BRECKSVILLE, OH 44141

and

Leo Mandelkern

DEPARTMENT OF CHEMISTRY AND INSTITUTE OF MOLECULAR BIOPHYSICS,
FLORIDA STATE UNIVERSITY, TALLAHASSEE, FL 32306

Introduction

The last 15 years have seen considerable excitement in the area of NMR spectroscopy. The development of Fourier transform techniques and high magnetic fields have changed NMR, as practiced by most chemists, from essentially a proton technique to a true multinuclear spectroscopy. Among all the additional nuclei, ^{13}C has dominated because of its central importance to organic chemistry. Nuclear magnetic resonance has proved quite powerful in polymer chemistry as well. The sensitivity of ^{13}C chemical shifts to subtle features of molecular bonding and stereochemistry makes it excellent for characterizing tacticity, copolymer sequence distribution, branching, and other structural irregularities in polymers. Pulsed Fourier-transform ^{13}C NMR also provides the opportunity to measure spin relaxation times of individual sites

© 1986 VCH Publishers, Inc.
Komoroski (ed): High-Resolution NMR Spectroscopy of Synthetic Polymers in Bulk

in an organic compound or polymer readily and to relate these times to the details of overall and segmental motions of the molecule.

More recently, three areas of NMR have seen rapid progress. One is the development of techniques for noninvasively imaging the interior of objects, in particular the human body. Another is the use of two-dimensional NMR, multipulse sequences, and double-quantum techniques to extract considerably more information, previously unavailable, from the NMR experiment. The last area of progress, which at this point is the most developed, is the ability to obtain high-resolution NMR spectra from rigid, solid materials. The coupling of the techniques of cross polarization, high-power proton decoupling, and magic angle sample spinning opened new areas of chemistry and physics to high-resolution NMR.[1] Although NMR experiments have been performed on solids and solid polymers since the beginning of NMR, the development of high-resolution capability has provided the hope of determining and then relating molecular structure to properties and solid-state chemistry in a straight-forward way.

Most polymer chemists are familiar with the capabilities of high-resolution NMR in characterizing the molecular structure of polymers in solution. In this regard, polymers are little different from low molecular weight organic molecules for which NMR has been a powerful analysis technique for 30 years. Hence the first expectation of many practitioners of polymer synthesis may be that high-resolution NMR in the solid state is the direct counterpart of the technique in solution. As such, the technique would provide the same structural information as solution-state studies, but with some advantages. These supposed advantages would be (1) an increase in sensitivity, because the sample need not be diluted; (2) the ability to study insoluble materials; and (3) the ability to observe species that are unstable or short lived because of their chemical reactivity or thermal instability.

On the other hand, those interested in the physical properties of polymers may be expected to view high-resolution NMR of solids differently. Generally they are more interested in properties intrinsic to the solid state, because polymers are used most often in their solid form. High-resolution, solid-state NMR, via both chemical shifts and relaxation times, offers the hope of providing a more direct link between polymer physical properties, even in end use, and the details of molecular structure. Hence it might be expected to occupy a place next to mechanical loss and dielectric relaxation methods. One goal of this book is to define the extent to which the above expectations have been or will be realized.

This book describes the application of high-resolution, solid-state NMR techniques to the study of polymers. This opening chapter provides some very general concepts pertaining to the solid state of polymers and some of the major areas of interest in polymer science. It suggests how high-resolution, solid-state NMR may be useful in studying these topics. Those areas where solids NMR is of unique application are described. Of particular interest is directing attention to specific polymer problems that can be attacked and better understood by these techniques. Areas where future development is likely to occur are mentioned also. More is said in Chapter 2 concerning instrumental advances and new techniques presently in development.

Structure and Morphology of Solid Polymers

Structure at the Molecular Level

It is appropriate at this early stage to define, in broad terms, what is meant by polymer structure, which exists at several levels. The most fundamental level concerns chemical composition—the identity and bonding arrangement of atoms in the molecule. For polymers, many possible variations of molecular structure exist. The chemical and stereochemical arrangement of an average chain needs to be specified for both homopolymers and copolymers. This includes monomer composition, sequence distribution, and tacticity. The molecular weight, its distribution, and the variation of chemical and stereochemical composition with molecular weight can be important factors. Branches, crosslinks, end groups, and chain defects, such as unsaturation or head to head linkages in vinyl polymers, even though in low concentration, can be very important aspects of molecular structure. The NMR spectrum represents the average polymer chain and is the sum of the responses from all the individual molecules in the sample.

Structure in the Condensed State

The next level of structure can be termed phase structure. For typical, high molecular weight polymers, this means either the solid or liquid (molten) states. At this point more precision is required in distinguishing between solid and liquid polymers. Polymers can crystallize partially into ordered, rigid lattices. The portions that do clearly are solids in the classical sense, although there may be molecular motion of some type in the crystal. The disordered portions that do not crystallize can be glassy and rigid, if the temperature of observation is below the glass transition temperature, T_g, or rubbery and molten if somewhat above T_g.

When a polymer is at a temperature well above T_g and is low enough in molecular weight ($\sim 10,000$ or lower), often it flows readily and behaves as a typical viscous liquid. From an NMR point of view, these materials are indeed liquids and for many purposes can be considered the same as polymers in solution. They are not discussed in this book, except as an adjunct to the discussion of higher molecular weight polymers or heterogeneous systems. All other cases—that is, all polymers that do not flow appreciably at the temperature of observation—are considered solids in this book. This definition includes completely amorphous polymers above T_g (except those of very low molecular weight). These latter polymers also are termed "rubbery," particularly at temperatures not too far above T_g. Rubbery polymers may or may not be crosslinked. The temperature of observation relative to T_g or to T_m, the crystalline melting temperature, has a strong influence on the NMR requirements necessary for obtaining high-resolution spectra.

Multiphase polymers present a more complicated situation, but nothing different in principle. For a semicrystalline polymer below T_m, the amorphous portion can be below or above T_g at the observation temperature and hence can be glassy or rubbery. Immiscible or partially miscible polymer blends require that the state of each component be defined. The same is also true of block copolymers that form phase-segregated systems. In multiphase situations the structure of the interphase between the major phases is also very important. For the crystalline phase, polymorphic forms can exist, usually at different temperatures or because of different crystallization conditions.

Above the level of the basic phase structure and chain mobility are higher order structures, termed supermolecular structures. This type of structure refers to higher levels of organization of the smaller lamellar crystallites. These include spherulitic and other comparable structures in semicrystalline polymers. This level of structure is characterized by randomly oriented regions of internal long-range order and results from crystallization in the absence of an external force. Block copolymers can form regular macroscopic structures made up of microscopic spherical, cylindrical, or lamellar phase-separated components. By processing or crystallization in the presence of an external force, oriented polymer samples exhibiting long-range molecular order and anisotropic mechanical properties are produced. Certain semiflexible polymers have been shown to exhibit the properties of liquid crystals, such as thermotropic or lyotropic phases, at temperatures above T_g. These phases, in which the polymer molecules exist in highly ordered conformations and are highly extended, can be oriented by application of magnetic or electric fields.

High-Resolution NMR: Solution versus Solid

Chemical Structure

The NMR chemical shift is sensitive to the detailed electronic environment of the nucleus being observed. The NMR spectrum can contain resolved peaks for each type of atomic site in the molecule and often can distinguish very similar sites, where the only difference is a substituent four or five bonds away. For copolymers, this results in a good sensitivity to sequence distribution. In addition, spin–spin coupling patterns can reveal unambiguously which atoms are neighbors in a molecule. Using chemical shifts and spin couplings, it is sometimes possible to deduce the structure of an organic molecule a priori, without resort to direct, detailed comparison to spectra of similar known compounds. In many cases peak positions can be predicted approximately from empirical rules derived from the observation of model compounds. Numerous ancillary techniques, such as spin decoupling, two-dimensional (2D) NMR, multipulse sequences, and relaxation-time measurements, exist as interpretational aids.

The result is that NMR can provide detailed information on chemical structure when the spectral resolution is sufficient to resolve the necessary shifts or couplings. Numerous examples presented in this book confirm that the quality of high-resolution, solid-state NMR is not as good as in solution, and hence solid-state NMR is generally not as useful as solution-state NMR for characterizing molecular structure. This derives in part from technique deficiencies, but chiefly from effects unique to the solid state. These solid-state effects, although important, are conveniently removed in solution for characterization of chemical structure.

Stereochemical Arrangement

In addition to the chemical bonding pattern, the, stereochemical arrangement of atoms often is reflected in the NMR spectrum. Chemical shifts and couplings are sensitive to conformation, configuration, and the disposition of neighboring molecules. Therefore, solution-state NMR is the best probe available for elucidating the configuration (tacticity) of the polymer backbone. Although resonances resulting from configurational isomers are usually highly resolved in solution, they are seldom resolved for atactic polymers in the solid state at the present state of the art in resolution. Configurational isomerism is often manifested as a peak broadening or asymmetry in the solid-state spectrum.

Individual conformations about single bonds usually are not seen in solution because of rapid averaging of all possible conformations by bond rotation. However, the average chemical shift or coupling can be sensitive to population changes among the various conformers. In the solid state (see Chapters 5 and 6) resonances from individual conformers sometimes can be resolved, usually for crystalline substances. The ability to observe resolved resonances for individual conformations of the polymer backbone is a major advantage of solid-state over solution-state NMR. The same influences that determine the dependence of the solution ^{13}C chemical shifts on microstructure also can determine the solid-state spectra.

In solution, chemical shifts can be influenced by what are generally called solvent effects. This catchall term includes effects that can arise from changes in the average conformation about a chemical bond with solvent, from specific interactions, such as hydrogen bonding, or from more diffuse shielding by electronically anisotropic groups on solvent or other molecules. Analogous behavior occurs in solid-state spectra, although in this case little or no averaging from molecular motion occurs. In addition to the conformational effects mentioned above, interactions from molecular packing in the crystal lattice can also be observed.

Sensitivity and Resolution

The major disadvantage of NMR for both solids and liquids is its low sensitivity relative to other spectroscopic techniques. For solids the situation for the ^{1}H nucleus is not severe, but the small chemical-shift range and the difficulty in performing the necessary multipulse experiments limit the practical utility at this time (see Chapter 9). For the ^{13}C nucleus, use of cross polarization places the sensitivity of the solids, high-resolution experiment into the range where it is readily feasible.

Unfortunately, overall, the solids ^{13}C NMR experiment is not as sensitive as the solution-state experiment because of the increased linewidths for the solid. Hence minor components, such as chain branches or defects, which can be observed in solution, may not be observable in the solid. Typically it is possible to observe minor components on the order of several percent in the solid state if the resonances are totally resolved. It is reasonable to expect at least modest improvements in sensitivity to be forthcoming in the next several years as instrument design improves and solid-state studies move to higher magnetic fields. The recently developed electron–nuclear polarization technique,[2] in which the carbon signal is generated by cross polarization from free radicals (directly or via protons), promises substantial increases (10–100 times) in sensitivity when the technique is applicable.

The poor resolution of the solid-state ^{13}C experiment relative to that in solution limits the ability to observe minor components, for quite often these are not resolved to the extent necessary (about several ppm). Modest improvement in spectral resolution may be forthcoming in certain cases with higher magnetic fields, improved high-power decoupling, and ways for setting the magic angle more reproducibly and accurately. Unfortunately, it appears that a large portion of the broadening observed in high-resolution, solid-state spectra, particularly for glassy materials, results from effects intrinsic to the solid and may not be readily removable. As in solution-state NMR, synthesis of isotopically labeled compounds often is necessary to circumvent specific problems in sensitivity or resolution.

Multinuclear Capability

The commercial availability of easy to use and sensitive multinuclear spectrometers has led to a much increased use of nuclei other than ^{13}C or ^{1}H for solution studies. The same is becoming true for the solid state. Table 1-1 lists the most important spin-1/2 nuclei that have been observed in the solid state in high resolution. A number of other spin-1/2 and quadrupolar nuclei also have been observed in the solid state under conditions categorizable as high resolution. The nuclei listed in Table 1-1 should prove useful for studying polymers or polymer additives containing those nuclei. An example of the advantage of NMR's multinuclear character is provided by Figure 1-1. Figure 1-1A is the ^{15}N spectrum of a solid, model polymer containing nitrogen,

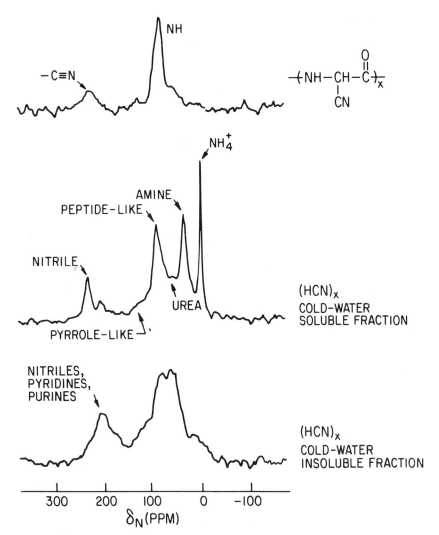

Figure 1-1. High-resolution, solid-state ^{15}N NMR spectra at 9.12 MHz of (A) poly(α-cyanoglycine); (B) the cold-water-soluble fraction of an HCN and NH$_3$ reaction mixture; and (C) the cold-water-insoluble fraction of the same reaction mixture. (Reprinted with permission from Schaefer, J.; Stejskal, E. O.; Jacob, G. S.; McKay, R. A. *Appl. Spectrosc.* **1982**, *36*, 179.)

whereas Figures 1-1B and C show the solid-state spectra of two fractions of a polymer produced by the reaction of HCN and NH$_3$. The spectral detail is more than sufficient for the identification of the types and relative amounts of organic nitrogen present.[3] The same information could not be obtained as readily from the ^{13}C spectra. The ability to observe different NMR nuclei independently provides the site specificity necessary to probe localized and subtle interactions in solid polymers.

TABLE 1-1. SOME SPIN-1/2 NUCLEI SUITABLE FOR HIGH-RESOLUTION STUDY OF SOLID POLYMERS

Nucleus	Natural abundance	Relative sensitivity	Approximate δ range
^{1}H	99.985	1.00	10
^{13}C	1.108	1.59×10^{-2}	200
^{15}N	0.37	1.04×10^{-3}	500
^{19}F	100	0.833	300
^{29}Si	4.70	7.84×10^{-3}	300
^{31}P	100	6.63×10^{-2}	300

Dynamic Processes and Polymer Motion

One of the major advantages of NMR for studying solid polymers is the sensitivity of the technique to various types of polymer motions (see, in particular, Chapter 7). Nuclear magnetic resonance interactions can be modulated or averaged by motions, greatly affecting the NMR spectrum and sometimes the experimental approach used to obtain the spectrum. Often the pertinent NMR parameters are the spin-relaxation times, although lineshapes determined by dipolar, quadrupolar, or chemical-shift interactions are becoming increasingly important. There are more potential motional probes in the solid than in the solution state.

The motional frequencies to which NMR can be sensitive cover a wide range, about $0.01-10^{10}$ Hz, and are selectable within broad limits. The motional selectivity of a particular NMR experiment is governed by the strength, and hence characteristic frequency, of the interaction being modulated by the motion (see Chapter 2). In some cases, equations can be derived directly relating NMR parameters to the actual frequencies of bond rotations in polymers, or bond rotations and overall molecular tumbling for small molecules.

The advantage of high-resolution NMR over standard pulsed NMR and other spectroscopic techniques for motion studies is that motional phenomena can be probed at all spectroscopically resolved sites in the molecule. This allows independent characterization of the primary backbone motions in the presence of motion of the side chains or end groups, for example, and vice versa. Once identified separately, these motions can then be related to transitions observed in mechanical spectroscopy or to macroscopic properties, in principle. In practice, NMR relaxation times, even in high resolution, can suffer from interpretational problems resulting from competing relaxation processes or the difficulty in modeling the often complex motions that are present. These difficulties are not insurmountable, however, and specific motional models can be tested. Chapter 7 presents a detailed analysis of the molecular motions in polycarbonate. One technique used to advantage in the polycarbonate study is specific ^{13}C labeling to observe the effects of motion on the lineshape.

For polymers available in deuterated form, recently developed deuterium NMR techniques for solids provide considerable information on polymer motion (see Chapter 10). Although it is not a high-resolution technique, ^{2}D NMR is included in this work because it nicely complements high-resolution techniques as a probe of polymer motion. The quadrupole interaction predominates over other interactions for the deuterium nucleus, greatly simplifying interpretation of the lineshape and relaxation times. Moreover, the quadrupole interaction is axially symmetrical about the C—D bond. This makes it easy to relate averaging of the ^{2}D spectrum to the geometry of motions in the molecule. Deuterium lineshapes are a sensitive probe of the geometry or amplitude of motion; relaxation times can provide information concerning motional rates. When the spin-alignment technique is considered in addition to lineshapes and relaxation times, ^{2}D NMR can probe motions over a wide range of frequencies.

Site specificity in ^{2}D NMR is achieved by specific labeling. Studies have been performed on a range of polymeric systems, including polyethylene, polycarbonate, polyesters, and rubbers. In multiphase systems, ^{2}D NMR can be a good probe of motion in each phase and can detect the degree of phase mixing.

In addition to the sensitivity to motion mentioned above, the NMR spectrum is sensitive to motions that cause exchange between chemically shifted sites. If a particular site exists in two different environments, two resonances are observed if the site does not switch between the two possibilities at a rate rapid relative to the frequency difference. If the site is exchanging rapidly between the two environments, a single resonance is seen, with a weighted average position determined by the residence times. Lineshape and temperature-dependence studies can yield information on exchange rates in these cases.

Directional Anisotropy

One important feature of NMR in the solid state (with no counterpart for isotropic liquids) is the directional nature of basic NMR interactions (see Chapter 2). The dependence of the chemical-shift and spin dipolar interactions on the direction relative to the magnetic field necessitates additional experimental procedures to eliminate this anisotropy and produce high-resolution spectra similar to those of liquid. On the other hand, the additional information resulting from this directional dependence can be of value for probing local anisotropies and motions in isotropic solids as well as anisotropic solids, such as oriented polymers or liquid crystalline polymers in their oriented phases. The manner in which this directional dependence is modulated by segmental motion can provide information on both the rate and the geometry of that motion.

Domains in Polymers

Through a process known as spin diffusion, NMR can be used in favorable cases to detect the existence of distinct domains and to measure domain sizes in heterogeneous systems (see Chapters 2, 5, and 9). Spin diffusion is not an actual molecular diffusion but is a transfer of spin magnetization over a distance among coupled spins. Certain pulse experiments can detect spin diffusion among different species and provide an estimate of the diffusive distance. The ability to determine a domain size depends on the existence of a substantial relaxation time difference between distinguishable regions. Depending on the relaxation time being used, characteristic sizes on the order of 20–1000 Å or more can be measured. Only a few applications of this type of NMR experiment have appeared,[4] possibly because many systems do not fulfill the necessary criteria to a sufficient extent.

Imaging

Recently NMR has assumed a major role as a technique for producing spatial images of the interior of the human body. This monumental success has spurred efforts to apply NMR imaging to solid, inanimate objects at a microscopic level. The basis of the NMR imaging technique is the labeling of the different spatial locations in a sample with different resonance frequencies by applying linear magnetic field gradients in various directions across the sample. The spatial resolution in the NMR imaging experiment is determined by (among other things) the strength of this gradient across the sample. Typically the spatial resolution in NMR imaging of the human body is about 1 mm. It is also possible to get a chemical-shift spectrum for a localized region in an object. Ideally, one would like to obtain the NMR spectra of regions corresponding to the domains in multiphase polymers or supermolecular structures in semicrystalline polymers.

At this point this goal is extremely far away. Very early experiments have been reported on glass fiber-reinforced epoxy resin composites.[5] These workers were able by NMR imaging to determine the spatial distribution of water in 2.5-cm diameter rods of the composites at a resolution of about 0.2 mm. Progress in NMR imaging of glassy or crystalline polymer samples will require technique improvements in several areas. First, line broadening in rigid solids will have to be mitigated, probably via multipulse line-narrowing techniques. Second, the spatial resolution will need to be improved well beyond that available now. Multiple quantum NMR is being investigated as a way to increase the effective magnetic field gradient across the sample,[6] and hence increase resolution perhaps by a factor of 10 or more. Last, as the resolution increases, the amount of sample per volume element decreases rapidly, with the obvious effect on signal-to-noise ratio considerations. Sensitivity increases will be necessary as resolution improves. It is clear that many years of instrumental and technique development are necessary before images can be obtained on solid polymers at a resolution beyond the several micron level.

Polymer Physical Chemistry: Current and Future Areas for Solid-State NMR Studies

There are a number of areas in polymer science where solid-state NMR can be expected to have an impact, and in some cases this will be unique. Here, the questions of a fundamental nature that have been or can be addressed by high-resolution solids NMR studies are reviewed.

Transitions

Much effort in the physical chemistry of polymers involves the study of transitions, that is, temperatures at which some thermal or physical property of the polymer undergoes a significant change. Transitions can be followed by observing changes in some thermodynamic quantity, such as enthalpy or heat capacity, or by some mechanical or spectroscopic property. Transitions often determine the range of use temperature of a polymeric material or they can be related to some desirable physical property of the polymer, such as the ability to withstand impact. It is desirable to understand the molecular basis for a given transition. For example, does it arise from localized or long-range motions of the backbone? of the side chain? Is the transition associated with the crystalline, amorphous, or perhaps interfacial, phases? What is the exact nature of the motion (for example, small- or large-angle jump between minima in a potential well) and how does it depend on temperature?

High-resolution NMR of solids appears well suited to answering these questions. A number of NMR parameters sensitive to motion and/or the electronic environment of the nucleus are available, in principle, for each resolvable site. One approach is exemplified by a recent study demonstrating that the correlation times determined by ^{13}C NMR for the backbone atoms of a series of amorphous polymers obey the Williams-Landel-Ferry (WLF) relation (see Chapter 4). Semicrystalline polymer transitions can be studied also, and work has appeared relating to both the β transition and the melt transition. As becomes evident in Chapters 3–5, one NMR technique may not be sufficient to address many of these subjects. For example, to investigate the nature of the glass transition fully it probably will be necessary to use techniques appropriate for glassy or crystalline and for rubbery solids.

Totally Amorphous Polymers

For this class of polymers, solid-state NMR techniques have already proved to be important. Fundamental research has centered and will continue to center on such topics as the nature of the glassy state and the glass transition, sub-T_g annealing and free volume effects, and chain motion in both glassy and rubbery states. Resolution of individual conformations in the glassy state, if possible, probably will depend on instrumental or technique advances or the choice of suitable polymers for study. In the rubbery state, the weighted average of the available conformations is represented by the observed chemical shift, as in solution.

Certain other topics of a more practical nature may be amenable to solid-state NMR analysis. For example, the effect of crosslinks on polymer chain motion and the motion at the crosslink site are topics of current interest. Of course, if crosslinks are to be observed, the sensitivity problem must be solved, either by using isotopic enrichment, in the case of ^{13}C, or by introducing a crosslink with an abundant NMR nucleus. The interaction of polymers, particularly rubbers, with fillers is also of interest. In this case it is necessary to observe molecules at the interface independently of the bulk of the polymer. This presents a problem of sensitivity and resolution that in general has not been solved.

Semicrystalline Polymers

The ability to observe resolved resonances for the crystalline and amorphous regions of semicrystalline polymers will make solids ^{13}C NMR techniques of great importance for these materials. Although resolved resonances may not always be observed in the same spectrum, the two phases may be observable independently by a multitechnique approach at the appropriate temperature. It may be possible to study independently a signal arising from the interfacial region of a semicrystalline polymer. Interfaces in polymeric systems will be difficult to observe by solid-state NMR techniques, even with the available approaches for resolution enhancement and phase discrimination. These approaches may not be sufficiently discriminating to yield a signal solely from the interface.

Chemical shifts for backbone carbons in polymers are determined in part by the relative populations of the various conformations via γ-gauche interactions. Chain conformation is the major factor that determines the chemical shift of carbons in the crystalline relative to the amorphous state and contributes to differentiation among different crystalline forms (see Chapter 6). Figure 1-2 shows the spectra of three crystalline forms of isotactic poly(1-butene).[7] The spectra are clearly different and mainly arise from the different chain conformations in the crystalline habitats. Such effects allow comparison of the X-ray and solid-state NMR results for crystalline polymers. It also holds out the hope of resolving individual conformations in glassy polymers or in the interfacial regions. Even for the same chain conformation, chain packing differences in the crystallites can produce different ^{13}C spectra. The distribution of certain chemical structures, such as comonomer units, branches, or chain defects, among the various regions may be obtainable in carefully controlled experiments.

Also of interest are questions pertaining to the nature of the phases themselves. How similar in motional or conformational properties is the amorphous component of a semicrystalline polymer to that of the same totally

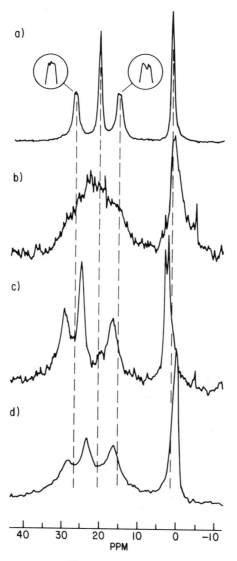

Figure 1-2. Solid-state 50.3-MHz ^{13}C spectra of poly(1-butene). (A) Form I at 20°C; (B) form II at −60°C; (C) form III at −10°C; (D) amorphous at 43°C. The vertical dashed lines represent the peak positions of form I. The chemical-shift scale is referenced to the amorphous methyl resonance at 0.00 ppm. (Reprinted with permission from Belfiore, L. A.; Schilling, F. C.; Tonelli, A. E.; Lovinger, A. J.; Bovey, F. A. *Macromolecules* **1984**, *17*, 2561. Copyright 1984 American Chemical Society.)

amorphous polymer? Such questions can be answered by chemical-shift and relaxation-time measurements. Is the interfacial region more like the amorphous or more like the crystalline portion? What happens upon annealing? What is the nature of the amorphous region for polymers of different supermolecular structures, or higher levels of organization? Solid-state NMR techniques have just begun to probe some of these questions (Chapter 5).

Polymer Blends

The situation for polymer blends is similar to that for semicrystalline polymers. The ability to observe resolved resonances for the different phases in incompatible blends depends on the structural differences between the two components. This usually is much easier than for semicrystalline polymers because of the chemically different nature of the blend components. If one phase is above T_g and the other below, then their spectra may be separable by applying several techniques. For fully or partially compatible blends, solid-state NMR studies should provide information on such features as the intimacy of mixing and specific interactions such as hydrogen bonding between the two components. If an interfacial region is present in sufficient quantity, it may be observable independently by solid-state techniques. For incompatible blends, spin-diffusion studies may provide estimates of domain sizes.

Oriented and Liquid Crystalline Polymers

The fundamental NMR interactions of the chemical shift and dipolar coupling depend on the direction of the tensor relative to the external magnetic field. Hence it is not surprising that these parameters are useful for studying anisotropic systems, such as oriented polymers (Chapter 8). Information concerning the extent and nature of the orientation is available. In addition, the range of studies available for isotropic polymers is also available for oriented systems. Hence one may expect to learn something about the nature of the amorphous regions in oriented semicrystalline polymers. Other areas of interest include the effect of drawing and annealing on oriented and unoriented regions as well as a comparison of chain motion in oriented and unoriented systems.

There is increasing interest in systems that combine the properties of polymers with those of liquid crystals. These systems can possess mesogenic groups on either the side chain or the backbone and can exist in both isotropic and liquid crystalline states. As mentioned earlier, these ordered phases can be oriented by magnetic or electric fields. The degree and type of order exhibited by these systems both above and below T_g greatly influence their properties. As for oriented, nonmesogenic polymers, solid-state NMR has a role to play. For example, it is possible by deuterium NMR with specific labeling to determine the degree of order of the mesogenic group and at different locations on the backbone or side chain. Motional information can be obtained also and compared to the corresponding isotropic phase.

Polymer–Small Molecule Interactions

In addition to studies of polymers in the pure, undiluted state, solid-state NMR techniques will be useful in studying the interaction of small molecules and polymer chains. One strength of NMR for this type of study is that the polymer and the diluent, such as a plasticizer, can be monitored independently. Spin-relaxation-time measurements probably will play a large role in such studies. An example of the type of information that will be forthcoming is provided by the work of Sefcik et al,[8] who studied the antiplasticization–plasticization effect of tricresyl phosphate on poly(vinyl chloride) (PVC). They found that the main-chain motions of PVC, as measured by ^{13}C spin–lattice relaxation times in the rotating frame, correlate with gas permeability on the same samples. Results such as these should provide considerable insight into the nature of plasticization, gas sorption, and the interaction of small molecules with polymers in general.

Analytical Applications, Biopolymers, and Surfaces

The use of high-resolution NMR techniques as an analytical tool for solid polymers has received considerable attention. Here analytical applications are considered to include those in which the chemical or stereochemical structure only is being identified. For most problems of identification of the chemical or stereochemical structure of polymers, solution-state NMR, if feasible, is to be preferred. The peak resolution attainable in solid-state spectra of amorphous polymers probably will not closely approach that of the solution state for the foreseeable future. Hence resolution, sensitivity, and experimental ease all currently favor solution-state studies.

High-resolution, solid-state NMR will have a large impact on the analysis of insoluble polymers, particularly those that are difficult to characterize by other techniques. This is true for optically opaque materials, which cannot be characterized easily by other spectroscopies. One area that has received considerable attention is that of electrically conducting polymers, such as polyacetylene and polypyrrole. Solids ^{13}C NMR has been able to provide detailed structural information not obtainable by any other method. For polyacetylene, cis and trans double bonds are resolved, as are sp^3-hybridized carbons. The relative amount of these structures is one factor governing the level of conductivity attainable. For polypyrrole and derivatives, solid ^{13}C NMR has confirmed the predominant α–α' backbone linkages, as well as the basic chemical structure. Figure 1-3A shows the cross-polarization, magic angle spinning (CPMAS) ^{13}C spectrum of neutral poly(β,β'-dimethylpyrrole), which is nonconducting.[9] The aromatic and methyl carbons are clearly identifiable. In addition, such polymers can be examined in both the conducting and nonconducting states, potentially providing fundamental information on the nature and mechanism of the conductivity. The spectrum obtained for

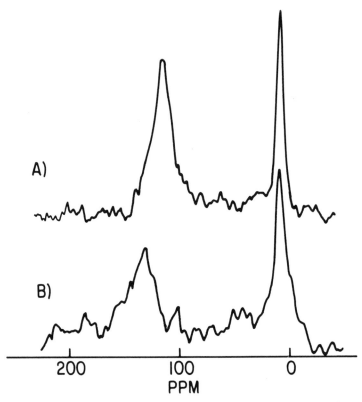

Figure 1-3. Carbon-13 spectra of (A) neutral poly(β,β'-dimethylpyrrole) and (B) poly(β,β'-dimethylpyrrole) perchlorate. (Copyright 1983 by International Business Machines Corporation; reprinted from reference 9 with permission.)

poly(β,β'-dimethylpyrrole) perchlorate is shown in Figure 1-3B. For this conducting material the aromatic peak shifts downfield as expected on oxidation of the pyrrole ring. More is said in Chapter 3 concerning solid-state NMR studies of conducting polymers.

Quite a few "intractable" systems, such as cured resins, have been studied by solid-state NMR. The analyses of cured phenol–formaldehyde resins and carbon black-filled rubbers are described in Chapters 3 and 4, respectively.

High-resolution, solid-state NMR techniques have been used with success in the area of biopolymers. Many of the principles outlined here for synthetic polymers also apply to biopolymers. The chemical complexity of many biopolymers, such as enzymes, make them difficult to study in any detail. This is not true of structural biopolymers, such as cellulose and other polysaccharides, the structural proteins elastin and collagen, and the nucleic acids. Early work centered on the structural proteins,[10] whereas more recently there has been considerable interest in cellulose and its derivatives.[11] For example, high-resolution, solid-state ^{13}C NMR is sensitive to the crystalline morphology of cellulose and can distinguish cellulose forms I and II, as well as celluloses from different natural sources. Although the topic is of great interest and importance, biopolymers are not considered further in this book.

The interaction of polymers with surfaces is an important and little understood topic at the molecular level. This is particularly true with regard to adhesion phenomena and the specific interaction of reinforcing fillers and fibers with polymers. The low sensitivity of NMR relative to other spectroscopic techniques makes it generally unattractive for surface studies, except in special cases. At this time it seems that solids NMR is likely to be valuable for studying polymer–surface interactions only under limited circumstances— when an abundant nucleus or isotopic enrichment is used, or when the experiment involves a very high surface area substrate, such as silica gel, which is also transparent to the technique. Success in this area will be a strong function of the ability to prepare suitable samples to circumvent problems of resolution and sensitivity. At the time of this writing, experiments along these lines for model systems are in progress.[12]

Summary

More applications of high-resolution, solid-state NMR undoubtedly will arise beyond those mentioned here. This short overview is intended to serve as a stimulus for these additional applications. The field is one that will experience both predictable and unpredictable advances in technique and instrumentation. At this relatively early stage in the application of solids NMR to polymers, only the initial, ground-breaking experiments have been performed in many cases.

References

1. Fyfe, C. A. "Solid State NMR for Chemists"; C. F. C. Press: Guelph, 1984.
2. Wind, R. A.; Anthonio, F. E.; Duijvestijn, M. J.; Smidt, J.; Trommel, J.; de Vette, G. M. C. *J. Magn. Reson.* **1983**, *52*, 424.
3. Schaefer, J.; Stejskal, E. O.; Jacob, G. S.; McKay, R. A. *Appl. Spectrosc.* **1982**, *36*, 179.
4. Caravatti, P.; Neuenschwander, P.; Ernst, R. R. *Macromolecules* **1985**, *18*, 119.
5. Rothwell, W. P.; Holecek, D. R.; Kershaw, J. A. *J. Polym. Sci., Polym. Lett. Ed.* **1984**, *22*, 241.
6. Garroway, A. N.; Baum, J.; Munowitz, M. G.; Pines, A. *J. Magn. Reson.* **1984**, *60*, 337.
7. Belfiore, L. A.; Schilling, F. C.; Tonelli, A. E.; Lovinger, A. J.; Bovey, F. A. *Macromolecules* **1984**, *17*, 2561.
8. Sefcik, M. D.; Schaefer, J.; May, F. L.; Raucher, D.; Dub, S. M. *J. Polym. Sci., Polym. Phys. Ed.* **1983**, *21*, 1041.
9. Clarke, T. C.; Scott, J. C.; Street, G. B. *IBM J. Res. Develop.* **1983**, *27*, 313.
10. Torchia, D. A.; VanderHart, D. L. *Top.* ^{13}C *NMR Spectrosc.* **1979**, *3*, 325.
11. Fyfe, C. A.; Dudley, R. L.; Stephenson, P. J.; Deslandes, Y.; Hamer, G. K.; Marchessault, R. H. *J. Macromol. Sci. Macromol. Revs.* **1983**, *C23(2)*, 187.
12. Komoroski, R. A.; Rahrig, D. B., unpublished results, 1984.

2

PRINCIPLES AND GENERAL ASPECTS OF HIGH-RESOLUTION NMR OF BULK POLYMERS

Richard A. Komoroski

B F GOODRICH RESEARCH AND DEVELOPMENT CENTER
9921 BRECKSVILLE ROAD, BRECKSVILLE, OH 44141

Introduction

This chapter describes the principles involved and some of the techniques used to obtain high-resolution ^{13}C NMR spectra of rigid, solid polymers. This necessitates a description of the various NMR interactions and their behavior in the solid state. Quite often the solution-state spectrum provides a convenient reference[1] because polymers span the range from ordered, crystalline solid, to glass, to rubber, to low molecular weight liquid. A knowledge of the basic principles of NMR spectroscopy is assumed here. Only aspects of the subject of interest for polymer studies are covered in this chapter, which is not intended to serve as a review of all the recently developed solid-state NMR techniques. The reader is referred to a recent review article[2] and text[3] on the subject for further details and nonpolymer applications.

Some of the ideas and terms used in this chapter and throughout the book may not be familiar to practitioners of high-resolution NMR. The excellent text by Fukushima and Roeder[4] presents many fundamental descriptions of concepts basic to pulse NMR. In-depth discussions on a number of topics appear in the appropriate chapters of the present work as needed. Many of the

Komoroski (ed): High-Resolution NMR Spectroscopy of Synthetic Polymers in Bulk

experimental requirements for performing both standard high-resolution, Fourier-transform nuclear magnetic resonance (FT NMR) and cross-polarization, magic angle spinning (CPMAS) NMR have been given elsewhere[3-7] and are not repeated here. Instead, this chapter is confined to the basic principles of the experiments as they relate to polymers and the effect of various NMR and instrumental parameters on the spectral intensities and resolution.

High-resolution NMR of solids can be divided into three classes of experiments, depending on the type of nucleus observed. In the first category are abundant spins ($I = 1/2$), which includes ^1H and ^{19}F. Principles specific to obtaining high-resolution ^1H or ^{19}F spectra of solid polymers are covered in Chapter 9.[8,9] Because most of this book is concerned with high-resolution NMR of ^{13}C in natural abundance, this chapter, for the most part, concentrates on the observation of the second class of nuclei, dilute spins ($I = 1/2$) in solids. These spins can be dilute because of their isotopic abundance, as for ^{13}C, ^{15}N, and ^{29}Si, or because of their relatively uncommon occurrence in the system of interest, as for ^{31}P. In the third category are the quadrupolar nuclei of nonintegral spin, such as ^{17}O, ^{11}B, and a number of metal nuclei. Although the quadrupolar interaction can be a severe broadening mechanism in the solid state, spectra categorizable as high resolution have been observed for some of these nuclei in solid samples. Because polymer applications have not yet appeared for these nuclei, they are not discussed further. Principles relating to the observation of deuterium ($I = 1$) in solid polymers are covered in Chapter 10. Although solid-state deuterium NMR is not a high-resolution technique, it is included in this book because of its importance in studying polymer motion and morphology.

Solid-State Interactions

There are several NMR interactions or effects that must be considered to understand ^{13}C NMR spectra of solids. All of the interactions that broaden NMR lines in solids are also operative in nonviscous liquids. In liquids the rapid, essentially isotropic motion of the molecules or molecular segments averages these interactions to zero or to a single, nonzero value representative of the average interaction. For high molecular weight polymers in both the solution and solid states, it is usually not the overall motion but the bond rotations in the chain backbone and side groups that provide this averaging. In general, this averaging is not isotropic, even in dilute solution. Polymers in their solid form have chain mobilities that vary over many orders of magnitude in time. For solid polymers the extent and nature of the averaging of broadening interactions determines to a large extent the techniques used to obtain the high-resolution spectrum and to some extent the resolution in that spectrum.

Dipolar Coupling

The Interaction. The major reason the NMR spectra (for $I = 1/2$) of solid compounds and polymers in their crystalline or glassy states are broad is the dipolar interaction among nuclear magnetic moments, or the spin–dipolar interaction. For protons, the major portion of this dipolar interaction occurs among the protons themselves. For ^{13}C in natural abundance, it is the interaction with the usually abundant and nearby protons. The ^{13}C atoms are sufficiently dilute to render the ^{13}C–^{13}C dipolar interactions weak and hence negligible in most cases. The spin–dipolar coupling operates through space, depends on the orientation of the interdipole vector relative to the magnetic field, and depends on the distance between the dipoles (Figure 2-1A). The dipolar interaction between two spins can be written in the general form:

$$\mathcal{H}_D = \bar{I}_1 \cdot \mathbf{D} \cdot \bar{I}_2 \tag{2-1}$$

where $\mathcal{H}_D$ is the magnetic dipolar Hamiltonian and $\mathbf{D}$ is the dipolar coupling tensor.[6] The trace of $\mathbf{D}$ is zero, which means that with fast isotropic averaging, as in solution, dipolar coupling does not affect the energy and hence the resonance frequency of the transition. The truncated dipolar Hamiltonian for two spin species I and S can be written[10]:

$$\mathcal{H}_D = \mathcal{H}_{II} + \mathcal{H}_{SS} + \mathcal{H}_{IS} \tag{2-2}$$

where:

$$\mathcal{H}_{II} = \tfrac{1}{2}\gamma_I^2\hbar^2 \sum_{i<j}^{N_I}\sum^{N_I} r_{ij}^{-3}(3\cos^2\theta_{ij} - 1)(\bar{I}_i \cdot \bar{I}_j - 3I_{iz}I_{jz}) \tag{2-3}$$

$$\mathcal{H}_{IS} = -\gamma_I\gamma_S\hbar^2 \sum_i^{N_I}\sum_k^{N_S} r_{ik}^{-3}(3\cos^2\theta_{ik} - 1)I_{iz}I_{kz} \tag{2-4}$$

Here γ_C and γ_H are the carbon and proton gyromagnetic ratios, which are proportional to the magnetic moments of the nuclei; r is the distance between the nuclei; and θ is the angle the C–H vector makes with the external magnetic field. The equation for $\mathcal{H}_{SS}$ is analogous to that for $\mathcal{H}_{II}$. For the situation of interest, when I is ^1H and S is ^{13}C in natural abundance, $\mathcal{H}_{SS}$ can be ignored because the dipolar interaction among the ^{13}C nuclei is weak owing to their large internuclear distances and small magnetic moments.

The dipolar coupling of a set of magnetically isolated ^{13}C–^1H pairs at a single angle relative to H_0 results in a splitting in the ^{13}C spectrum (corresponding to the two allowable proton spin states) given (in Hz) by:

$$D = \hbar\gamma_C\gamma_H(3\cos^2\theta - 1)/2\pi r^3 \tag{2-5}$$

This is shown schematically in Figure 2-1B. Obviously, the dipolar coupling is independent of the strength of H_0. It is a strong function of the internuclear distance. For example, a 25% increase in r causes the dipolar interaction to be reduced by about a factor of 2. Table 2-1 lists some representative static dipolar couplings involving ^{13}C and ^1H.

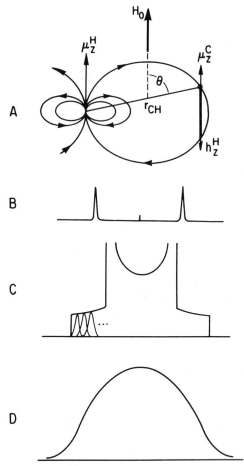

Figure 2-1. (A) Dipolar interaction between a ^{13}C and proton spin. The μ are the z components of the magnetic moments, and h_z^H is the z component of the proton dipolar field at the ^{13}C nucleus. (B) Dipolar splitting of isolated CH pairs at one angle relative to the magnetic field. (C) Pake pattern expected for isolated C–H pairs distributed at all angles as in polycrystalline or glassy materials. Several components from specific angles are illustrated schematically. (D) Approximate Gaussian lineshape observed for non-isolated C–H pairs, where all dipolar interactions are operative.

Isolated C–H pairs are encountered rarely in organic compounds. Hence the total C–H dipolar interaction seen by an individual ^{13}C nucleus involves a sum over all protons at their respective distances and angles. For very simple compounds observed as single crystals this produces additional splittings in the ^{13}C spectrum, with some broadening from peak overlap or dipolar coupling to more distant protons on the same or other molecules. Because the dipolar couplings are angle dependent, rotating the single crystal sample with respect to the magnetic field changes the values of the various splittings in a predictable way, providing geometric information, at least in principle.

TABLE 2-1. STATIC DIPOLAR COUPLINGS IN SELECTED MOLECULES[a]

Spin pair	Compound	r (Å)	$\langle \Delta\omega^2 \rangle^{1/2}$ (kHz)
$^1H–^1H$	H_2	0.746	194
	H_2O	1.51	23.4
	C_2H_4	1.82	13.4
$^{13}C–^{13}C$	C_2H_2	1.20	2.92
	C_2H_4	1.34	2.12
	C_2H_6	1.54	1.40
$^1H–^{13}C$	C_2H_2	1.06	11.4
	C_2H_6	1.11	9.9

[a] From reference 2. Reprinted by permission of the copyright holder, Elsevier (North Holland).

However, for polycrystalline organic compounds and glassy or crystalline, unoriented polymers, the various C–H vectors assume all possible angles with respect to the external magnetic field. For a single type of magnetically isolated C–H vector in such a material, a range of dipolar couplings is seen, producing the characteristic Pake doublet (Figure 2-1C). Without magnetic isolation of the individual C–H vectors, many additional dipolar interactions occur. This produces a so-called inhomogeneously broadened line tending toward a Gaussian shape (Figure 2-1D). Inhomogeneous broadening refers to the fact that the various parts of the broad line (at different resonance frequencies) arise somewhat independently from C–H dipoles at various internuclear separations and at various angles with respect to the magnetic field. This is to be contrasted with homogeneous broadening, where all nuclei in the sample give rise to the entire line, which is broad because of spin-relaxation or spin-diffusion effects (see Signal Generation and Sensitivity).

At this point it is appropriate to mention a recently developed technique called zero-field NMR,[11] which has circumvented part of this inhomogeneous broadening problem. This technique relies on removing the sample from the main magnetic field at certain times during the experiment and allowing the local dipolar (and other) interactions alone to determine the spectrum. The angular dependence is removed. This produces spectra for polycrystalline or amorphous solids that are similar to those expected for single crystals. Zero-field NMR has not yet seen wide application but holds promise for providing detailed structural information for amorphous solids.

Dipolar Decoupling. As a result of the dipolar interactions with protons, the ^{13}C NMR linewidth in an organic solid is on the order of 20 kHz. The fine structure associated with the various chemically shifted resonances occurs over a range much less than this. Hence it is necessary to remove the effects of this dipolar interaction if high-resolution spectroscopy is to be performed. Because the trace of the dipolar coupling tensor is zero, its effect can be removed from the spectrum if it can be suitably averaged. Nature has done this for us in nonviscous liquids. The very rapid and for the most part random rotational and translational motions occurring in the liquid state totally remove the effect of dipolar coupling by averaging the angle-dependent term to zero. For polymers in solution or in the rubbery state, only a small residual component of the dipolar interaction may remain.

In glassy or crystalline solids there may be little or no molecular motion at the proper frequencies to average the heteronuclear dipolar interaction $\mathscr{H}_{IS}$. Very slow and highly anisotropic motions do occur, but in most cases these are not effective at averaging the dipolar interaction to an extent sufficient for high-resolution NMR. Equation (2-4) for $\mathscr{H}_{IS}$ has been written to show the spin and spatial parts separately. If either of these parts can be averaged to zero by some coherent process, the effect of $\mathscr{H}_{IS}$ can be removed from the spectrum. One possibility is physically spinning the sample at 54.74°, the so-called magic angle, which reduces the $(3 \cos^2 \theta - 1)$ term to zero.[12] This approach, which is described in detail later in this Section, is not technically feasible for the typical C–H or H–H dipolar interactions of the strength encountered in organic solids. Magic angle spinning (MAS) can reduce weak dipolar interactions, as explained later in this Section.

The other option available is to average $I_z S_z$, the spin part of $\mathscr{H}_{IS}$, rather than the spatial part. The ^{13}C magnetic moments experience the z component of the local field h_z^H from the proton moments (Figure 2-1A). If this local field can be averaged to zero, for example, by inverting it half the time on average, the ^{13}C spin sees zero net field from the protons and the effect of heteronuclear dipolar coupling is removed. In the typical ^{13}C experiment in rigid solids, radiofrequency (rf) irradiation of sufficient power at the proton frequency forces the proton spins to precess rapidly. This averages the effective field from ^{1}H that is seen by the ^{13}C nucleus.[13] For the averaging to be effective, the irradiation obviously must be powerful enough to average $\mathscr{H}_{IS}$, the C–H dipolar coupling. It also must be powerful enough to average $\mathscr{H}_{II}$, which is usually larger than $\mathscr{H}_{IS}$. This latter requirement derives from proton spin diffusion (see Signal Generation and Sensitivity), which can disrupt the forced precession of the ^{1}H spins. Dipolar decoupling is analogous to the proton decoupling normally done in ^{13}C NMR of liquids to remove C–H spin–spin (scalar) couplings, which range from 0 to about 200 Hz. For liquids the decoupling field strength is usually about 1 G (4 kHz) and is modulated in some manner to cover the ^{1}H chemical-shift range. For solids, decoupling fields of 10 G (40 kHz) or more are needed to reduce the much stronger dipolar coupling.

At this point a review of how the various "fields" in the NMR experiment are described is needed. The large, static field that permits NMR absorption is given in units of Tesla or Gauss (1 Tesla = 10,000 G) and lies between 14 and 141 kG. This field determines the Larmor frequency (5–600 MHz) for a given nucleus via the gyromagnetic ratio ($v_0 = (\gamma/2\pi)H_0$). The situation is different for the rf or alternating fields used for excitation (H_1) or decoupling (H_2). Of course, the frequency of the rf field is at or near the Larmor frequency of the nucleus being observed or decoupled. The strength of these rf fields can be given in magnetic field units or, for a particular nucleus, in frequency units (Hz), the two being interconvertible via γ. The rf field strength in frequency units is convenient because it allows a direct comparison with other quantities, such as linewidth, chemical-shift range, and spinning rate, which are in frequency units.

Decoupling is most effective when the decoupler frequency is exactly on resonance. For a typical organic sample, the proton chemical-shift range (including anisotropies) is about 10 ppm or more, and the decoupler frequency is not exactly on resonance for all protons. Residual broadening from incomplete decoupling, $\Delta\langle\omega_{CH}^2\rangle^{1/2}$, is approximately given by[14,15]:

$$\Delta\langle\omega_{CH}^2\rangle^{1/2} \simeq \langle\omega_{CH}^2\rangle^{1/2} \cos\psi \qquad (2\text{-}6)$$

where $\tan^{-1}\psi = \omega_{1H}/\Delta\omega_{0H}$, the ratio of the decoupling field strength to the frequency offset from the proton being decoupled and $\langle\omega_{CH}^2\rangle^{1/2}$ is the second moment from heteronuclear dipolar coupling. For optimum decoupling, $\psi = \pi/2$. Generally, decoupling field strengths of 25–75 kHz are needed to average the typical C–H dipolar interaction effectively. The effect of incomplete decoupling on the linewidth is treated in more detail in Chapter 5. The extent to which incomplete decoupling contributes to the residual linewidths for polymers depends on the system of interest. Because of the strong coupling among protons in the crystalline regions of semicrystalline polymers, these systems are more likely to display residual broadening from incomplete decoupling than are amorphous, glassy polymers.

Anisotropic Chemical Shift

The Interaction. The power of high-resolution NMR spectroscopy in organic and polymer chemistry can be attributed to its ability to resolve peaks arising from chemically or stereochemically different types of atoms in a molecule. The resonance frequency of a particular nucleus depends on both the type of nucleus (eg, ^1H, ^{13}C) and the electronic environment of that nucleus (ie, the arrangement of electrons and other nuclei). The surrounding electrons shield the nucleus from the external magnetic field. Because the electronic environment is usually different in different directions in the molecule, the nucleus sees a different shielding and hence has a different chemical shift in different directions. The single-valued chemical shifts observed in solution are in reality isotropic averages of anisotropic shieldings resulting from rapid molecular rotation.

Like other solid-state interactions, the chemical shift can be represented by a second-rank tensor[6]:

$$\mathcal{H}_\sigma = \hbar\bar{I} \cdot \sigma \cdot \bar{H}_0 \qquad (2\text{-}7)$$

where σ is the chemical shift tensor and $\mathcal{H}_\sigma$ is the chemical-shift Hamiltonian. Unlike the dipolar interaction tensor, which is traceless and so does not affect the position of the NMR resonance in solution, σ is not traceless. Because the components of σ are small compared to unity, Eq. (2-7) can be written:

$$\mathscr{H}_\sigma = \gamma\hbar\sigma_{zz}H_0 \qquad (2\text{-}8)$$

The quantity σ_{zz}, the projection of the chemical shift tensor along the z axis, can be expressed in terms of the principal values of σ, σ_i, and the respective direction cosines, l_i, with respect to H_0:

$$\sigma_{zz} = \sum_{i=1}^{3} \sigma_i l_i^2 \qquad (2\text{-}9)$$

The principal values represent the magnitude of the tensor in three mutually perpendicular directions in the molecule, and the direction cosines specify the orientation of the tensor. Because the isotropic average of each l_i^2 is $1/3$:

$$\sigma_{zz}^{avg} = \tfrac{1}{3}\mathrm{tr}\sigma = \sigma_{iso} \qquad (2\text{-}10)$$

where σ_{iso} is the isotropic chemical shift seen in solution.

For a single crystal in the absence of other interactions, a single, relatively narrow signal is observed. The position of this signal depends on the orientation of the principal axes of σ, and hence of the crystal, with respect to H_0. Two such signals are shown schematically in Figure 2-2A for two orientations of a single crystal of solid carbon monoxide. Similarly to the dipolar interaction, a study of the chemical shift as a function of orientation of the single crystal can map out the shift anisotropy and its orientation in the molecule.

For a polycrystalline solid or an unoriented, glassy or semicrystalline polymer, an envelope of signals resulting from all possible orientations is seen. For carbon monoxide this envelope looks like Figure 2-2B because of the axial symmetry of the molecule.[16] In this case each portion of the envelope arises uniquely from molecules oriented at one angle relative to the magnetic field. Figure 2-2C shows a generalized chemical-shift anisotropy (CSA) powder pattern for the more common nonsymmetrical case. The full equations describing the CSA lineshape are given in Chapter 7, along with a treatment of the effect of certain molecular motions on the CSA. The principle values of the CSA can be obtained directly from the pattern as indicated, whereas the weighted-average position in this envelope corresponds to the isotropic chemical shift in solution. However, no absolute determination of the orientation of the tensor in the molecular frame can be made from the powder spectrum. Sometimes the orientation can be inferred from symmetry (for axially symmetrical tensors) or by comparison to similar functional groups in molecules where the orientation is known. For rigid, unoriented polymers the powder lineshape always is encountered because polymer single crystals are rarely available.

One of the original motivations for the development of high-resolution NMR techniques for solids was the observation of these CSA patterns. Observing the full anisotropy yields more electronic information with which to

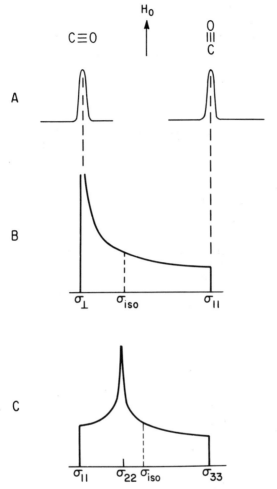

Figure 2-2. (A) Chemical shifts observed for two orientations of a carbon monoxide single crystal relative to the static magnetic field. (B) Axially symmetrical chemical-shift anisotropy powder pattern as would be observed for polycrystalline carbon monoxide. The components of the shift tensor parallel and perpendicular to the symmetry axis are shown. The weighted-average value, or isotropic chemical shift, observed in solution is given by σ_{iso}. (C) Fully anisotropic CSA powder pattern with tensor components indicated.

probe electronic structure and on which to base theories of the chemical shift than does the average chemical shift. Changes in molecular structure may produce large changes in one or more of the CSA principle values while affecting σ_{iso} only slightly. The correlation of CSAs with chemical structure and reactivity is just beginning. In addition, CSA patterns can be sensitive to polymer motions in the solid state (see Chapter 7). Table 2-2 lists some typical CSA values for some common nuclei in different chemical environments.

TABLE 2-2. SOME TYPICAL CHEMICAL SHIFT ANISOTROPIES (CSA)[a]

Environment	CSA[b]
^{13}C	
CH_3CH_2OH	24
CH_3OH	63
$(CH_2)_n$	33
(CH_2CH_2O)	58
Olefinic, trans-polybutadiene	187[c]
Aromatic, PBT	201–215[d]
Ester, PBT	127[d]
^1H	
Aliphatic	5
Olefinic, aromatic	5–7
Carboxylic	18–32
^{15}N	
Pyridine	782
CH_3CN	489
L-Histidine · HCl · H$_2$O	
Imidazole	216, 222[e]
NH_3^+	< 12[e]
Glycylglycine · HCl · H$_2$O	
Peptide	150[f]
^{19}F	
C_6F_6	155
$(CF_2CF_2)_n$	90
$CFCl_2CFCl_2$	240
$[(CF_3)CO]_2O$	69
^{29}Si	
$CH_3Si(OCH_3)_3$	45
$[(CH_3)_3Si]CH$	26
^{31}P	
$P(OR)_3$	~ 450[g]
KH_2PO_4	15
Ph_3PO	210[g]
Ph_3P	53[g]

[a] Data from reference 6 except as noted.
[b] $|\sigma_{33}-\sigma_{11}|$, in ppm.
[c] From reference 48.
[d] From Jelinski, L. W. *Macromolecules* **1981**, *14*, 1341.
[e] From Harbison, G.; Herzfeld, J.; Griffin, R. *J. Am. Chem. Soc.* **1981**, *103*, 4752.
[f] From Harbison, G. S.; Jelinski, L. W.; Stark, R. E.; Torchia, D. A.; Herzfeld, J.; Griffin, R. G. *J. Magn. Reson.* **1984**, *60*, 79.
[g] Komoroski, R. A., unpublished results, 1984.

Magic Angle Spinning.[12] For a molecule with more than one or a few different types of carbons, the CSA patterns, particularly for unsaturated carbons, will overlap extensively, preventing their individual definition. Figure 2-3A shows the ^{13}C spectrum of poly(ethylene terephthalate) without magic angle spinning and its decomposition into overlapping, individual CSA patterns.[17] It is clear that direct observation of ^{13}C CSAs is difficult or impossible in all but the simplest molecules. Ideally, one would like a spectrum with the simplicity of the solution spectrum but with the CSA information available. This goal has been realized recently by implementation of certain two-dimensional techniques for solids (see Peak Assignments and Two-Dimensional NMR). However, a more limited and easily attained goal is to have the same situation as in solution, observing just the average chemical shifts in a single spectrum, while sacrificing the information concerning the anisotropy.

Inspection of the chemical-shift Hamiltonian shows that there exists only the option of averaging a spatial term to zero. This is done by physical spinning of the sample at the proper rate and angle.[12] When the rigid array of

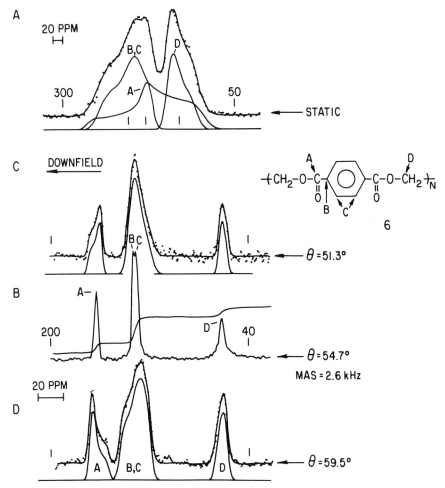

Figure 2-3. Carbon-13 NMR spectra of solid poly(ethylene terephthalate). (A) static, (B) at the magic angle, (C) and (D) off the magic angle as indicated. (Reprinted with permission from reference 17.)

nuclei is rotated about an axis making an angle α with H_0 and angle β_i with the ith principal axis of the chemical-shift tensor, the direction cosine becomes:

$$l_i = \cos \alpha \cos \beta_i + \sin \alpha \sin \beta_i \cos (\omega_r t + \varphi_i) \qquad (2\text{-}11)$$

where ω_r is the rate of rotation and φ_i is the azimuthal angle (Figure 2-4A). The chemical shift Hamiltonian can then be divided into time-independent and time-dependent parts:

$$\mathscr{H}_\sigma^{avg} = \hbar H_0 \gamma I_z \left[\tfrac{1}{2} \sin^2 \alpha \, \text{tr}\sigma + \tfrac{1}{2}(3 \cos^2 \alpha - 1) \sum_i \sigma_i \cos^2 \beta_i \right] \qquad (2\text{-}12)$$

$$\mathscr{H}_\sigma(t) = \hbar H_0 \gamma I_z \left[\tfrac{1}{2} \sin 2\alpha \sum_i \sigma_i \sin 2\beta_i \cos (\omega_r t + \varphi_i) \right.$$
$$\left. + \tfrac{1}{2} \sin^2 \alpha \sum_i \sigma_i \sin^2 \beta_i \cos 2(\omega_r t + \varphi_i) \right] \qquad (2\text{-}13)$$

The time-dependent portion is periodic in ω_r and gives rise to side bands at multiples of ω_r. At spinning rates greater than about half the width of the CSA pattern, the intensities of these side bands go to zero, leaving only the time-independent portion. Each component of the shielding tensor, regardless of initial orientation, is forced to undergo an oscillatory motion (as illustrated in Figure 36 of reference 18). The average of each of these motions is the isotropic shift when α equals the magic angle 54.74°, for which $\cos^2 \alpha = 1/3$, $\sin^2 \alpha = 2/3$. The magic angle admits of a simple physical picture because it is just the angle that the diagonal of a cube makes with a coincident edge (Figure 2-4B), ie, the [1,1,1] direction. Clearly this direction does not favor orientation with respect to any of the three coordinate axes.

Even at exactly the magic angle, the linewidth of the remaining peak usually does not reduce to a value typical of the solution state, because residual interactions remain that are not eliminated by MAS or by dipolar decoupling. For ^{13}C NMR of solids at typical magnetic fields, values of ω_r of 1–6 kHz are necessary to collapse the CSA pattern into a single line. Because nuclear shielding is a field-dependent phenomenon (Eq. 2-7), the width of the CSA pattern in frequency units increases linearly with the static field. Higher magnetic fields therefore require higher MAS speeds, a technological problem that is touched on later in this section.

Any solid-state interaction that displays the $3 \cos^2 \alpha - 1$ dependence can be removed by MAS, in principle.[12] The speed requirements are different for homogeneous and inhomogeneous interactions, however. In the homogeneous case, as for H–H dipolar interactions in a solid, ω_r must exceed the full strength of the interaction. This limits the practical utility of MAS in such cases. For inhomogeneous interactions, such as the CSA, MAS at less than the full width of the interaction yields a relatively narrow center band and a spectrum of side bands.[19] Thus a high-resolution spectrum cluttered with side bands is obtained. Weaker dipolar interactions of spin-1/2 nuclei, on the order

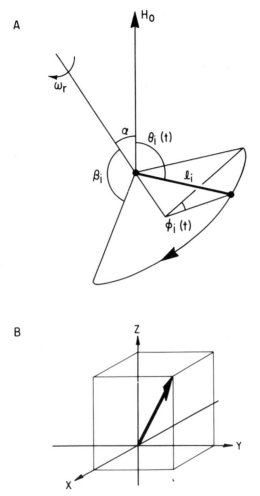

Figure 2-4. (A) Angles used in the analysis of the effect of sample spinning on the chemical-shift anisotropy as in the text; (B) the diagonal of a cube ([1,1,1] direction) is at the magic angle relative to the coincident edges.

of 5 kHz or less, can be eliminated totally by MAS. These include the natural abundance C–C dipolar interaction, as well as long-range C–H interactions.

An exception to the above is the dipolar coupling of a spin-1/2 nucleus to a quadrupolar nucleus. In the analysis of the effect of MAS on the dipolar inter-action, it is assumed that the spins are quantized along the direction of the static magnetic field, as is usually the case. For quadrupolar nuclei this is not so. The spins actually are quantized along an effective field resulting from the main field and the quadrupole interaction.[20] For nuclei with substantial quad-rupole couplings, this effective field can be very different from the main field. When the presence of both interactions is taken into account, terms occur in the Hamiltonian that do not display the $3 \cos^2 \alpha - 1$ dependence. The result is that MAS fails to remove completely the dipolar coupling of the spin-1/2 nucleus to the quadrupolar nucleus.

Figure 2-5[21] shows the example of crystalline glycine for carbon bonded to ^{14}N ($I = 1$), a common occurrence. A single, narrow resonance is seen for the α-carbon in the ^{15}N-enriched sample, indicating that MAS has removed the relatively weak ^{15}N–^{13}C dipolar coupling. A more complicated pattern is seen for the natural abundance ^{14}N sample. This effect has been investigated in some detail and the theory described.[20,22] This behavior also has been seen for coupling to $^{35,37}Cl$ in PVC[23] and Kel-F.[24] For example, in PVC at 22.6 MHz the CHCl resonance was broader than the CH_2 resonance, particularly for a highly crystalline PVC.[23] Although the possibility exists for obtaining useful information from these residual couplings,[22] for such systems as chlorine-, bromine-, and nitrogen-containing polymers they probably constitute another unwanted line-broadening interaction. Higher magnetic fields reduce this effect because the Zeeman component of the effective field increases while the quadrupolar component remains constant. However, high magnetic fields may not be desirable because of the higher MAS speeds necessary to average the CSA.

Broadening from variations of bulk magnetic susceptibility within the sample can be eliminated by MAS insofar as it exhibits the $3 \cos^2 \alpha - 1$

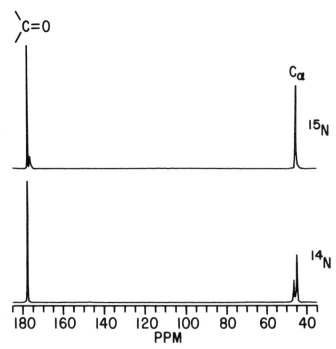

Figure 2-5. Carbon-13 MAS spectrum of solid glycine, showing the effect of residual dipolar coupling to the quadrupolar nucleus ^{14}N. (Reprinted with permission from Hexem, J. G.; Frey, M. H.; Opella, S. *J. Am. Chem. Soc.* **1981**, *103*, 224. Copyright 1981 American Chemical Society.)

dependence.[25] Broadening by susceptibility contributions can be an important factor for filled elastomers and for the amorphous regions of semicrystalline polymers above T_g.

From a practical point of view, rapid sample rotation at the magic angle often has been the most difficult aspect of obtaining high-resolution, ^{13}C NMR spectra of solid polymers. Several types of MAS deficiencies can affect spectral quality dramatically. If the rate of spinning is less than about half the width of the CSA, side bands are seen at multiples of ω_r, as mentioned above. The number, intensity, and position of these side bands strongly depend on the CSA and ω_r. From the intensity distribution of the side bands it is possible to trace out the full CSA pattern.[26] Figure 2-6 shows the high-resolution ^{13}C spectrum of solid poly(methyl methacrylate) at a number of different MAS speeds.[27] Obviously, the presence of side bands complicates both qualitative and quantitative interpretation of ^{13}C CPMAS spectra. Techniques have been developed for removing spinning side bands from slow-spinning MAS spectra.[28] Although these techniques have not yet been developed to the point of routine use, progress is being made with regard to their effectiveness.

Spinning side bands are primarily a problem for unsaturated carbons, which have large anisotropies, generally 150–200 ppm (Table 2-2). The currently available maximum spinning rate of about 6 kHz can cover the full sp^2 carbon anisotropy at an observation frequency of about 37 MHz for ^{13}C, far from the maximum of 125 MHz available today. For aliphatic carbons ($\Delta\sigma \lesssim 50$ ppm), side bands can be avoided at observation frequencies of about 100 MHz or lower, which includes the overwhelming majority of all high-resolution, ^{13}C instruments available today. Because many polymers studied so far contain aromatic or carbonyl carbons, some workers have preferred to operate at the lower magnetic fields characteristic of electromagnet-based spectrometers. The widespread use of MAS techniques for ^{13}C at very high magnetic fields will probably depend on the future success of quantitative side-band-elimination techniques.[28] For other nuclei, the problem of side bands may be more or less severe than for ^{13}C, depending on the nucleus and its electronic environment (Table 2-2).

Beyond the question of spinning rate is the question of the deviation of the spinning angle from the magic angle, and angle stability. Setting the spinning angle sufficiently close to $54.74°$ is an important factor in spectral resolution, because a gross missetting can broaden the lines severely. A missetting of the angle results in a reduced powder pattern, the scaling factor being $1/2(3\cos^2\alpha - 1)$.[29] Therefore, the angle setting is particularly critical for nuclei with large CSAs. In fact, deliberate off-magic-angle spinning can be used to determine the CSA patterns of multiple-line spectra.[17,27] Figures 2-3B to D show the effect of off-axis spinning for poly(ethylene terephthalate). Note the inversion of the CSA pattern when traversing the magic angle.

A number of techniques are available for accurately setting the magic angle.[17,30,31] The angle stability from sample to sample of the common double-bearing rotor geometries is good enough that the angle only

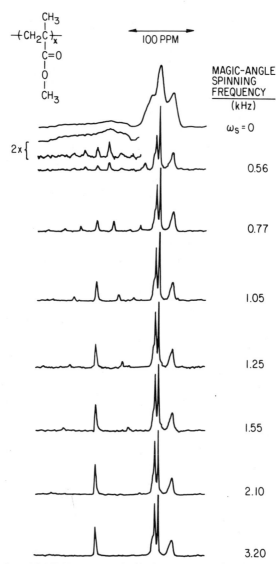

Figure 2-6. Carbon-13 NMR cross-polarization spectra of poly(methyl methacrylate) at 15.1 MHz as a function of spinning frequency. (Reprinted with permission from reference 27.)

occasionally needs adjustment. This is not true for the Andrew-Beams geometry, where frequent resetting of the angle is often required. This can be performed easily and directly on most analytical samples using the KBr method.[31]

Given the sensitivity of the MAS spectrum to the spinning angle, stability of that angle during spectral accumulation is critical. Single-bearing MAS rotors can undergo other, slower motions, such as nutation or precession, superimposed on the high-speed spinning. Because these additional motions result in other angles being visited, peak broadening or low-frequency side bands can be observed.[15] The cure for this problem is the use of well-balanced rotors.

Scalar Coupling

An interaction of considerable importance for high-resolution NMR studies in solution is the spin–spin or scalar coupling. Multiplet splittings caused by scalar coupling provide information on the number and identity of nearest-neighbor atoms in organic molecules and polymers[1] and are a critical feature of ^1H NMR spectra. Although for ^{13}C the C–H scalar couplings are removed by broadband decoupling to provide spectral simplicity, the lost information can be recovered by additional experiments.

Scalar couplings are of little value for high-resolution NMR studies of solid polymers. Proton–proton couplings have not been resolved in high-resolution, solid-state proton spectra (Chapter 9). Carbon–hydrogen scalar couplings are removed, along with dipolar couplings, from solid-state spectra by proton decoupling. As for liquids, one-bond C–H couplings are recoverable by additional experiments. Occasionally the scalar coupling to other spin-1/2 nuclei, such as ^{31}P, may be resolvable in the solid-state spectrum if the magnitude of J is sufficiently large (about 15 Hz or greater).

Signal Generation and Sensitivity

Relaxation Times

As in conventional FT NMR of liquids, spin relaxation times are extremely important for obtaining high-resolution NMR spectra of solid polymers. However, for solids there are several additional important relaxation times. In this section the various relaxation times for high-resolution ^{13}C NMR of solid polymers are described briefly. Typical values for various types of polymers are given in Chapters 3–5.

The most fundamental relaxation times, and the ones that are important for both liquid-state NMR and high-resolution NMR of solid polymers at any temperature, are the spin–lattice (T_1) and spin–spin (T_2) relaxation times. These time constants govern the return of the z and the x,y components, respectively, of the magnetization or NMR signal back to their equilibrium values. When the magnetization is perturbed from along the direction of the static field, the signal grows back to its equilibrium value along this direction with time constant T_1. In this process the spin system gives its excess energy to its surroundings, historically called the "lattice" because of origins in solid-state physics. The nonradiative processes of spin relaxation occur via the modulation of local magnetic fields by molecular motion of the proper frequency spectrum. For the ^{13}C nucleus in polymers, the local fields arise from the proton magnetic moments. Spin–lattice relaxation is most efficient when the correlation frequencies of these motions are near the Larmor frequency, typically 5–500 MHz depending on nucleus and magnetic field.

When an x,y component of the magnetization is created, as by an rf pulse, this component decays to zero with time constant T_2. There is no energy associated with this component; it is purely an entropy relaxation. This decay occurs because all the spins do not precess at exactly the same rate and tend to get out of phase. The linewidth $v_{1/2}$ is related to T_2 by:

$$v_{1/2} = 1/(\pi T_2) \qquad (2\text{-}14)$$

for a Lorentzian line. Molecular motion at the Larmor frequency affects T_2; low-frequency processes, on the order of 100–1000 Hz, can also affect T_2. Inhomogeneity in the static field affects the apparent T_2, because it causes the spins to see different fields and hence get out of phase more rapidly. However, for high-resolution studies of solids this last factor usually is negligible in a high-resolution magnet. For nonviscous liquids at typical Larmor frequencies, $T_1 = T_2$ (see Chapter 4). For high molecular weight polymers in solution and for solids, $T_1 \gg T_2$.

Each isotope or molecular environment has its own T_1 and T_2. These relaxation times are different for different nuclei because of differences in Larmor frequencies, motions, and available relaxation mechanisms. Because different sets of nuclei can be coupled by various interactions, their relaxation times may not be independent. For isotopically abundant, nonisolated nuclei (^{1}H and ^{19}F) in solids, the relaxation times of different chemical types often are averaged by spin diffusion, or mutual spin flips among strongly coupled nuclei. More is said about the important process of spin diffusion later in this section and in Chapter 5. Techniques for measuring T_1 and T_2 have been covered in detail elsewhere[4,5] and are not repeated here.

Many important motional processes in solids, particularly for polymers, have characteristic frequencies in the range of tens of kHz, a range not effective in producing T_1 relaxation under normal circumstances. Because the frequencies effective for T_1 relaxation are determined by the size of the static magnetic field, one alternative is to perform the T_1 measurement in a very small static field, corresponding to a Larmor frequency in the tens of kHz range. Because NMR sensitivity approximately goes as the three-halves power of the static field, this approach is not a practical one. The problem is circumvented by performing the T_1 measurement in the rotating frame.[32] The procedure used is shown in Figure 2-7. A 90° rf pulse in the y direction brings the magnetization along the x direction. Because the pulse is on or close to resonance, in a reference frame rotating at the Larmor frequency the rf field and the magnetization each point in a fixed direction. The phase of the rf pulse is then shifted by 90°, ie, it is moved from the y to the x direction and is now aligned with the magnetization. This procedure is called "spin locking." In the rotating frame the magnetization behaves as if it is only in a field H_1 with a

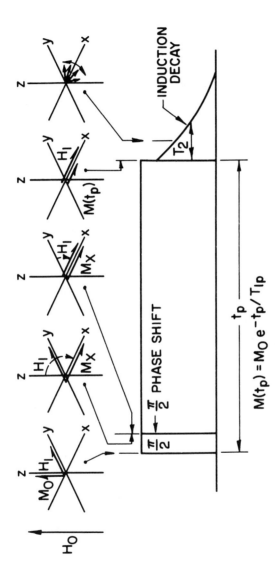

Figure 2-7. Pulse sequence used for spin locking and measurement of $T_{1\rho}$, the spin–lattice relaxation time in the rotating frame. (A) Initial magnetization and direction of exciting field. (B) Tipping of M_0 to M_x by 90° rf pulse (H_1). (C) Phase shift of 90° to lock M_x along H_1. (D) Decay of spin-locked signal after t_p. (E) Removal of spin-locking field and subsequent FID. (Reprinted with permission from reference 32.)

strength in the tens of kHz range. Of course, the size of the magnetization, and hence the sensitivity, was determined by the much larger static field. The decay constant for the magnetization in this lower rf field is called $T_{1\rho}$, the spin-lattice relaxation time in the rotating frame. It is determined as in Figure 2-7, by varying the length of the phase-shifted rf pulse.

The time $T_{1\rho}$ has been used extensively to study solid polymers because of its sensitivity to motions in the tens of kHz range[7,32] (see Chapters 3 and 7). Its quantitative interpretation in terms of motion can be complicated by the contribution of non-spin–lattice processes to the spin lock decay[14] (see Chapter 3). These additional contributions depend on the existence of another relaxation time, the dipolar relaxation time, T_{1D}, of the protons. To understand T_{1D} it is necessary to consider the concept of dipolar order.[4] The occupation of spin energy levels in the static magnetic field according to the Boltzmann distribution creates a state of so-called Zeeman order. This is nothing more than the ordinary magnetization. If a sample in a state of Zeeman order and with a long T_1 is removed slowly from the magnetic field, the Zeeman order disappears. A state of dipolar order is created in which the spins are aligned according to the local dipolar fields. If the sample is slowly returned to the magnetic field (total time $\ll T_1$), almost all of the original signal reappears. The initial ordering in the main field was stored in the dipolar field—in each small region of the sample the signal became aligned with the local dipolar field. Because the local fields are oriented randomly, there is no total signal. These local fields are much smaller than the static field, and hence the signal aligned along the local fields decays to its equilibrium value in that field. The decay time for this process is T_{1D}. Because the characteristic frequencies of these local fields are similar to those in the $T_{1\rho}$ experiment, they can compete with those of spin–lattice processes in the spin-lock decay when they are coupled into the spin-locked system (ie, the ^{13}C nuclei) (see Chapter 3).

The general behavior of the various relaxation times with temperature is shown schematically in Figure 2.21 of reference 3 (see also Figure 4-2, this volume). This figure shows the relative ordering of the various relaxation times with frequency, lower frequencies shifting the minimum (where relaxation is most efficient) to lower temperatures. More is said in Chapters 3 and 4 concerning the explicit functional form of these curves.

In addition to the problem of obtaining narrow lines for solid polymers, the problem of adequate sensitivity must be considered. In conventional FT NMR spectroscopy of liquids, the spectrum is obtained as the free-induction decay (FID) following a 90° rf pulse.[5] Usually it is necessary to acquire a number of such decays and add them coherently in a computer to improve the signal-to-noise ratio. The rate at which the rf pulses can be applied is limited by the rate at which the carbon magnetization returns to its equilibrium value, ie, by the spin–lattice relaxation time T_1^C. For polymers in solution, typical T_1^C are several hundred milliseconds for protonated carbons and several seconds for nonprotonated carbons.[33] For bulk polymers well above T_g, T_1^C can be even shorter, approaching the theoretical minimum (for a C–H dipolar mechanism) of tens of milliseconds at ^{13}C Larmor frequencies of 15–25 MHz.[34] In glassy and crystalline polymers, T_1^C can range from about 5 to 20 s for glassy polymers and up to 4000 s for the crystalline regions of some polymers (see Chapter 5). This is because the high-frequency motions necessary for efficient spin–lattice relaxation usually have very low spectral densities in rigid solids. The long T_1^C encountered for rigid bulk polymers make acquisition of the standard FID a very time-consuming process relative to solution-state ^{13}C NMR. The situation is compounded by the broader lines encountered in high-resolution spectra of solid polymers. It is offset only slightly by the fact that the sample concentration is higher in the bulk solid than in solution.

On the other hand, the proton magnetization in a solid polymer returns to equilibrium much faster than the carbon magnetization.[18] Because $T_1^H \gg T_2^H$, the proton spin system equilibrates internally much faster than it does with the lattice. Nonequilibrium magnetization in any part of the proton spin system is transferred to surrounding protons in times on the order of 100 μs by spin diffusion.

Spin diffusion is not an actual molecular diffusion but is the transport of spin energy within the spin system by mutual, energy-conserving spin flips. Consider two adjacent protons in a solid coupled by the dipolar interaction. Even if chemically different, the protons are at essentially the same resonance frequency because of their large linewidths. If the two have antiparallel magnetic moments, it is an energetically favorable process for both of them to change orientations simultaneously or "flip," again yielding an antiparallel pair but with moments reversed. Such flips can occur rapidly among neighboring, coupled protons in a solid and serve to distribute excess energy or magnetization among all the coupled spins. The process, like any diffusion process, is driven by a concentration gradient, in this case the spatial gradient in the magnetization. It is spin diffusion that insures that the ^{1}H linewidth is homogeneous.

Because of spin diffusion, the whole proton spin system is coupled to the lattice via the most efficient portions of the system. Energy transfer to the lattice is very efficient near paramagnetic impurities, lattice defects, and molecular segments in rapid motion, such as methyl side chains or possibly end groups. Even low levels of these relaxation sites can have a large effect on T_1^H. The presence of spin diffusion and such strongly relaxing sites has complicated interpretation of proton spin-relaxation behavior in standard, low-resolution pulsed NMR experiments.

Cross Polarization

The basic idea behind cross polarization (CP)[35,36] is the generation of the ^{13}C NMR signal by using the energy and relaxation properties of the protons. To understand cross polarization, it is necessary to introduce the concepts of an isolated spin system and its spin temperature.[10] Consider a simple organic solid whose only spin-possessing nuclei are ^{1}H and ^{13}C. In a magnetic field H_0 the ^{1}H spins populate their set of Zeeman levels (Figure 2-8A, left), separated by $\Delta E = \gamma \hbar H_0$, according to the Boltzmann distribution:

$$n_u/n_l = \exp\left(-\Delta E/kT_s\right) \qquad (2\text{-}15)$$

where T_s is the temperature of the spin system and n_u and n_l are the populations of the upper and lower states, respectively. Equation (2-15) says that a larger polarization (greater population difference) corresponds to a lower spin temperature. Via spin diffusion the ^{1}H system equilibrates at a single T_s in a very short time (< 100 μs) and can be considered an isolated spin reservoir (Figure 2-9). After a much longer time the ^{1}H system equilibrates with the surroundings (the lattice), and T_s equals T_1, the lattice temperature. The ^{1}H magnetization is given by the Curie law:

$$M_{0H} = C_H H_0/T_1 \qquad (2\text{-}16)$$

where

$$C_H = N_H \gamma_H^2 \hbar^2/4k \qquad (2\text{-}17)$$

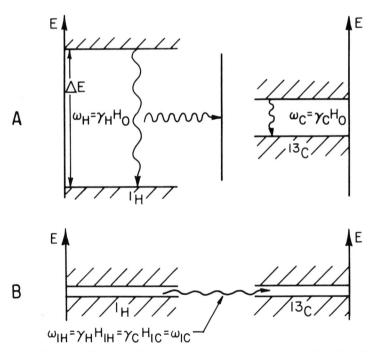

Figure 2-8. Schematic diagram of the cross-polarization process. (A) In the laboratory frame, the ^{1}H levels have four times the separation of the ^{13}C levels, making energy transfer from H to C an unfavorable process. (B) When both spin systems are in frames rotating at the same rate, the energy levels are matched (the Hartmann-Hahn condition is satisfied) and energy transfer is allowed. The level separation in B is not drawn to the same scale as in A.

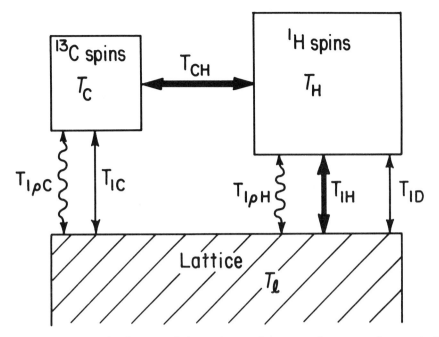

Figure 2-9. Schematic diagram of the carbon and proton spin systems in a typical organic solid. The italic Ts indicate spin temperatures. The Ts indicate various relaxation processes that may occur among the carbon, proton, and lattice systems.

The ^{13}C spins also equilibrate among their levels, which have only about 1/4 the separation of the proton levels at the same temperature because $\gamma_H \cong 4\gamma_C$ (Figure 2-8A, right). The equilibration times for the ^{13}C spin reservoir are longer than for ^1H because of the larger internuclear distances and different relaxation processes involved. The fact that the separation of the proton levels is greatly different from that of the ^{13}C levels (the nuclei have greatly different resonance frequencies) means that these two sets of nuclei cannot undergo mutual, energy conserving spin flips efficiently and hence are isolated from each other (Figure 2-8A). This allows them to have different relaxation times.

The proton reservoir is taken to a low T_s by one of several techniques. The most common one is the spin-locking technique, shown in Figure 2-7, which is part of the process used for measuring $T_{1\rho}$. By consideration of Eqs. (2-15) and (2-16), it can be shown that in the rotating frame the protons now are characterized by a very low spin temperature, given by:

$$T_R = (H_{1H}/H_0)T_1 \tag{2-18}$$

This was brought about not by an actual lowering of the temperature, but by a lowering of the effective field seen by the protons.

The cross-polarization process occurs when the smaller, hotter ^{13}C reservoir is brought into contact with the larger, cooler ^1H reservoir. One way this is accomplished is by irradiating the carbons with an rf field H_{1C} of a strength such that the so-called Hartmann-Hahn condition[35] is satisfied:

$$\gamma_H H_{1H} = \gamma_C H_{1C} \qquad (2\text{-}19)$$

When this condition is met the original proton and carbon levels, which did not match, are brought to lower, matching levels (Figure 2-8B). The energy-conserving spin flips can now occur between the carbon and proton spins instead of just within the individual reservoirs. In NMR jargon, the rotating frames of the two spin systems are made to rotate at the same rate.

The larger, cooler proton reservoir cools the carbon reservoir; that is, a carbon signal is created along H_{1C}. Assuming ideal conditions, the carbon signal created by cross polarization (CP) is four times the original equilibrium carbon magnetization.[36] After a suitably long contact, H_{1C} is turned off and the carbon signal is detected and stored in the computer as in conventional FT NMR. For high-resolution studies dipolar decoupling is performed during signal acquisition to remove the C–H dipolar interaction. Magic angle spinning can be performed if desired. Moreover, depending on the size, spin temperature, and decay rate of the spin-locked proton signal, the carbons can be brought into contact one or more additional times, the carbon signal acquired and coadded to previously acquired signal. Figure 2-10A shows the cross-polarization sequence utilizing proton spin locking.

The efficiency of the CP process is determined by the relative values of several NMR relaxation times and the CP time, t_c. Figure 2-10A shows schematically the buildup of the carbon magnetization. The CP rate under spin-lock conditions ($1/T_{CH}$) is determined by the effective strength of the dipolar interaction (as determined by molecular motion and interatomic C–H distances) and the degree to which the Hartmann-Hahn condition is fulfilled. It also depends on the initial polarization of the proton reservoir. For protonated carbons in rigid solids, most of the signal buildup occurs over about the first 100 μs. For nonprotonated carbons most of the signal appears in about the first millisecond. Considering the protons as an infinite reservoir, the ^{13}C signal grows as in Eq. (2-20)[14]:

$$M(t) = M_0 \lambda^{-1}[1 - \exp(-\lambda t/T_{CH})] \exp(-t/T_{1\rho}^H) \qquad (2\text{-}20)$$

where

$$\lambda \equiv 1 + T_{CH}/T_{1\rho}^{C'} - T_{CH}/T_{1\rho}^H \qquad (2\text{-}21)$$

Here $T_{1\rho}^{C'}$ is the carbon rotating-frame relaxation time with proton decoupling.

Once signal buildup proceeds, the decay of the proton magnetization in the spin-locked state, $T_{1\rho}^H$, tends to deplete the signal. If the $T_{1\rho}^H$ is very short, as can occur at a minimum in the $T_{1\rho}^H$–temperature curve, the cross-polarization process can be short circuited, preventing buildup of a substantial carbon magnetization. When $T_{1\rho}^H > t_c$, T_{CH}, the C–H cross-polarization experiment can be used to measure the $T_{1\rho}^H$ of local protons by following the decay of the signals of the carbons of interest as a function of t_c (see Chapter 3). The time

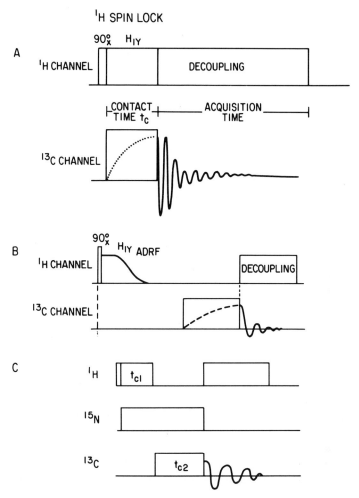

Figure 2-10. (A) A C–H cross-polarization sequence utilizing proton spin locking. Carbon-13 magnetization buildup during the contact time t_c is illustrated schematically. The ^{13}C signal is acquired after turning off the matching carbon rf field, but with high-power proton decoupling; (B) ADRF cross-polarization sequence; (C) simple double-cross-polarization sequence.

T_{CH} can be measured by observing the growth of the carbon signal for short t_c, and fitting the data to Eq. (2-20).

Several assumptions were made in the above analysis of the cross-polarization process. First, the rf fields H_{1H} and H_{1C} must be larger than their respective linewidths. Also, $T_1^C > T_1^H \geq T_{1\rho}^H > t_c > T_{CH}$. When these conditions are not met, then cross polarization is less than optimum.[37] An in-depth theoretical treatment of cross-polarization dynamics has been given elsewhere.[6]

In addition to a larger single-shot carbon magnetization, the repetition rate of the process is limited by the proton T_1, which is usually less than 1 s for polymers, and not by the carbon T_1 of 10–4000 s. The net result is a substantial increase in signal to noise relative to acquisition of the FID by repetition of the standard one-pulse experiment. This is true only for those cases where cross polarization is an efficient process, ie, where there is a sizable static dipolar interaction. Cross polarization can occur in more mobile systems, such as elastomers,[38] and even in liquids via the J coupling,[39] but it is not particularly efficient in these cases. More is said in Chapter 4 concerning the relative efficiency of CP and one-pulse FT NMR for rubbery polymers.

The T_1^H of some crystalline solids can be very long, necessitating long pulse repetition times.[40] In these cases it may be that acquisition of the standard ^{13}C FID with dipolar decoupling is the best way to obtain the spectrum. This usually is not the case for polymers, either glassy or crystalline. It can be the case for pure, crystalline, low molecular weight, polymer models.[40] From a practical point of view, long T_1^H sometimes can be shorted by deliberate introduction of paramagnetic impurities (as by grinding the sample with $CuSO_4$).[41]

Quantitative analysis of NMR spectra requires that a resonance intensity be proportional to the number of nuclei contributing to that resonance. Because CP rates do vary depending on the number of attached or nearby protons as well as on molecular motion, it is necessary that all carbons in a sample be given sufficient time to equilibrate with the proton reservoir during CP contact. In this regard, the contact time t_c acts somewhat like the pulse repetition time in the one-pulse experiment.[5] Figure 2-11 shows a set of ^{13}C spectra for poly(hydroxybenzoic acid) (PHBA) taken by CPMAS with different contact times.[42] The nonprotonated carbons cross polarize more slowly than the protonated carbons because of the weaker C–H dipolar interaction that they experience. A contact time of 1–3 ms is typical for quantitative studies on rigid polymers.

Figure 2-12 demonstrates the advantage of combining line-narrowing techniques and CP relative to more limited possibilities for a typical glassy polymer, poly(methyl methacrylate) (PMMA). Comparison of C and D illustrate the value of cross polarization.[43]

An additional experiment is described next that is important for relating ^{13}C $T_{1\rho}$ to molecular motion. It was mentioned earlier that nonmotional processes could contribute to the ^{13}C $T_{1\rho}$ decay. The pulse sequence shown in Figure 2-10B provides a measurement of T_{CH}^D, the cross-relaxation time linking the proton dipolar state with the ^{13}C spin-lock state (Figure 3-24, this volume). Using the sequence of Figure 2-10B, a state of proton dipolar order is created by a process called adiabatic demagnetization in the rotating frame (ADRF). As shown, this consists of slowly (compared to the ^1H T_2) removing the proton spin-locking field and is conceptually similar to the process of removing the sample from the field described earlier. This dipolar state relaxes with time constant T_1^D. Cross polarization then occurs from the proton dipolar state to the carbon spin-lock state at a rate $1/T_{CH}^D$.

It is this process that can compete with motional or spin–lattice processes in the $T_{1\rho}$ experiment.[7] It is more efficient when the carbon H_1 field is less than

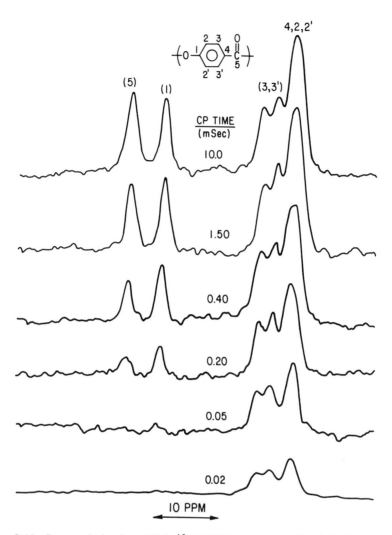

Figure 2-11. Cross-polarization MAS ^{13}C NMR spectrum of poly(hydroxybenzoic acid) as a function of contact time (in ms). All spectra were obtained from 2 K FTs of 1000 FID accumulations at a 3.5-s repetition time. (Reprinted with permission from Fyfe, C. A.; Lyerla, J. R.; Volksen, W.; Yannoni, C. S. *Macromolecules* **1979**, *12*, 757. Copyright 1979 American Chemical Society.)

the local dipolar field from protons. The quantity T_{CH}^D is not to be confused with T_{CH}, the cross-polarization time under proton spin-lock conditions. The Hartmann-Hahn condition is not satisfied for T_{CH}^D, and hence this quantity is much longer than T_{CH} (but not necessarily much longer than $T_{1\rho}^C$). Increasing H_1 makes T_{CH}^D longer relative to the motional contribution to $T_{1\rho}^C$. With the knowledge of T_{CH}^D it is then possible to assess the importance of nonmotional contributions to $T_{1\rho}^C$ as demonstrated in Chapter 3.

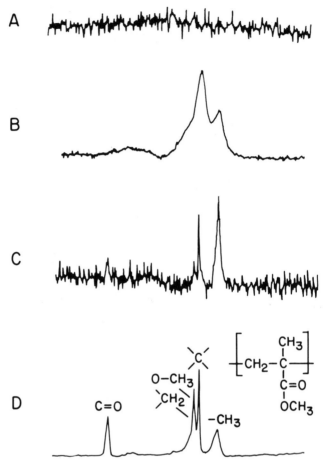

Figure 2-12. Solid-state NMR spectra of poly(methyl methacrylate) obtained under different conditions. (A) Stationary sample, no cross polarization, low-power decoupling; (B) stationary sample, cross polarization, high-power decoupling; (C) magic angle spinning, no cross polarization, high-power decoupling; (D) magic angle spinning, cross polarization, high-power decoupling. (Reprinted from reference 43, p. 194, by courtesy of Marcel Dekker, Inc.)

Line-Narrowing Schemes and Molecular Motion

Backbone and side-chain rotational motions in solid polymers occur at rates spanning about 12 orders of magnitude (from about 0.01 to about 10^{10} Hz). Because of their very high molecular weight and/or the presence of crystallinity, polymers can retain their macroscopic shape even while possessing rapid rotational motion about backbone chemical bonds. As has been seen, the techniques needed to obtain high-resolution ^{13}C spectra of solid organic compounds depend to a large degree on molecular motion in the solid. At this point it is appropriate to examine the effect of molecular motion on the line-narrowing schemes described above. If an interaction in the solid is narrowed partially by molecular motion, is it easier to narrow the line by artificial line-narrowing methods? This topic has been discussed clearly by Waugh.[44]

For a homogeneous line-broadening NMR interaction to be averaged effectively by an artificially imposed coherent motion [as in dipolar decoupling (DD) and MAS], the frequency of the motion v_r must exceed the frequency width arising from the interaction. This width is determined by the strengths of the interactions present as modulated by molecular motion.

There are four cases that can be distinguished for ^{13}C NMR of solid polymers. This discussion is restricted to the C–H dipolar interaction because it is usually the predominant broadening mechanism.

1. A glassy or crystalline solid displaying the full rigid-lattice dipolar line-width. For DD (or MAS) to be effective in this case, the decoupling strength v_r must exceed the strength of the strongest dipolar coupling affecting the line-width. This is the dipolar coupling among the protons, D_{HH}, which is usually larger than the carbon–hydrogen dipolar coupling D_{CH} and, in addition, is a homogeneous interaction.[4] This generally requires v_r to be at least about 10 kHz (Figure 2-13), but more typically about 40 kHz. The backbone carbons of glassy polymers and crystalline regions of all semicrystalline polymers studied to date fall into this category. A typical example is the backbone of poly(methyl methacrylate).[45]

2. The dipolar interaction is partially, but for the most part isotropically, averaged by molecular motion. In this case, the undecoupled signal exhibits a width reduced from the full static width. However, it is not sufficient to average coherently at a frequency (decoupling field strength) corresponding to either the reduced or the static linewidth. Additional averaging must exceed the frequency v_c of the molecular motion responsible for the partial narrowing. Figure 2-13 shows this for two cases of partial narrowing by molecular motion. The motional frequency v_c in principle can range from the static dipolar linewidth (about 4×10^4 Hz) to the motional narrowing region seen for nonviscous liquids (about 10^{10} Hz). Rubbery polymers fit into this category, although the linewidth also may contain a reduced static component. An example is bulk polyisobutylene, with backbone motion on the order of 10^8 Hz at 45°C.[46] The only feasible option for narrowing the width in this case is changing the temperature. Increased molecular motion further narrows the lines at higher temperature, whereas a sufficient lowering of temperature slows molecular motion to the vicinity of the static linewidth, where coherent averaging techniques can work. Polyisobutylene is discussed in more detail in Chapter 4.

What can be regarded as a special instance of this case has been described by VanderHart and co-workers.[15] When the sample has significant molecular motions at the Larmor frequency of the decoupling v_{1H}, these motions compete with the decoupling process, resulting in a line broadening. In this case, either the decoupling power can be increased, although this is feasible only within relatively narrow limits, or the temperature can be changed to shift the spectral density of motions away from the decoupling Larmor frequency. An example of this case is poly(ethylene oxide) at room temperature.[15] This phenomenon is described in more detail in Chapters 3 and 5.

3. Very rapid and anisotropic motion. Here the resonance is narrowed to a reduced value by a rapid, highly anisotropic rotation. Because the motion is extremely rapid, its spectral density at v_r is very small. Hence the reduced width is described by an effectively static interaction, which is susceptible to further averaging by a coherent process. In this case the characteristic frequencies of the molecular motion are weak at the frequency of the coherent motion. Examples of this situation are the amorphous regions of polyethylene[47] (see Chapter 5) and filled rubber (see Chapter 4). Figure 2-14 shows the distinction between this case and case 2.

4. A superimposed, zero-frequency interaction not related to polymer motion. This is broadening by magnetic susceptibility differences in the sample and is included here because it is affected by MAS. It is more likely to occur

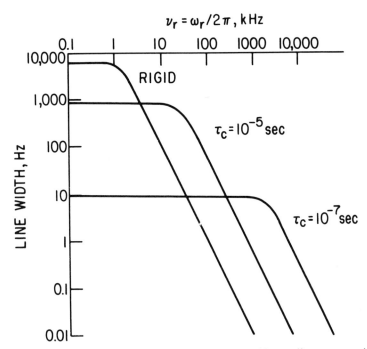

Figure 2-13. Effect of a coherent averaging process (decoupling or magic angle spinning) on the linewidth (homogeneous interaction) for three different motional situations in a solid. The quantity v_r is the strength of the decoupling field or the spinning rate, and τ_c is the correlation time of the motional process. (Reprinted from reference 44, p. 207, by courtesy of Marcel Dekker, Inc.)

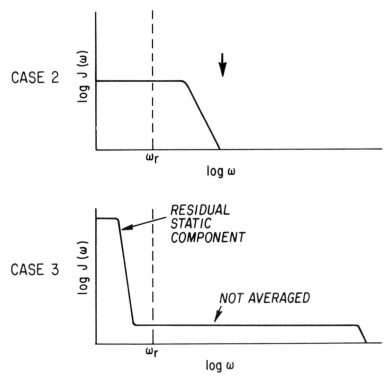

Figure 2-14. Plots of spectral density versus frequency for the two motional processes described in the text. In case 2, the coherent averaging frequency ω_r is not high enough to supercede the effect of the motional process. In case 3, two components are present. One is a residual static component, which can be averaged by the coherent process. (Reprinted from reference 44, p. 208, by courtesy of Marcel Dekker, Inc.)

for semicrystalline or filled polymers. In any event, it is never more than a few parts per million in magnitude. Portions of the broadening that exhibit a $3 \cos^2 \alpha - 1$ dependence can be eliminated by MAS at a rate exceeding the broadening.[25] To the extent that the susceptibility broadening contains anisotropic contributions, it cannot be spun away by MAS. An example of this case is the amorphous portion of trans-polybutadiene.[48]

Peak Assignments and Two-Dimensional NMR

Peak Assignments

General Methods. Peak assignment in the ^{13}C NMR spectra of solid organic compounds is in many respects the same as for solution-state spectra. The usual options of comparison to model compounds, empirical chemical-shift calculations, and isotopic enrichment are still available. It is usually wise to compare the solid-state spectrum to that in solution to ascertain, if possible, those factors responsible for the inevitable loss in spectral resolution and to assess the extent of peak overlap. For example, if a resonance is broadened by tacticity effects, this becomes apparent from the solution spectrum. Also, if multiple resonances are seen in the solid-state spectrum because of frozen conformations or the presence of both crystalline and amorphous phases, the solution spectrum does not show these effects.

Spin–lattice relaxation times have been used in selected cases to assign carbon resonances in solution.[49] For the most part this method has been supplanted by the more reliable and rapid multipulse-decoupling methods. In the glassy or crystalline solid state, spin relaxation times are seldom useful for assignment of carbon multiplicities because they do not depend directly on the number of attached protons[18] (see Chapter 3). Relaxation times are useful for distinguishing resonances from different phases in mixed-phase systems (see Chapter 5).

Another general approach for making ^{13}C peak assignments is isotopic substitution. In the solid state, specific ^{13}C labeling sometimes may be necessary to make peak assignments. Specific ^{13}C enrichment is more likely to become a method for observing a single carbon site free of other peaks that interfere in natural abundance. Examples of this are described in Chapters 3 and 7. Such overlap is much more a feature of solid-state than solution-state spectra. Deuterium substitution also can be of value because, when protons are replaced by deuterons, the dipolar coupling to a directly bonded ^{13}C atom is reduced by a factor of about 7. Hence deuterated carbons behave like quaternary carbons in the CP experiment and in the multipulse experiments described next.[50]

Multipulse Experiments. For insoluble samples or when it is desired to assign the solid-state spectrum without recourse to the solution spectrum, methods comparable to those for solution-state spectra have been developed. A simple technique called "delayed decoupling" or "dipolar dephasing" yields a spectrum from only nonprotonated carbons.[51] In this technique a small delay is inserted between the time when the ^{13}C signal is created by cross polarization and the acquisition period. The protons are not decoupled during this delay. Because the dipolar coupling depends on the inverse cube of the C–H distance, the coupling is strong for carbons with directly bonded protons and less so for nonprotonated carbons. The delay is chosen such that signals from protonated carbons totally dephase during this time, whereas those from nonprotonated carbons for the most part remain in phase. Rapidly rotating methyl groups are not suppressed to the same extent as other protonated carbons because of their motionally reduced dipolar interaction. Examples of applications of this technique for mixed-phase systems are given in Chapter 5, and for cured resins in Chapter 3.

Techniques called "J-resolved NMR spectroscopy" have been developed to distinguish carbon types and observe directly bonded C–H couplings in solid-state ^{13}C spectra.[52,53] These techniques rely on using homonuclear decoupling (see Chapter 9) to remove the effect of $\mathscr{H}_{II}$ above, leaving only $\mathscr{H}_{IS}$. One version[52] yields a spectrum similar to that of the Attached Proton Test (APT) technique used for liquids,[54] where CH and CH_3 carbon peaks are 180° out of phase relative to CH_2 and C_q carbon peaks. Figure 2-15 shows an example. Another version[53] yields the fully coupled ^{13}C spectrum. The experiment also can be performed in two dimensions.[55] Because of certain, relatively stringent experimental requirements, these methods have not yet seen much application.

Double Cross Polarization. Schaefer, McKay, and Stejskal[56] have developed an elegant method for assigning peaks in solid-state NMR spectra. The method consists of the sequential cross polarization from protons to a second

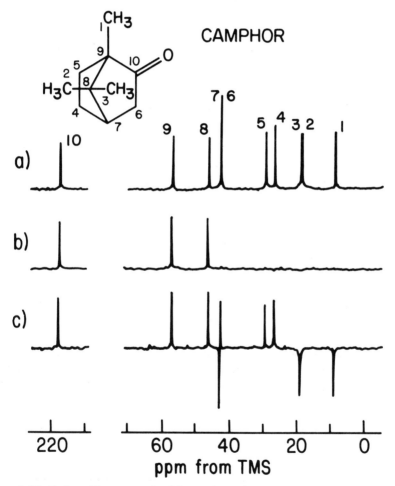

Figure 2-15. Carbon-13 spectra of solid camphor. (A) Normal CPMAS spectrum with assignments shown. (B) Nonprotonated carbons selectively observed. (C) Signals of nonprotonated and methylene carbons are positive, whereas those of methine and methyl carbons are negative. (Reprinted with permission from Terao, T.; Miura, H.; Saika, A. *J. Am. Chem. Soc.* **1982**, *104*, 5228. Copyright 1982 American Chemical Society.)

nucleus, say ^{15}N (isotopically enriched), then from the second to a third nucleus, such as ^{13}C, which is dipolar coupled to the second. The dipolar coupling for the $^{15}N-^{13}C$ case is significant only for directly bonded pairs. The spectrum observed can be that of the second or the third nucleus. The essential features of the double-cross-polarization (DCP) sequence are shown in Figure 2-10C, with observation of the third nucleus. For the example mentioned, this spectrum is only from those ^{13}C nuclei directly bonded to ^{15}N; ^{13}C atoms not bonded to ^{15}N do not contribute. Hence, in addition to being a peak-assignment method, DCP is a method for following the fate of doubly labeled chemical bonds in molecules. If the doubly labeled pairs remain intact, a DCP signal is observed. Broken pairs give no DCP signal. Any two spin-1/2 nuclei can be used, although the $^{13}C-^{15}N$ pair is an important one from the biologic point of view. The method has been used to study the complex products of hydrogen cyanide polymerization[57] as well as plant metabolism.[58]

Two-Dimensional NMR

The idea of two-dimensional Fourier transform NMR (2D NMR) was originated by Jeener in 1971.[59] In the last few years 2D NMR techniques have been used to a rapidly increasing extent, primarily because of computer advances. Although most 2D NMR techniques have been developed for liquids, applications to solids have appeared with increasing frequency. The reader is referred to the recent text for an in-depth treatment of the principles of 2D NMR.[60]

In the normal, 1D FT NMR experiment, an FID is elicited from the spin system by an rf pulse and is sampled as a function of time. This time domain signal is then transformed to give the standard frequency spectrum. The 2D NMR experiment involves the introduction of a second time variable or dimension. The key characteristic of this second time variable is that during this time the spin system is allowed to evolve under the influence of some NMR interaction or state of the experimenter's choosing.

All 2D NMR experiments can be illustrated diagrammatically as in Figure 2-16.[60] During the preparation period, a nonequilibrium spin system is created by rf pulses. This state evolves during the evolution period for a time t_1, which

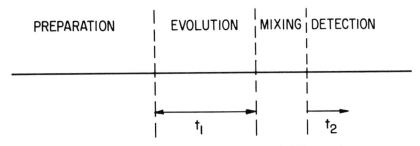

Figure 2-16. Generalized diagram of the 2D NMR experiment.

gives the second time dimension (called first by convention). The mixing period, which is not always used, allows for transfer of spin information among different sets of spins, as required by certain types of experiments. Detection is performed during the final period t_2 as in the 1D experiment.

The experiment is repeated for various t_1 values to fill in the second dimension. This yields a series of FIDs. After Fourier transformation in just the t_2 dimension, a series of spectra with phase or amplitude distortions depending on t_1 is obtained. These distortions carry the additional information relating to the interaction operating during the evolution period. This interaction can be the chemical shift, scalar coupling, dipolar coupling, or other NMR interaction.

The 2D NMR experiments for solutions usually are directed toward unraveling complex spectra, as in the experiment to separate chemical shifts from scalar couplings, or toward determining geometrical or chemical relationships among spins, as in various correlation experiments.[60] The same is true for solids. Some 2D experiments have been devised to recover information concerning the dipolar and anisotropic chemical shift interactions that was sacrificed to produce the high-resolution spectrum. Correlation of resonances associated with different regions of a solid polymer by spin diffusion can be studied by 2D experiments. Sequences have also been designed to observe C–H scalar couplings[55] and to correlate carbon and proton chemical shifts.[61]

Recovery of Dipolar Patterns. As described above, the dipolar interaction of an isolated ^{13}C–^1H pair is known if the internuclear distance is known, as is the case for directly bonded pairs. The reduction in the dipolar interaction by molecular motion at a rate greater than the strength of the interaction provides a measure of the amplitude of the motion. Two-dimensional NMR techniques have been devised for recovering information concerning the dipolar interaction[62,63] and have been applied to polymers by Schaefer and co-workers.[64] The pulse sequence used to obtain the so-called dipolar rotational echo spectra is shown in Figure 2-17 along with a typical result. One key feature of the sequence is the H–H multipulse decoupling, which removes interproton interactions to leave only "isolated" C–H dipolar interactions. Another is that the sequence is synchronized to the magic angle spinning rate. This is necessary to obtain coherent acquisition of the dipolar modulation, which is affected by sample spinning. The dipolar interactions are followed by varying the number of decoupling pulses in the t_1 period.

The bottom of Figure 2-17 shows the dipolar patterns for three carbons of poly(p-isopropylstyrene). These patterns represent slices parallel to the t_1 axis of the 2D spectrum. The side bands occur at multiples of the spinning frequency. The patterns are considerably different, reflecting differences in molecular motion among the three carbons. The use of these patterns to obtain information on molecular motion in glassy polymers is described in more detail in Chapter 3.

Recovery of CSAs. Several 2D NMR techniques have been developed for recovering chemical shift anisotropy (CSA) patterns[65]. Figure 2-18 shows the

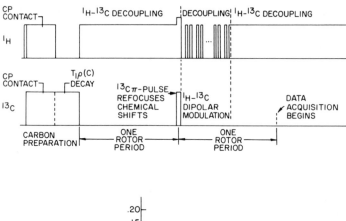

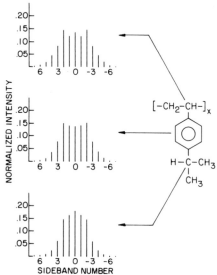

Figure 2-17. Top: Pulse sequence for the dipolar rotational spin-echo ^{13}C NMR experiment. Bottom: Dipolar Pake patterns for three carbons of solid poly(p-isopropylstyrene). Each pattern is broken up into spinning side bands separated by 1.894 kHz. Each spinning side band is represented by a single point in the frequency (F_1) domain. (Adapted from Figures 1 and 3 of reference 64. Reprinted with permission.)

pulse sequence and results of one method, which involves mechanical flipping of the sample spinning axis between the evolution and detection periods.[65] The CSA information is encoded in the spectrum during t_1, when the sample is spun at an angle of 90°. The signal is then "stored" along the z axis by the 90(y) rf pulse until the angle can be changed to 54.7° for high-resolution acquisition. Figure 2-18 (bottom) shows the isotropic spectrum of solid p-dimethoxybenzene and the four cross sections from the 2D spectrum, giving the appropriate CSA patterns. Through this and similar experiments it is possible to realize the full value of solid-state NMR. That is, the anisotropic shift information is obtained without sacrificing the high-resolution character of the experiment.

Spin Diffusion and Domains in Solids. Spin diffusion between individual sites in solids is analogous to chemical exchange in liquids and can be examined by 2D NMR techniques.[66] Because it is usually confined to neighboring molecules, spin diffusion between two species can occur only if they are mixed

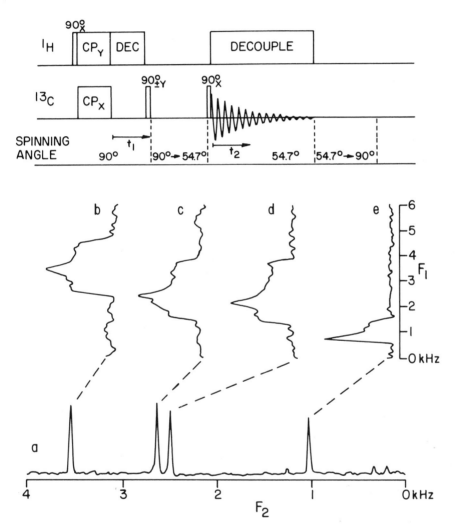

Figure 2-18. Top: 2D NMR experiment for mapping out the CSA pattern in the F_1 dimension. In the present case a sample spinning at 2.5 kHz undergoes the angle change between t_1 and t_2 in 0.5 s. Dec. = decouple. Bottom: Projection of the 2D spectrum of p-dimethoxybenzene with cross sections showing the CSA patterns. Because of the way the experiment was performed, the CSA patterns are half as wide as in the static case. (Adapted from Figures 1 and 2 of reference 65. Reprinted with permission.)

intimately. Figure 2-19 shows the ^{1}H NMR results for two blends of poly-styrene (PS) and poly(vinyl methylether) (PVME).[67] At the top of the figure is the 1D ^{1}H spectrum with peak assignments. In the 2D contour plots, diagonal peaks represent spin exchange among the same type of spin, and hence there is a diagonal peak for each peak in the spectrum. Off-diagonal peaks represent spin diffusion among chemically different types of spins. For the PS–PVME blend precipitated from chloroform (Figure 2-19A), off-diagonal peaks occur for peaks from PS and PVME individually, but not for peaks between the two polymers. For the blend precipitated from toluene, cross peaks are seen between the two polymers, as indicated by the arrows in Figure 2-19B. The above results confirm that although mixed domains do occur in the blend obtained from toluene, they do not occur in the blend obtained from chloro-form.[67] From quantitative measurements of cross-peak intensities and spin-diffusion rates, it may be possible to estimate the fraction of mixed domains and domain sizes. Analogous experiments also can be performed for ^{13}C and other nuclei.[68]

Slow Motions in Solids. It was explained earlier (see Signal Generation and Sensitivity) that the various NMR relaxation times were sensitive to molecular motions in different frequency ranges. However, the spin relaxation times T_1, $T_{1\rho}$, and T_2 are not sensitive to very slow molecular motions in the range of approximately 0.01–10 Hz. These very slow motions can be studied by the spin-alignment technique developed for deuterium NMR (see Chapter 10). Of course, this last technique requires deuterium enrichment, ideally only at a particular chemical site in the molecule.

Veeman and coworkers[69] have proposed a 2D, ^{13}C NMR experiment, similar to that just described for spin diffusion measurements, for studying very slow molecular motions. The pulse sequence is shown in Figure 5-27A (Chapter 5). The ^{13}C signals are created in the x,y plane as usual by cross polarization. During t_1 these signals are allowed to precess, which labels them according to precession frequency; they are then taken to the z-axis by a 90° ^{13}C pulse. At the end of the mixing period (which can be as long as the ^{13}C T_1), a second 90° pulse samples the signal. If any spin packet experienced a change in chemical shift during the mixing time (by either spin diffusion or slow molecular motion), this will produce an off-diagonal peak in the 2D spectrum.

One additional feature of the experiment permits separation of slow molecu-lar motion from spin diffusion. The experiment is conducted with slow MAS (at a rate smaller than the CSA), with the mixing time being set to an integral number of rotor periods. This insures that any off-diagonal intensity from spinning side bands is returned to the diagonal. The slow MAS should elimi-nate the contribution of spin diffusion to off-diagonal intensity, although this latter point needs to be verified. As a result, all off-diagonal intensity should arise from slow motion.

Figure 2-20 shows the results for polyoxymethylene at 50°C.[69] Any side-bands, and hence off-diagonal intensity if present, arise from the crystalline

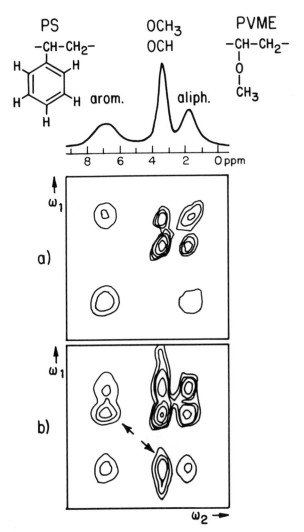

Figure 2-19. Two-dimensional proton spin-diffusion spectra of blends of polystyrene (PS) and poly(vinyl methylether) (PVME): (a) cast from chloroform; (b) cast from toluene. Note the absence of cross peaks between signals belonging to different polymers when cast from chloroform in (a), whereas strong cross peaks, indicated by arrows, appear in (b) for the blend cast from toluene. The MAS frequency was 2.8 kHz, the temperature was 328 K, and the mixing time was 100 ms. The spectra were recorded on a Bruker CXP-300 spectrometer. (Reprinted with permission from Caravatti, P.; Neuenschwander, P.; Ernst, R. R. *Macromolecules* **1985**, *18*, 119. Copyright 1985 American Chemical Society.)

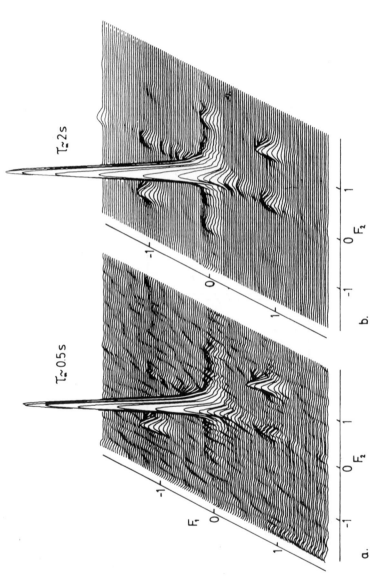

Figure 2-20. The 2D ^{13}C exchange NMR of polyoxymethylene at 50°C. (A) Mixing time of 750 rotor periods with a spinning frequency of 1475 Hz (64 × 870 FIDs) (31 h); (B) Mixing time of 3000 rotor periods with a spinning frequency of 1520 Hz (64 × 750 FIDs) (47 h). (Reprinted with permission from Kentgens, A. P. M.; de Jong, A. F.; de Boer, E.; Veeman, W. S. *Macromolecules* **1985**, *18*, 1045. Copyright 1985 American Chemical Society.)

fraction of this polymer, which displays the full CSA powder pattern at room temperature. At a mixing time t_m of 0.5 s, off-diagonal peaks are barely visible. They become more pronounced at $t_m = 3$ s. These results indicate the presence of a very slow molecular motion at less than about 1 Hz. The authors associate this slow motion with the α-relaxation seen by dielectric relaxation for this material. These early results indicate that the technique should be good for studying motions in the range $t_1 < \tau_c < T_1$, if the spin diffusion contribution is effectively eliminated.

Some Practical Aspects of NMR of Solid Polymers

In this section several aspects are discussed that, although not involving principles, are considered important enough for inclusion here.

Sample Considerations for MAS

Samples can be examined in essentially any form that can be spun successfully. A number of MAS rotor designs are currently in use, and all can be used for solid polymers. If feasible, the entire rotor itself can be machined out of the polymer of interest. The Beams-Andrew and Lowe geometries allow this. Because of the close tolerances involved, it may not be feasible for some double-bearing geometries. This option requires that the mechanical properties of the polymer be sufficiently high to withstand alone the intense centrifugal force exerted with MAS. This author has injection-molded Beams-Andrew geometry rotors of PVC, polycarbonate, and acrylonitrile-butadiene-styrene (ABS) that spin at up to 3 kHz with little or no additional machining. This option allows for the direct study of the effect of processing on polymer properties. Of course, when processing aids are used, their effect must be considered. It may not be desirable to perform relaxation time studies on full polymeric rotors when some of the rotor is far removed from the rf coil. Spin diffusion effects between the regions in and out of the coil may alter the observed T_1, as sometimes seen in liquid-state studies.[4]

Of course, the polymeric sample can be studied in any form that fits and spins in a hollow sample rotor. This may be a machined, molded, or extruded cylinder, a powder, a film or sheet, or pellets if they can be balanced to the extent that they spin stably. Balancing can be accomplished by filling the void volume with a powder such as talc. Films and sheets can be wound around a mandrel and inserted into the MAS rotor, or they can be punched into disks that stack in the rotor. Even very rubbery materials, such as uncured, unfilled elastomers 100°C above T_g, can be spun at reasonably high speeds with double-bearing geometries. For elastomers, much slower speeds suffice to give all the narrowing that is possible given the motional properties of the polymer chain.

The materials available for use as hollow MAS rotors are limited. Delrin (polyoxymethylene) has good mechanical properties for ambient temperature work but has a very strong resonance at about 89 ppm. Hence it is not well suited for some polymers containing oxygen or chlorine. It is also not well suited for observing weak, broad signals, as obtained for conducting polymers or possibly for highly degraded materials. Deuterated PMMA also has been used for MAS rotors, although the residual ^{1}H content is high enough to give a significant background signal in many cases. Kel-F [poly(chlorotrifluoro-ethylene)] is transparent to proton-decoupled solids NMR techniques because the large C–F dipolar coupling remains, broadening the Kel-F signal beyond high-resolution detection. However, it is not rigid enough to spin at speeds greater than about 3 kHz or at temperatures much above ambient. Double-bearing and similar designs generally use NMR-transparent inorganic materials, such as aluminum oxide or boron nitride. Vespel (polyimide), which gives a weak signal, can be used for high-temperature work.[70]

Except for the obvious effect of sample dilution, the presence of additional materials, such as inert fillers, is of little direct, practical concern. Of course, any additive that affects the molecular properties, such as crosslinking agents and plasticizers, affects the solid-state spectra. The presence of metallic or ferromagnetic particles is to be avoided because they degrade resolution and may prevent sample spinning. Electrically conducting samples, such as polyacetylene, can detune the probe circuitry and can be subject to severe rf heating. These samples must be examined by grinding to a fine powder and dilution with an inert filler, such as powdered glass.[71]

Variable-Temperature Studies

Because relaxation times in the solid state are very dependent on temperature, the capability to perform solids ^{13}C NMR studies over a range of temperature is highly desirable. There are several reasons for this. If transitions such as T_g are to be studied, it is essential to be able to monitor the spectra and relaxation times well above, well below, and at the transition temperature. If phase-discrimination studies are to be performed, it may be necessary to operate at a temperature well above the T_g of one phase to perform the separation. Interpretation of relaxation-time data in terms of various models of chain motion relies heavily on being able to predict the temperature dependence. In a more practical vein, cooling of rubbery materials to T_g or below facilitates their study by cross polarization, whereas their mechanical properties are improved with regard to high-speed MAS. Variable-temperature capability for high-resolution NMR studies of solids has become commercially available in the last several years. The value of variable temperature CPMAS capability is well demonstrated in Chapters 3 and 7.

Summary

This chapter has explained the basic interactions and experiments pertinent to high-resolution NMR of solid polymers, with emphasis on rigid systems. The influence of molecular motion on coherent line-narrowing schemes was outlined briefly. Finally, some advanced experiments were described that permit the extraction of much additional information from solid-state NMR spectra. The treatment was at a level necessary for understanding most of the material in the following chapters. Additional aspects of both theory and experiment appear throughout the book as necessary.

References

1. Bovey, F. A. "High Resolution NMR of Macromolecules"; Academic Press: New York, 1972.
2. Duncan, T. M.; Dybowski, C. R. Surface Sci. Repts. 1981, 1, 157.
3. Fyfe, C. A. "Solid State NMR for Chemists"; C. F. C. Press: Guelph, 1984.
4. Fukushima, E.; Roeder, S. B. W. "Experimental Pulse NMR—A Nuts and Bolts Approach"; Addison-Wesley: Reading, 1981.
5. Farrar, T. C.; Becker, E. D. "Pulse and Fourier Transform NMR"; Academic Press: New York, 1971.
6. Mehring, M. "High Resolution NMR in Solids"; Springer-Verlag: Berlin, 1983.
7. Schaefer, J.; Stejskal, E. O. Top. ^{13}C NMR Spectrosc. 1979, 3, 283.
8. Haeberlen, U. In "High Resolution NMR in Solids: Selective Averaging, Adv. Magn. Reson. Ser., Waugh, J. S., Ed., Suppl. 1"; Academic Press: New York, 1976.
9. Gerstein, B. C.; Dybowski, C. R., "Transient Techniques in the NMR of Solids: Introduction to the Theory and Practice"; Academic Press: New York, 1985.
10. Abragam, A. "The Principles of Nuclear Magnetism"; Clarendon Press: Oxford, 1961.
11. Weitekamp, D. P.; Bielecki, A.; Zax, D.; Zilm, K.; Pines, A. Phys. Rev. Lett. 1983, 50, 1807.
12. Andrew, E. R. Phil. Trans. R. Soc. London 1981, A299, 505.
13. Sarles, L. R.; Cotts, R. M. Phys. Rev. 1958, 111, 853.
14. Garroway, A. N.; Moniz, W. B.; Resing, H. A. ACS Symp. Ser. 1979, 103, 67.
15. VanderHart, D. L.; Earl, W. L.; Garroway, A. N. J. Magn. Reson. 1981, 44, 361.
16. Beeler, A. J.; Orendt, A. M.; Grant, D. M.; Cutts, P. W.; Michl, J.; Zilm, K. W.; Downing, J. W.; Facelli, J. C.; Schindler, M. S.; Kutzelnigg, W. J. Am. Chem. Soc. 1984, 106, 7672.
17. Murphy, P. D.; Taki, T.; Gerstein, B. C.; Henrichs, P. M.; Massa, D. J. J. Magn. Reson. 1982, 49, 99.
18. Lyerla, J. R. Meth. Exp. Phys. 1980, 16A, 241.
19. Maricq, M. M.; Waugh, J. S. J. Chem. Phys. 1979, 70, 3300.
20. Zumbulyadis, N.; Henrichs, P. M.; Young, R. H. J. Chem. Phys. 1981, 75, 1603.
21. Hexem, J. G.; Frey, M. H.; Opella, S. J. J. Am. Chem. Soc. 1981, 103, 224.
22. Hexem, J. G.; Frey, M. H.; Opella, S. J. J. Chem. Phys. 1982, 77, 3847.
23. Komoroski, R. A. J. Polym. Sci., Polym. Phys. Ed. 1983, 21, 1569.
24. Fleming, W. W.; Fyfe, C. A.; Kendrick, R. D.; Lyerla, J. R.; Vanni, H.; Yannoni, C. S. ACS Symp. Ser. 1980, 142, 193.
25. Doskocilova, D.; Schneider, B. Macromolecules 1972, 5, 125.
26. Herzfeld, J.; Berger, A. E. J. Chem. Phys. 1980, 73, 6021.
27. Stejskal, E. O.; Schaefer, J.; McKay, R. A. J. Magn. Reson. 1977, 25, 569.
28. Raleigh, D. P.; Olejniczak, E. T.; Vega, S.; Griffin, R. G. J. Am. Chem. Soc. 1984, 106, 8302, and references therein.
29. Murphy, P. D.; Stevens, W. C.; Cheung, T. T. P.; Lacelle, S.; Gerstein, B. C.; Kurtz, Jr., D. M. J. Am. Chem. Soc. 1981, 103, 4400.

30. Bodenhausen, G.; Caravatti, P.; Deli, J.; Ernst, R. R.; Sauter, H. *J. Magn. Reson.* **1982**, *48*, 143.
31. Frye, J. S.; Maciel, G. E. *J. Magn. Reson.* **1982**, *48*, 125.
32. Connor, T. M. *NMR, Basic Princ. Prog.* **1971**, *4*, 247.
33. Komoroski, R. A.; Mandelkern, L. In "Applications of Polymer Spectroscopy", Brame, E. G. Jr., Ed., Academic Press: New York, 1978, p. 57.
34. Doddrell, D.; Glushko, V.; Allerhand, A. *J. Chem. Phys.* **1972**, *56*, 3683.
35. Hartmann, S. R.; Hahn, E. L. *Phys. Rev.* **1962**, *128*, 2042.
36. Pines, A.; Gibby, M. G.; Waugh, J. S. *J. Chem. Phys.* **1973**, *59*, 569.
37. Yannoni, C. S. *Acc. Chem. Res.* **1982**, *15*, 201.
38. Komoroski, R. A. *Rubber Chem. Technol.* **1983**, *56*, 959.
39. Chingas, G. C.; Garroway, A. N.; Moniz, W. B.; Bertrand, R. D. *J. Am. Chem. Soc.* **1980**, *102*, 2526.
40. Murray, D. P.; Dechter, J. J.; Kispert, L. D. *J. Polym. Sci., Polym. Lett. Ed.* **1984**, *22*, 519.
41. Schaefer, J. Personal communication, 1984.
42. Fyfe, C. A.; Lyerla, J. R.; Volksen, W.; Yannoni, C. S. *Macromolecules* **1979**, *12*, 757.
43. Fyfe, C. A.; Dudley, R. L.; Stephenson, P. J.; Deslandes, Y.; Hamer, G. K.; Marchessault, R. H. *J. Macromol. Sci. Rev. Macromol. Chem.* **1983**, *C23*, 187.
44. Waugh, J. S. In "NMR and Biochemistry"; Marcel Dekker: New York, 1979, Ch. 13, p. 203.
45. Schaefer, J.; Stejskal, E. O.; Buchdahl, R. *Macromolecules* **1977**, *10*, 384.
46. Komoroski, R. A.; Mandelkern, L. *J. Polym. Sci., Polym. Symp.* **1976**, *54*, 201.
47. Dechter, J. J.; Komoroski, R. A.; Axelson, D. E.; Mandelkern, L. *J. Polym. Sci., Polym. Phys. Ed.* **1981**, *19*, 631.
48. Komoroski, R. A. *J. Polym. Sci., Polym. Phys. Ed.* **1983**, *21*, 2551.
49. Allerhand, A.; Doddrell, D.; Komoroski, R. *J. Chem. Phys.* **1971**, *55*, 189.
50. Barker, P.; Burlinson, N. E.; Dunell, B. A.; Ripmeester, J. A. *J. Magn. Reson.* **1984**, *60*, 486.
51. Opella, S. J.; Frey, M. H. *J. Am. Chem. Soc.* **1979**, *101*, 5854.
52. Terao, T.; Miura, H.; Saika, A. *J. Am. Chem. Soc.* **1982**, *104*, 5228.
53. Terao, T.; Miura, H.; Saika, A. *J. Chem. Phys.* **1981**, *75*, 1573.
54. Patt, S. L.; Shoolery, J. N. *J. Magn. Reson.* **1982**, *46*, 535.
55. Mayne, C. L.; Pugmire, R. J.; Grant, D. M. *J. Magn. Reson.* **1984**, *56*, 151.
56. Schaefer, J.; McKay, R. A.; Stejskal, E. O. *J. Magn. Reson.* **1979**, *34*, 443.
57. McKay, R. A.; Schaefer, J.; Stejskal, E. O.; Ludicky, R.; Matthews, C. N. *Macromolecules* **1984**, *17*, 1124.
58. Schaefer, J.; Skokut, T. A.; Stejskal, E. O.; McKay, R. A.; Varner, J. E. *J. Biol. Chem.* **1981**, *256*, 11574.
59. Jeener, J. Ampère International Summer School, Basko Polje, Yugoslavia, 1971, unpublished.
60. Bax, A. "Two Dimensional Nuclear Magnetic Resonance in Liquids"; Delft University Press/ Reidel: Dordrecht, 1982.
61. Roberts, J. E.; Vega, S.; Griffin, R. G. *J. Am. Chem. Soc.* **1984**, *106*, 2506.
62. Munowitz, M. G.; Griffin, R. G.; Bodenhausen, G.; Huang, T. H. *J. Am. Chem. Soc.* **1981**, *103*, 2529.
63. Munowitz, M. G.; Griffin, R. G. *J. Chem. Phys.* **1982**, *76*, 2848.
64. Schaefer, J.; McKay, R. A.; Stejskal, E. O.; Dixon, W. T. *J. Magn. Reson.* **1983**, *52*, 123.
65. Bax, A.; Szeverenyi, N. M.; Maciel, G. E. *J. Magn. Reson.* **1983**, *55*, 494, and references therein.
66. Szeverenyi, N. M.; Sullivan, M. J.; Maciel, G. E. *J. Magn. Reson.* **1982**, *47*, 462.
67. Caravatti, P.; Neuenschwander, P.; Ernst, R. R. *Macromolecules* **1985**, *18*, 119.
68. Caravatti, P.; Deli, J. A.; Bodenhausen, G.; Ernst, R. R. *J. Am. Chem. Soc.* **1982**, *104*, 5506.
69. Kentgens, A. P. M.; de Jong, A. F.; de Boer, E.; Veeman, W. S. *Macromolecules* **1985**, *18*, 1045
70. "NMR Solids Accessories," Doty Scientific, Inc., Columbia, S.C., 1983.
71. Clarke, T. C.; Scott, J. C.; Street, G. B. *IBM J. Res. Develop.* **1983**, *27*, 313.

3

HIGH-RESOLUTION NMR OF GLASSY AMORPHOUS POLYMERS

James R. Lyerla

IBM RESEARCH LABORATORY, SAN JOSE, CA 95193

Introduction

The properties of amorphous polymers below their respective glass transition temperatures (T_g) are of great technological importance; nonetheless, the molecular basis for most sub-T_g properties is not well understood.[1] Indeed, the elucidation of these molecular structure–physical property relationships is emerging as an important area of basic polymer science for such understanding is fundamental to the design of materials with specific properties as well as to advancing our knowledge of the amorphous state. For example, the mechanical properties, such as toughness or impact strength of glassy polymers, are often key to their utility, yet development of basic connectivity between the local structure and type, amplitude and time scale of molecular motions in glassy polymers and the observed mechanical properties is still in its infant stages. Likewise, the reaction chemistry during cure of thermosetting resins controls their ultimate physical properties, yet detailed understanding of that chemistry is often lacking. High-resolution, solid-state NMR experiments offer the potential to probe, at the molecular level, these and many other questions relating to the primary structure and motional properties of glassy polymers. The purpose of this chapter is to illustrate several such applications of NMR techniques.

The utility of multipulse NMR and deuterium NMR for inquiring into the

© 1986 VCH Publishers, Inc.
Komoroski (ed): High-Resolution NMR Spectroscopy of Synthetic Polymers in Bulk

properties of glassy polymers is discussed in Chapters 9 and 10, respectively; this chapter focuses on ^{13}C cross-polarization magic angle spinning (CPMAS) experiments (as outlined in Chapter 2). This chapter is not intended to be an exhaustive review of CPMAS NMR applications to amorphous polymers but is directed at describing certain spectral features characteristic of high-resolution ^{13}C spectra of amorphous polymers and illustrating the use of such spectra (and variations thereof) in obtaining structural, chemical, and dynamical information on these polymers.

NMR Structural Studies of the Glassy State

Spectral Features

The ^{13}C spectra[2] of four states of the resin from diglycidyl ether of bisphenol A (DGEBA) presented in Figure 3-1 illustrate many of the spectral features of glassy amorphous polymers and form a basis to discuss the types of information available on such systems from NMR studies. The solution spectrum of the prepolymer is to be contrasted to the CPMAS spectrum of the crystalline form in both the number of resonances and the linewidths of the resonances. In the solid state, rapid interconversion between many molecular conformations usually is precluded, and resonance lines that are chemically equivalent in solution often are split unless there is crystallographic equivalency. The spectrum of the polycrystalline form of the DGEBA prepolymer displays a significant number of these "solid-state" splittings of resonance lines—eg, the methyl carbon resonance in solution (peak i) gives rise to two resonance lines separated by 4.9 ppm in the crystalline phase, whereas the proton-bearing ring carbons (peaks c and d) give rise to multiple resolved lines, some significantly shifted (6.5–10.1 ppm) from the resonance positions found in solution. The polycrystalline spectrum suggests that there is more than one monomer conformation within the unit cell. This result is consistent with the X-ray analysis, which finds one stereoisomer present[2] but that one end of the monomer is slightly disordered, thereby yielding two possible conformations for the epoxide ring. Based on the X-ray analysis and a model of steric hindrance effects on chemical shifts,[3] Garroway and co-workers[2] have attempted to explain the precise origin of each of the resonance lines in Figure 3-1B. Although the steric hindrance model predicts solid-state splittings of the right order of magnitude, the detailed agreement is not good. However, irrespective of a detailed model of the origin of the shifts, the major point to be made from spectrum 3-1B is that unaveraged conformations in the solid state can increase the multiplicity of resonance lines over that found in a counterpart liquid-state spectrum.

The second contrasting feature between the solution and polycrystalline ^{13}C spectra of the DGEBA prepolymer is the greater than tenfold difference in resonance linewidth. Despite the regularity of crystal habit and the line-

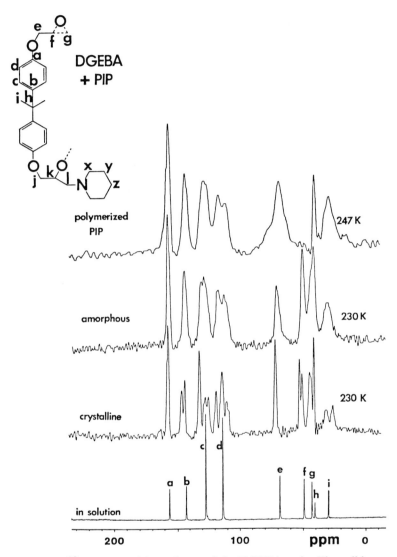

Figure 3-1. The ^{13}C spectra of four phases of the DGEBA resin. The solids are compared well below room temperature to highlight the solid-state splittings in the cured polymer. The solvent peaks have been eliminated in the solution-state spectrum. Loosely speaking, the amorphous resin spectrum is a blurred version of that for the crystalline state. Moreover, the polymerized and amorphous resins are virtually superposable in the aromatic regions, and the difference in chemical structure is readily apparent in the aliphatic region. (Reprinted with permission from Garroway, A. N.; Ritchey, W. M.; Moniz, W. B. *Macromolecules* **1982**, *15*, 1051. Copyright 1982, American Chemical Society.)

narrowing schemes employed, the resonance lines in the CPMAS spectrum do not approach those of the liquid-state spectrum. VanderHart and co-workers[4] have examined, in detail, the possible sources of line broadening, eg, dispersions of isotropic chemical shifts (arising from distributions of anisotropic sources of magnetic susceptibility, bond angles, or nonaveraging conformations), motional modulation effects, and experimental imperfections (such as missetting of the magic angle, nonuniform spinning rate, off-resonance proton-decoupling effects, insufficient proton-decoupling power). Many combinations of these sources of line broadening can account for the observed linewidths from polymer solids. The caveat for the polymer scientist is not to expect the same resolution from "high-resolution" solid-state and solution-state NMR spectra. This distinction is further compounded for the glassy amorphous state of polymers as evidenced in the ^{13}C CPMAS spectrum (Figure 3-1C) of the DGEBA prepolymer. The resonance linewidths in the glassy spectrum are ca. 50–100 Hz or 5–10 times those in the polycrystalline spectrum. Here, the discrete solid-state splittings of 5–10 ppm found for the crystalline phase are "blurred out," leading to distributions of isotropic chemical shifts that limit the achievable resolution. The breadth of the chemical shift distribution is affected significantly by the nonregularity of the molecular packing and conformations in the glassy state.

Chemistry of Polymer Resins

The ^{13}C CPMAS spectrum of the DGEBA prepolymer cured with piperidine (5% by weight) at 393 K for 16 h is shown in Figure 3-1D. In this polymerized state, the epoxy is no longer soluble. When compared to the spectrum of the amorphous prepolymer, the spectrum of the cured resin is very similar in the aromatic region but, in the aliphatic region, clearly reflects the chemical changes occurring during polymerization. The opening of the epoxide ring results in significant downfield chemical shifts for resonance lines f and g. These resonances, along with resonance e, all fall under the broad resonance between 65 and 75 ppm in the cured polymer. A crude estimate of the degree of polymerization can be obtained based on the disappearance of peaks f and g and indicates that about 90–95% of the rings are opened. Thus, despite the broad lines encountered for CPMAS ^{13}C spectra of glassy polymers, it is clear that the capability of obtaining spectra sufficient to resolve individual carbons in an insoluble system presents the potential to elucidate the curing chemistry of intractable amorphous polymers. For an overview of further work on the epoxies, the reader is referred to the literature.[2,5,6]

The chemistry of phenol–formaldehyde resins, both novolac and resole types, has been investigated by ^{13}C CPMAS techniques.[6–9] The study by Fyfe et al[7,8] on resole resins demonstrates how selective ^{13}C enrichment of reactants can be used to enhance the information from NMR spectra on polymerization and solid-state reactions. The cured resin was prepared via the scheme outlined below[8]:

Scheme 1 Preparation of Resole

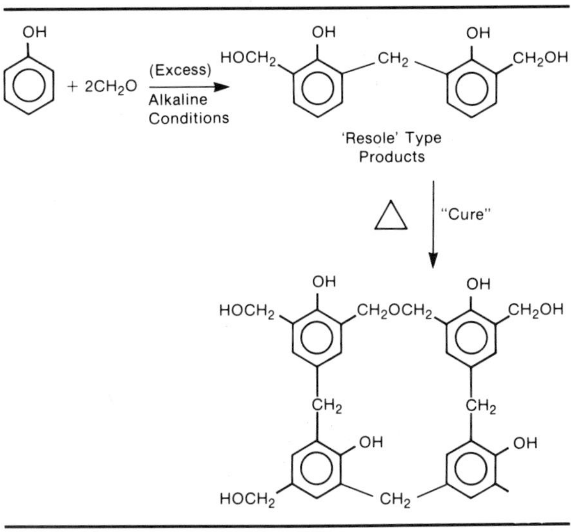

The ^{13}C CPMAS spectrum of the cured resin with reactants unenriched in the ^{13}C isotope is compared in Figure 3-2 to that in which the formaldehyde reactant was ^{13}C enriched to ca. 5%. In the spectrum of the ^{13}C-enriched resin, methylene resonances are assigned to free paraformaldehyde, methylol groups, and bridging methylenes according to Scheme 1. The spectrum provides a direct measure of the nature and the extent of cross linking in the polymer network.

The resonance lines in the case of the resole are even broader (150–250 Hz) than for the epoxy (Figure 3-1). The significant broadening arises from the fact that in the phenolic resin prepolymer (B stage resin) there are a variety of closely related isomeric forms and so the resonance position of a given type of carbon depends on the number, type, and phenolic substitution pattern of proximate linkages. The combination of the usual chemical-shift dispersion, which arises from the heterogeneity in the three-dimensional arrangements of the glassy state, and the isomer content accounts for the very broad lines. Despite these linewidths, the enriched resin can be used to provide insight into the thermal degradation chemistry of these resins.[7] Before the results are examined, an NMR pulse sequence is introduced that allows differentiation of carbons with directly bonded protons from those carbons without protons in a CPMAS spectrum. The experiment[10] consists of a normal cross-polarization period followed by insertion of a short delay (25–100 μs) in which both ^{13}C and ^{1}H rf fields are turned off before the ^{1}H field is turned back on (to decouple the protons) and data acquisition begun (see Figure 2-10A, Chapter 2). During the delay period, the ^{13}C spins precess in their local proton dipolar fields. For carbons subject to strong C–H dipolar couplings (eg, carbons with directly bound protons and not subject to rapid motion), the magnetization

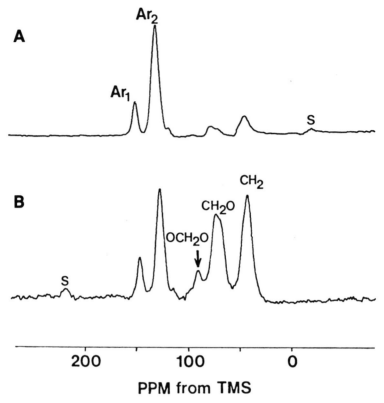

Figure 3-2. Solid-state ^{13}C NMR spectra obtained from a solid rotor of a cured resole-type material (phenol–formaldehyde–sodium hydroxide = 1 : 2 : 0.01, cured at 110°C for 24 h) under the following experimental conditions: (A) magic angle spinning at 3.6 kHz, 1-ms cross-polarization time, 2-s recycle time, 2000 FIDs; (B) conditions as in (A) except that 500 FIDs were averaged and the sample was prepared using formaldehyde ^{13}C enriched to ~5%. The small peaks marked s denote spinning side bands. (Reprinted with permission from Fyfe, C. A.; Rudin, A.; Tchir, W. J. *Macromolecules* **1980**, *13*, 1322. Copyright 1980, American Chemical Society.)

quickly decays because of rapid dephasing of the ^{13}C spins. The resulting spectrum therefore consists mainly of signals arising from carbons without directly attached protons. Methyl carbons often appear but at reduced intensity because of their particular relaxation characteristics. The sequence of spectra shown in Figure 3-3 illustrates the experiment. Figure 3-3A is the "normal" CPMAS spectrum of a resole, whereas Figure 3-3B demonstrates the effect of interjecting a 40-μs delay before data acquisition is begun. Figure 3-3B represents primarily the spectrum of carbons without attached protons. Figure 3-3C is the difference spectrum and hence corresponds to the spectrum of only the proton-bearing carbons. The aromatic carbons in spectrum 3-3B arise from the phenolic hydroxyl carbon (Ar$_1$) and the ortho- and para-substituted ring carbons (Ar$_3$). In spectrum 3-3C, the aromatic carbon resonances are

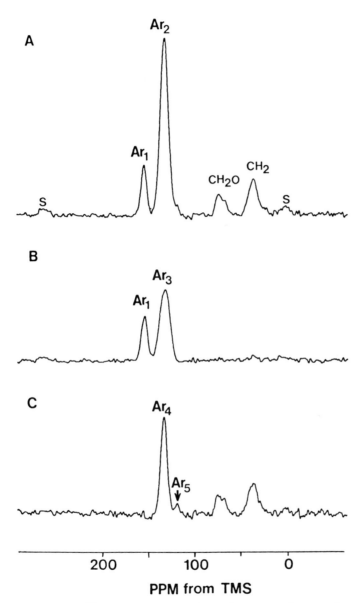

Figure 3-3. Solid-state ^{13}C NMR spectra of a solid rotor of a cured resole-type material. The small peaks marked s denote spinning side bands. (A) Magic angle spinning, 1-ms cross-polarization time, 2-s recycle time, 500 FIDs; (B) conditions as in (A) except that during the 50-μs dwell before acquisition, the proton decoupler was gated off for the first 40 μs; (C) result of subtracting 1.15 × spectrum (B) from spectrum (A). (Reprinted with permission from Fyfe, C. A.; Rudin, A.; Tchir, W. J. *Macromolecules* **1980**, *13*, 1322. Copyright 1980, American Chemical Society.)

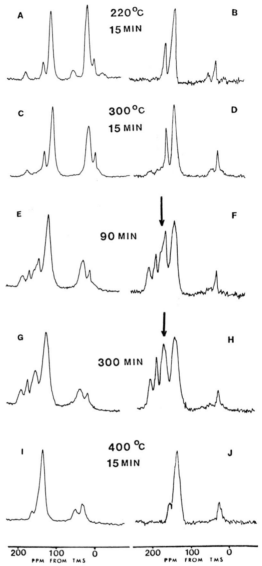

Figure 3-4. Carbon-13 CPMAS spectra of various heat-treated resoles. Resins were prepared with 5% ^{13}C-enriched formaldehyde P : F : NaOH = 1 : 2 : 0.01 and cured at 180°C for 160 min under 15-psi nitrogen. Cross-polarization spectra obtained with 1-ms contact and 1-s recycle delay. Nonprotonated spectra obtained with 1-ms contact, 1-s recycle delay, and decoupler gated off for 50 μs before acquisition. (A) Spectrum of cured resole heated in air to 220°C for 15 min, 3600 scans; (B) same sample as (A), only nonprotonated carbon selection, 3600 scans; (C) spectrum of cured resole heated in air to 300°C for 15 min, 3600 scans; (D) same sample as (C), only nonprotonated carbon selection, 3600 scans; (E) spectrum of cured resole heated in air to 300°C for 90 min, 5200 scans; (F) same sample as (E), only nonprotonated carbon selection, 5200 scans; (G) spectrum of cured resole heated in air to 300°C for 300 min, 2200 scans; (H) same

from the meta ring carbons (where no reaction takes place) (Ar_4) and small amounts of unsubstituted ortho and para carbons. Overall the results demonstrate the high degree of ring substitution occurring in the reaction of the phenol and formaldehyde moieties.

The CPMAS spectra displayed in Figure 3-4 correspond[7] to a series of spectra of the ^{13}C-enriched cured resin subjected to various thermal cycles in air. For each experimental condition, a pair of spectra is reported, one corresponding to a standard CPMAS experiment (A, C, E, G, I) and the other to a delayed decoupling experiment (B, D, F, H, J). Comparison of spectra 3-4A and 3-4B with that of Figure 3-2B indicates that the major spectral changes induced by heating the polymer at 220°C for 15 min (which yields a 3% weight loss of the resin) are a marked decrease of methylol groups (70 ppm) and an increase in the methylene absoption. Most likely, the CH_2O carbons are converted to CH_2 by the normal curing process utilizing the residual (unreacted) ortho and para sites. In addition, there is a small amount of oxidation, as indicated by the low-intensity resonance at 190 ppm, which on the basis of comparison of 3-4A and 3-4B is found to bear a proton and hence is assigned to an aldehyde formed by direct oxidation of a methylol grouping as in Scheme 2. Finally, there is the formation of some methyl carbons as revealed by the resonance at ca. 15 ppm.

Scheme 2

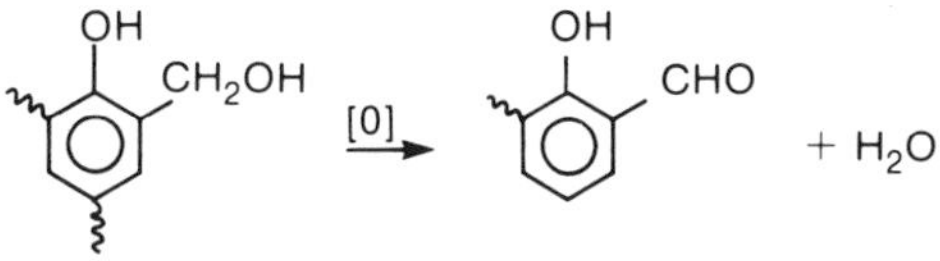

At 300°C, there is complete loss of CH_2O groups in only 15 min but only a 7% weight loss, pointing to a small degree of oxidation. Prolonged heating of the resin at 300°C results in a substantial decrease in the CH_2 resonance intensity, with simultaneous formation of a number of carbonyl resonances (150–195 ppm) that can be assigned to aldehydes, ketones, carboxylic acids, and anhydrides. At 400°C, there is rapid degradation and decomposition (43% weight loss after 15 min) and the ^{13}C spectra (3-4I,J) of the product are significantly different than those of the initial resole. As at 300°C, there has been almost complete loss of the CH_2 resonance, but now there is no carbonyl functionality present. This result suggests a conversion to an essentially all-aromatic structure, which can arise via schemes such as 3.

sample as (G), only nonprotonated carbon selection, 2200 scans; (I) spectrum of cured resole heated in air to 400°C for 15 min, 3600 scans; (J) same sample as (I), only nonprotonated carbon selection, 3600 scans. (Reprinted with permission from Fyfe, C. A.; McKinnon, M. S.; Rudin, A.; Tchir, W. J. *Macromolecules* **1983**, *16*, 1216. Copyright 1983, American Chemical Society.)

Scheme 3

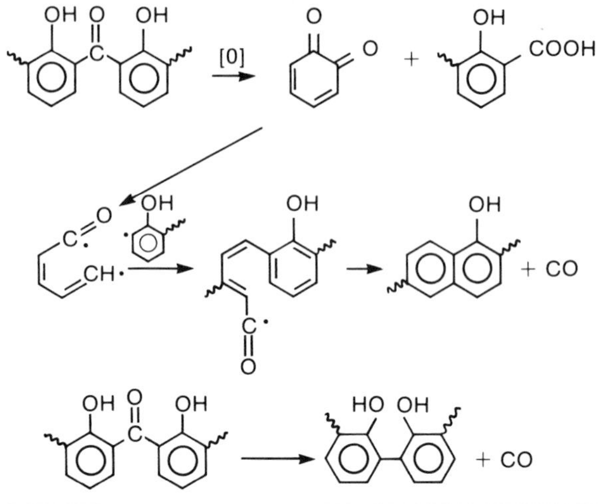

When the 400°C thermolysis is carried out under vacuum, the product gives rise to the ^{13}C CPMAS spectra displayed in Figure 3-5. Again, there is loss of CH_2O functionality but formation of considerable methyl groups, probably via the route in Scheme 4. However, there are no other significant structural differences in the polymer, verifying the oxidative nature of the degradation process. Indeed, when the sample used to obtain the spectra in Figure 3-5 was subjected subsequently to 300°C thermolysis conditions in air, spectral changes similar to those in Figure 3-4 were observed. This result confirms the oxidative degradation mechanism and the fact that the oxidized functionalities do not arise from CH_2O groups. Collectively, the results on the resoles demonstrate the power of CPMAS experiments when combined with selective ^{13}C enrichment to follow polymer chemistry.

Maciel and co-workers have studied both resoles[9] and novolac resins.[11] In addition, these workers have focused attention on the curing of furfuryl alcohol resins.[12] In this last case, results from ^{13}C CPMAS experiments have

Scheme 4

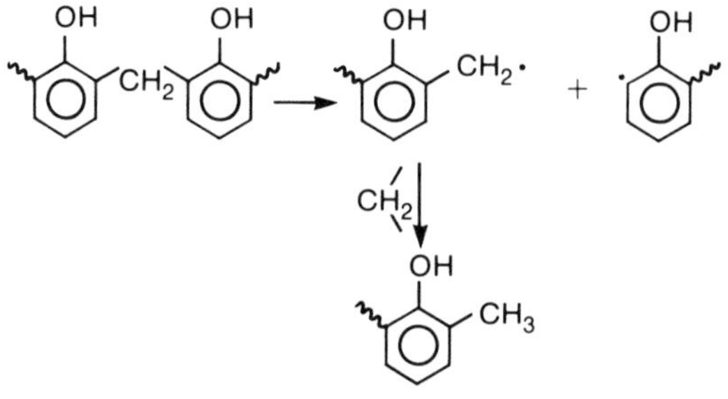

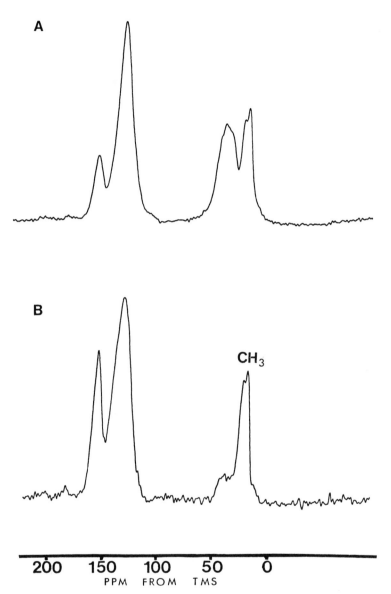

Figure 3-5. Carbon-13 CPMAS spectra of cured resole heated under vacuum to 400°C for 60 min. Resins prepared with 5% ^{13}C-enriched formaldehyde P : F : NaOH = 1 : 2 : 0.01. (A) Spectrum using 1-ms contact and 1-s recycle delay, 3600 scans; (B) same sample as (A), only nonprotonated carbon selection, 3600 scans. (Reprinted with permission from Fyfe, C. A.; McKinnon, M. S.; Rudin, A.; Tchir, W. J. *Macromolecules* **1983**, *16*, 1216. Copyright 1983, American Chemical Society.)

Scheme 5

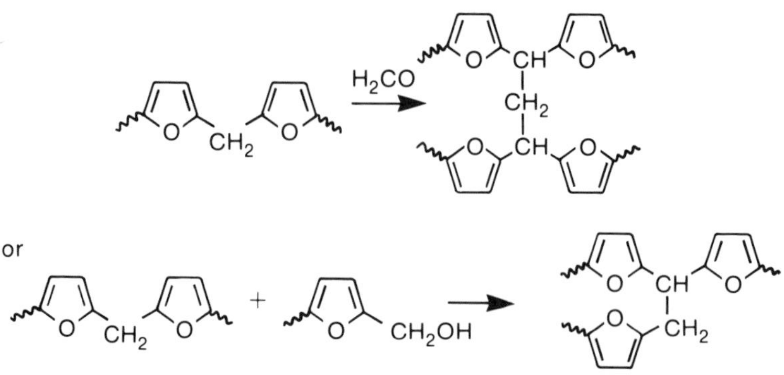

been employed to reveal that the main curing process is the formation of crosslinking branches through methylene linkages according to the Scheme 5. Crosslinking through the 3 and 4 positions of the furan ring was found to be unimportant. More recent CPMAS studies[13] involving the prepolymer formation have demonstrated the lack of appreciable concentration of formaldehyde, methylol groups, or dimethylene ether linkages in the very early stages of the curing process. This establishes that the condensation process in Scheme 6 (as opposed to Scheme 7) is the dominant route in the polymerization of furfuryl alcohol.

Scheme 6

Scheme 7

Other examples of the utilization of CPMAS NMR to follow polymer chemistry can be found in the recent literature,[6] including studies on acetylene-terminated polyimides[14] and sulfones,[6] polyacrylonitrile,[15] polyimides,[16] urea–formaldehyde and melamine–formaldehyde resins,[6] HCN-derived polymers,[17] and plasma-polymerized polymers.[18] In this last case, that of the intractable films from plasma polymerization, Kaplan and Dilks[18] have utilized specific ^{13}C labeling of the ring and methyl carbons of toluene to trace reaction pathways of this compound in the plasma process. From relative intensities in spectra such as those in Figure 3-6, obtained as a function of the

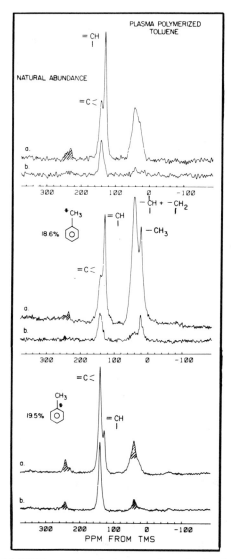

Figure 3-6. Carbon-13 NMR spectra of plasma-polymerized toluene and isotopically labeled materials. (From reference 19.)

amount of ^{13}C-enriched toluene[19] in the vapor injected into the inductively coupled rf system, it is possible to determine the polymer sites into which specific toluene carbons are incorporated (see Table 3-1). Approximately 20% of the C-1 carbons become saturated carbons in the plasma polymerization process, whereas 24% of the methyl carbons end up as unsaturated carbons in the final polymer film. The complete ^{13}C data require net saturation of ca. 30% of the toluene double bonds and a net displacement of hydrogen by carbon on ca. 20% of the toluene ring carbons. These results suggest that the

TABLE 3-1. DISTRIBUTION OF TOLUENE CARBONS IN THE PLASMA POLYMER[a,b]

| Toluene carbons | Structural type in polymer | | | | |
	=C	=CH	C, CH, CH$_2$	All Aliphatic	CH$_3$
All	27	33		40	
CH$_3$	10	14	51		25
C-1	73	7		20	
C-2–C-6[c]	21	42		37	

[a] From reference 18. Reprinted with permission of John Wiley and Sons, Inc., copyright holder.
[b] Numbers represent the percentage of a particular toluene carbon (or average percentage of several carbons) ending up in the specified structural type in the polymer. Data accuracy ± 15%.
[c] The average destination of the toluene-protonated ring carbons is determined from the appropriate weighted averaging of the preceding three rows of data.

major precursors to the polymer are the benzyl radical and toluene itself. When combined with the elemental analysis results, the ^{13}C data suggest a structure of plasma-polymerized toluene such as 1:

Structure 1

Molecular Geometry and Polymerization Mechanisms by CP-Nutation NMR

There is a novel NMR experiment that can provide insight into polymer structure and features such as polymerization mechanisms. The sources of this information are homonuclear and heteronuclear dipolar couplings. The magnitude of such couplings represents a source of geometric information on molecules because information on internuclear distances is contained in the relationship[20]:

$$D = \frac{3\gamma^2\hbar}{4\pi} \langle r^{-3}(3\cos^2\theta - 1)\rangle \qquad (3\text{-}1)$$

where D is the motionally averaged dipolar coupling between two like (homonuclear case) spins, γ is the magnetogyric ratio, θ is the angle between the internuclear vector and the applied magnetic field (B_0), and r is the internuclear distance. In general, it is difficult to extract the information on internuclear distances without samples having some degree of long-range order and without resorting to a multiparameter fitting procedure to resolve overlap of couplings and to deconvolute chemical-shift effects. For a nonoriented polycrystalline or amorphous material, the dipolar splitting from an isolated pair of nuclei is manifested as a Pake doublet[20] with outer shoulders (see Figure 3-7). Unfortunately, the doublet maxima usually are masked or distorted by

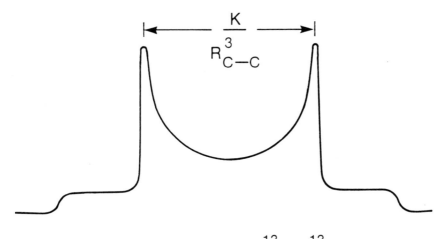

$$\underset{\text{C--C}}{R^3}$$

K

Pake doublet from $^{13}C - {}^{13}C$
dipolar coupling in an amorphous material

Figure 3-7. Idealized Pake doublet from $^{13}C-{}^{13}C$ dipolar coupling in an amorphous material. The splitting is given by K/r^3_{C-C}, where $K = 3\gamma^2\hbar/4\pi$.

chemical-shift effects. Recently, Yannoni et al[21,22] have introduced a cross-polarization nutation experiment that allows the measurement of homonuclear dipolar splittings in amorphous solids unencumbered by chemical-shift effects. In the ^{13}C version of this experiment (shown schematically in Figure 3-8), carbon magnetization is generated along the carbon rf field by CP methods. Following a 90° phase shift of the carbon rf field, a nutation (ie, a forced precession of nuclear magnetization) is excited by the application of a train of carbon rf pulses. The ^{13}C signal is observed (with continuous proton irradiation to eliminate heteronuclear couplings) in receiver windows between the pulses, allowing the entire time evolution of the magnetization to be recorded (thus avoiding a time-consuming point by point method).[22] The nutation spectrum, obtained by Fourier transformation of the acquired signal, has frequencies $\gamma H_1/2\pi$, ie, in the kHz range. The proton-decoupled ^{13}C nutation NMR spectrum[23,24] of doubly ^{13}C-labeled, unoriented cis-polyacetylene at 77 K is shown in Figure 3-9, demonstrating the well-resolved Pake powder pattern. (The well-resolved peak in the center of the powder pattern arises from isolated ^{13}C nuclei, ie, from singly labeled monomer in the sample.)

Although a highly crystalline polymer itself, polyacetylene provides a useful illustration of the utility of the CP-nutation technique as it may be applied to amorphous polymers. Polyacetylene and its analogs have attracted widespread attention during the past decade owing to the fundamental interest in the bonding in linear polyenes and the high conductivity $[>10^3 \; (\text{ohm-cm})^{-1}]$ observed when the polymer is doped with certain oxidizing and reducing agents.[25] Generally, the polymer is prepared by growth of films on the surface of a Ziegler-Natta catalyst solution and typically synthesized as the cis-transoid isomer (**2**). Conversion to the thermodynamically more stable trans-

Structure 2

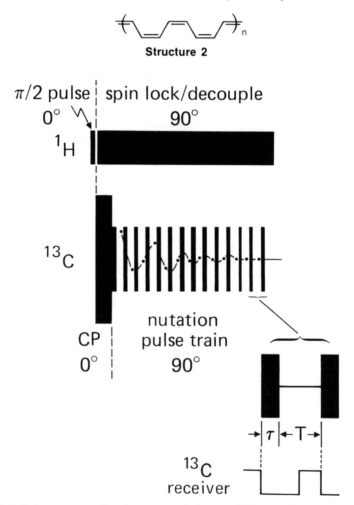

Figure 3-8. Pulse sequence for the cross-polarization (CP)–nutation experiment. For the spectra shown herein, τ and T were 8 and 17 μs, respectively, and the receiver was turned on 6 μs after the trailing edge of the transmitter pulse. The dots between pulses signify the ^{13}C signal captured in the receiver windows shown in the expanded view. (From reference 22. Reprinted with permission.)

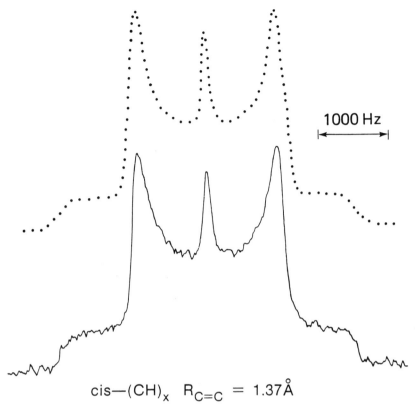

$$\text{cis—(CH)}_x \quad R_{C=C} = 1.37\text{Å}$$

Figure 3-9. Solid line: proton-decoupled ^{13}C nutation NMR spectrum at 77 K of doubly labeled cis-(CH)$_x$. Dotted curve: simulation of nutation spectrum of cis-(CH)$_x$ using a 1.37-Å C—C bond length. The peak in the middle is from isolated ^{13}C nuclei. (From reference 23. Reprinted with permission.)

transoid form (**3**) is accomplished thermally and can be followed by ^{13}C CPMAS spectroscopy[26,27] (Figure 3-10).

Structure 3

The highly disordered nature of the as-grown polyacetylene film prevented the direct determination of the bonding parameters in this material by usual X-ray methods. Application of CP-nutation NMR spectroscopy has resulted in bond length measurements in both isomers as well as providing information on the polymerization mechanism. For these studies, cis-(CH)$_x$ was prepared by the method of Ito et al[28] using a titanium tetra-n-butoxide–triethylaluminum catalyst mixture in toluene. Specifically, a mixture of 4% of doubly ^{13}C-enriched acetylene ($\geq$99% ^{13}C) in double-^{13}C-depleted acetylene ($>$99.9% ^{12}C) was polymerized. This procedure produced a material tailored

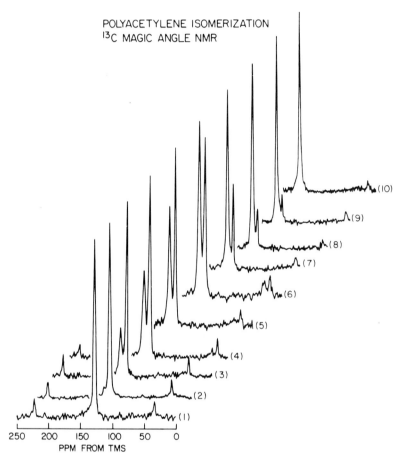

Figure 3-10. The 22.63-MHz ^{13}C NMR spectra of a 12-mg, initially high-cis content polyacetylene as isomerization progresses to the trans form. Proton-enhanced cross-polarization conditions: repetition time = 1 s; contact time = 1 ms, with resonant radio frequency field of 8 and 32 G for protons and carbons, respectively; proton decoupling power = 12 G; number of scans ~ 50,000 per spectrum; spinning speed = 2.2–2.5 kHz. Spectra are normalized arbitrarily to the larger of the two peaks. Percentage trans determined from IR absorption ratio (740 versus 1010 cm^{-1}) and the corresponding thermal history of each spectrum: (1) 32% (25°C, 1.5 days); (2) 47% (25°C, 9 days); (3) 47% (25°C, 11 days; 78°C, 1 h; 100°C, 1 h); (4) 52% (25°C, 11 days; 78°C, 1 h; 100°C, 2 h); (5) 51% (25°C, 11 days; 78°C, 1 h; 100°C, 4 h); (6) 48% (25°C, 11 days; 78°C, 1 h; 100°C, 7.5 h); (7) 59% (25°C, 11 days; 78°C, 1 h; 100°C, 22.8 h); (8) 54% (25°C, 11 days; 78°C, 1 h; 100°C, 47 h); (9) 85% (25°C, 11 days; 78°C, 1 h; 100°C, 64 h); (10) 94% (25°C, 11 days; 78°C, 1 h; 100°C, 64 h; 200°C, 21 min). (Reprinted with permission from Gibson, H. W.; Pochan, J. M.; Kaplan, S. *J. Am. Chem. Soc.* **1981**, *103*, 4619. Copyright 1981, American Chemical Society.)

for the CP-nutation experiment in that the spin pairs were relatively isolated, thereby avoiding interference from longer range (adjacent, etc.) dipole–dipole splittings, whereas the high enrichment produces good signal to noise in the spectra.

As indicated in Figure 3-9, the Pake powder pattern from cis-$(CH)_x$ is simulated via Eq. (3-1) with a C—C bond length of 1.37 Å. Although only one Pake pattern is observed for the polymer, this does not necessarily imply that there is complete bond equalization. Indeed, such a finding is contrary to the finding in the trans (see Figure 3-11) spectrum (where bond equalization would be more favorable) and to theoretical arguments.[23] Instead, the observation of only one bond length indicates that the carbons of the labeled pair end up

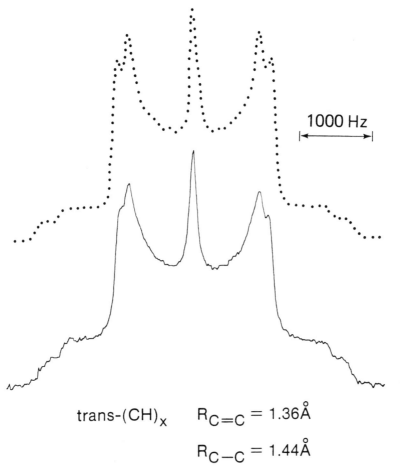

$$\text{trans-}(CH)_x \qquad R_{C=C} = 1.36 \text{ Å}$$

$$R_{C-C} = 1.44 \text{ Å}$$

Figure 3-11. Solid line: proton-decoupled ^{13}C nutation NMR spectrum of doubly labeled trans-$(CH)_x$ at 77 K. Dotted curve: simulation resulting from superposition of ^{13}C nutation spectra using bond lengths of 1.36 and 1.44 Å. Again, the central peak is from isolated ^{13}C nuclei. (From reference 23. Reprinted with permission.)

either singly or doubly bonded to one another, but not both. The bond length of 1.37 Å strongly suggests that the polymerization results in the labeled pairs being doubly bonded in cis-$(CH)_x$. This result carries the further implication that, for this particular Ziegler-Natta catalyst, the polymerization proceeds by a four-center acetylene insertion mechanism (Figure 3-12) and not a metallacycle mechanism,[29] where the labeled pairs would be singly bonded.

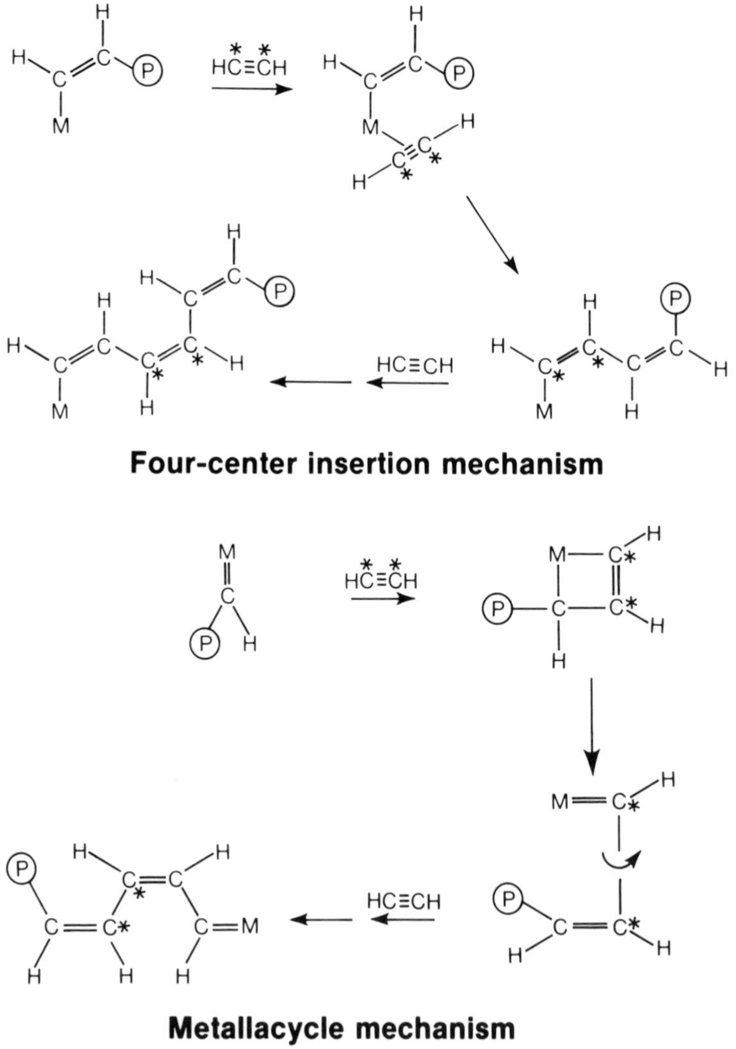

Four-center insertion mechanism

Metallacycle mechanism

Figure 3-12. Possible mechanisms for acetylene polymerization. (A) The direct four-center acetylene insertion mechanism; (B) a metallacycle mechanism proposed by Katz (Katz, T. J.; Lee, S. J. *J. Am. Chem. Soc.* **1980**, *102*, 422). (Reprinted with permission from Clarke, T. C.; Yannoni, C. S.; Katz, T. J. *J. Am. Chem. Soc.* **1983**, *105*, 7787. Copyright 1983, American Chemical Society.)

The CP-nutation spectrum of trans-$(CH)_x$ [prepared by heating a cis-$(CH)_x$ sample in an evacuated sealed tube at 160°C for 1 h] displays (Figure 3-11) evidence for two overlapping Pake powder patterns of comparable intensity. Both the splittings of the doublet maxima and the well-defined steps in the wings of the spectrum are simulated faithfully by ^{13}C homonuclear dipolar couplings arising for C—C bonds of 1.44 Å and 1.36 Å. These bond lengths, which are in close agreement with theory,[23] are accorded to C—C labels in single and double bonds, respectively, and confirm a bond-alternated structure. Based on the powder-pattern intensities, there are approximately equal populations of singly and doubly bonded labeled carbons in the trans isomer. These arise by thermal transformation of cis material in which all labeled pairs are doubly bonded. Because motion of neutral solitons [a mechanism invoked to explain the mobile neutral defect observed in undoped $(CH)_x$[30]] along a polymer chain would lead to an interchange of single and double bonds, the observation of a nonaveraged NMR spectrum places a limit on the dynamic interchange at less than ca. 10^3/s (the observed nonaveraged splitting between the doublets in Figure 3-11 being less than 1 kHz).[24] Such a slow rate of interchange appears contrary to a soliton model or, if applicable, suggests that only a small fraction of the sample sees such neutral solitons. The mechanism of the scrambling of the enriched carbon pairs in $(CH)_x$ upon cis–trans isomerization remains an area of continuing interest, as does the nature of the semiconductor to metal transition, which occurs when the polymer is doped with oxidizing and reducing agents. With respect to this latter question, CPMAS studies are being utilized in attempts to gain insight into the proposed models.[25] The interested reader is referred to the recent literature for details.[25,27] As stated, the purpose here was to introduce the CP-nutation experiment and illustrate how it might be used to study aspects of polymerization and determine geometries in amorphous materials.

Molecular Motions in the Glassy State

Lineshape Analysis

Cross-polarization MAS NMR spectroscopy has utility in studies of rate processes in solids. One of the important applications of NMR spectroscopy is that chemical rate processes with lifetimes of ca. 10^{-6}–10 s can be studied directly by observing lineshape changes as a function of temperature. An important situation arises when the solid state induces magnetic inequivalence in a pair or pairs of carbon atoms that are equivalent in solution because of rapid molecular reorientation. As has been seen, such a situation is found at 247 K for the ortho and meta aromatic ring carbons of an epoxy resin cured with piperidine (Figure 3-1D). If molecular motion, which renders these carbons equivalent, occurs in the glassy state at higher temperature, coalescence of the signals is observed. The ^{13}C CPMAS spectra as a function of temperature (Figure 3-13) of the same cured epoxy demonstrate the coalescence phenomenon and the continued line narrowing as temperature is increased.[2] The result indicates the motional process involving the polymer backbone (aromatic rings) occurs on a time scale faster than 10 ms at room temperature and above (100–200 Hz being the approximate chemical-shift separation for the two ortho carbons). (Before this motion is discussed in more detail, note that the methyl carbon resonance displays an increasing broadening as the temperature is lowered in the experiment. This behavior is discussed in more detail later in this section.) The coalescence of the aromatic resonances in the piperidine-cured epoxy is suggestive of a classical chemical exchange process in liquids: ie, as the exchange frequency τ_e between two nuclear sites

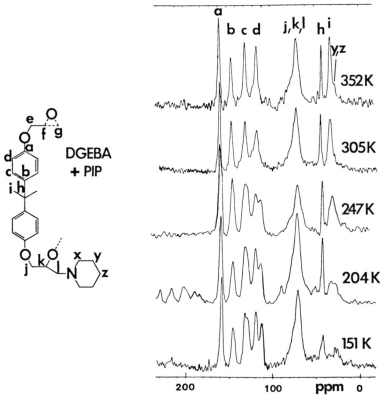

Figure 3-13. The ^{13}C spectra over a 200 K temperature range of the epoxy resin diglycidyl ether of bisphenol A (DGEBA) cured with piperidine (PIP). In the legend, one of the epoxide groups of the monomer is unreacted, whereas the other is shown in one possible structure with the piperidine molecule. (The spinning side bands in the 204 K spectrum arise from a slower rotation rate and should be ignored.) (Reprinted with permission from Garroway, A. N.; Ritchey, W. M.; Moniz, W. B. *Macromolecules* **1982**, *15*, 1051. Copyright 1982, American Chemical Society.)

increases, each line of the doublet of resonances (representing the two magnetically inequivalent sites) begins to broaden; then the lines merge into a single
line that narrows as the exchange rate continues to increase (narrowing continues to be observed to the point that other factors dominate the linewidth or
spin–spin relaxation process). However, this simple classical model fails to
account for the features of Figure 3-13; eg, a much wider temperature range is
required to complete the coalescence process than would be predicted by the
model when applied to the resonance separation (ca. 7 ppm) and observed
coalescence temperature (ca. 273 K). Furthermore, the ortho-carbon doublet
appears to coalesce by filling in its center rather than the two members
merging into one.[2] The classical picture fails for the glassy polymer because
the motion is not described well by a single correlation time model based on
an exponential decay of correlation for the local field.[31] Indeed, the dielectric
and mechanical relaxations of glassy polymers are generally not Debye-like
but require a distribution of correlation times for their description.

Garroway et al[2] have examined the NMR results on the epoxies employing
a Williams-Watts distribution function[32] to a motional process of 180° phenyl
ring flips. Two cases have been examined: (1) an inhomogeneous distribution,
where the distribution of exponential correlation times arises from a spatial
variation of single-exponential processes, and (2) a homogeneous distribution
which is a parameterization of an unknown nonexponential autocorrelation
function that describes the phenyl group motion. In an amorphous polymer
the inhomogeneous distribution could have its origin in the local structural
heterogeneity, whereby phenyl groups near regions of crosslinking or in
regions of unfavorable frozen-in chain conformations would have different
activation energies for ring flips than in other regions of the sample.[31] The
homogeneous distribution describes all common molecular processes as
having the same rate, but one which varies in time. The NMR lineshape distinguishes the two classes of distributions, because in the inhomogeneous case
spectra from broad and narrowed lines can coexist. Unfortunately, the
resolution in the epoxy case is not sufficient to allow the distinction and Garroway et al[2] have carried out a full analysis using both distributions and the
Williams-Watts function with a width parameter, α, inferred from the β-
relaxation process observed in mechanical loss spectroscopy.

The results from analysis of the NMR lineshapes and mechanical relaxation
data are summarized in Table 3-2. Whereas the NMR results are consistent
with the activation energy ($E = 63$ kJ/mol) and width parameter ($\alpha = 0.28 \pm$
0.02) found from mechanical spectroscopy, the 180° ring flips are detected to
be $3000 \times$ (inhomogeneous distribution) or $20 \times$ (homogeneous distribution)
slower by NMR. The data can be reconciled by allowing less than 180°
motions of the phenyl ring, ie, diffusive motion. A reorientation of ca. 3° per
root-mean-square step length (inhomogeneous distribution) or 40° per step
(homogeneous distribution) brings the NMR and mechanical data into accord.
Unlike 180° jumps, these smaller reorientations can couple to the applied
stress field to produce a mechanical relaxation and indeed, may be motions

TABLE 3-2. WILLIAMS-WATTS AUTOCORRELATION FUNCTION PARAMETERS FOR EPOXY RING MOTION[a]

	α^b	E (kJ/mol)	αE (kJ/mol)	τ_o (s)^c	τ_p(295 K) (μs)^d
β-Transition (mechanical relaxation)	0.28	63	18	7×10^{-18}	1
NMR lineshape analysis (180° phenylene reorientation)					
Inhomogeneous distribution	—	—	18^e	2×10^{-14f}	3000
Homogeneous distribution	0.28^g	60	17	4×10^{-16}	20

[a] Reprinted with permission from Garroway, A. N.; Ritchey, W. M.; Moniz, W. B. *Macromolecules* **1982**, *15*, 1051. Copyright 1982, American Chemical Society.
[b] Determines shape of the Williams-Watts distribution.
[c] Arrhenius prefactor for temperature dependence of τ_p.
[d] Determines position of Williams-Watts distribution on time axis.
[e] Determined only as the product.
[f] Assumed mechanical relaxation activation energy of 63 kJ/mol to calculate τ_o.
[g] Assumed. Reasonable fit for $\alpha = 0.28 \pm 0.05$.

superimposed on ring flips for glassy polymers [eg, refer to the data on polystyrene later in this chapter (Relaxation Measurements in the Solid State) or the data on polycarbonate (PC) in Chapter 7]. The important results from the epoxy study are that: (1) phenyl ring motion does take place; (2) the distribution of relaxation times for NMR and mechanical loss data are virtually the same when small-step diffusion is allowed; and (3) small-step ring diffusion accompanying 180° ring flips may be one of the molecular processes responsible for the β relaxation in epoxies. Clearly, this type of study indicates the potential of high-resolution, solid-state NMR to provide insight into local characteristics of glassy polymers and their possible relationship to observed mechanical properties.

Motional Information from Averaging of the Chemical Shielding Tensor

In principle, observation of the carbon chemical shielding tensor (CST) also provides a powerful means to elucidate low-frequency (kilohertz and below) motions in polymers. The collapse of the CST from its rigid lattice value accompanies the onset of motions on a time scale (order of milliseconds) that is roughly the inverse of the shielding anisotropy. If the principal elements of the CST can be assigned to specific directions in the molecular coordinate system, then the manner in which the chemical-shift anisotropy pattern changes as a function of such variables as temperature or diluent may provide insight into the specifics of the accompanying molecular motion (eg, motion along a chain axis, motion of an aromatic ring). In Chapter 7, Jones gives an excellent example of the use of ^{13}C enrichment in polycarbonate (the ring carbon ortho to the carbonate group) to define specific motions from changes of shielding patterns.[33] The ring motions accompanying a temperature change of 260°C (-160 to 100°C) can be defined from the change of anisotropy pattern from nonaxial to near axial symmetry. Full lineshape analysis is consistent with ring motion consisting of π flips plus restricted rotation, which is also in accord with the findings from deuterium NMR studies.[34]

It has been seen in Chapter 2 that overlap of the shielding anisotropies usually precludes detailed analysis of orientational effects in multiline spectra. Fortunately, because the chemical-shift anisotropy is totally static in origin, sample spinning at a rate much less than the anisotropy is sufficient to displace the spectral density completely from zero frequency and remove the line broadening from this source.[35] Side bands are prominent in slow-spinning experiments when the chemical shift anisotropies are large. However, the intensity pattern of the spinning side bands reflect the magnitude of the shift anisotropy and by proper analysis this information can be recovered by measuring spectra at a few spinning frequencies.[36] An alternative approach to obtaining chemical-shift anisotropy data in complex systems is to enrich isotopically with ^{13}C at sites of interest (as in the study on PC). Enrichment levels are required that produce a signal-to-noise ratio (S/N) sufficient to reduce the signal intensity from nonenriched sites to the noise level yet do not introduce problems from ^{13}C–^{13}C dipolar broadening.

In less complex systems, variable-angle spinning can be used to determine the elements of the shielding tensor. For example, in the static ^{13}C spectrum of glassy cis-1,4-polybutadiene (CPBD) at $-150°C$ only two elements (Figure 3-14A) of the shielding tensor σ are resolved. The third overlaps the —CH$_2$— pattern, thus making it impossible to assign the tensor fully. However, variable-angle spinning allows the full assignment to be made through the angular functional dependence of the spectrum (see Figure 3-14). With the sample spinning at an angle of 87°, the $P_2(\cos\theta)$ functional dependence (see Chapter 2) predicts a spectrum, relative to the static case, reduced in width by a factor of 0.492 and with the principal elements of the shielding tensor reversed with respect to the isotropic shift. On this basis, the principal values of σ calculated from the data of Figure 3-14, are $\sigma_{11} = 236 \pm 4$, $\sigma_{22} = 115 \pm 4$, $\sigma_{33} = 35 \pm 4$. Assuming a similar orientation of the shielding tensor as in ethylene ($\sigma_{11} = 236$, $\sigma_{22} = 124$, $\sigma_{33} = 27$),[37] these elements correspond to shielding approximately perpendicular (but in plane) to the double-bond direction, parallel to the bond direction, and perpendicular to the plane of the bond, respectively. The isotropic shift obtained from these values is 128.7 ppm as compared to 129.7 determined from the spectrum in the elastomeric state (Figure 3-14C) and the MAS spectrum at -150 (Figure 3-14B).

Even in spectra of overlapping anisotropy patterns, it is possible by use of multiple techniques to gain limited insight into motions in glassy systems. For example, Edzes[38] has utilized results of lineshape-fitting, slow sample spinning at the magic angle, and relaxation differences between carbons to determine the principal elements of each CST in glassy poly(methyl methacrylate) (PMMA). In Figure 3-15, the ^{13}C magic angle spinning spectrum of PMMA at 23°C is contrasted with a nonspinning spectrum. The elements of the carboxyl carbon shielding tensor are resolved and are indicated in the nonspinning

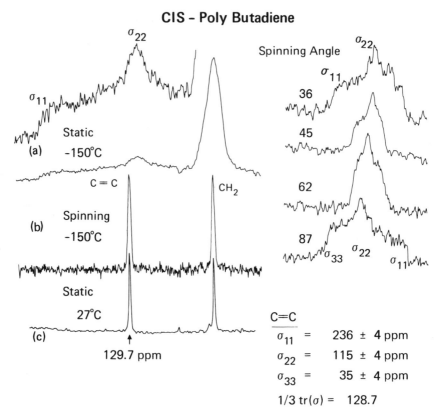

CIS - Poly Butadiene

Figure 3-14. Cross polarization ^{13}C NMR spectra of cis-1,4-polybutadiene at $-150°$C (a) static; (b) spinning at the magic angle; (c) static at 23°C. For the static spectrum at $-150°$C two elements of the shielding tensor are indicated for the vinyl carbon in the $4\times$ vertical expansion (inset). Also CP spectra as a function of spinning angle at $-150°$C are shown at the right and tensor elements are indicated. (Reprinted with permission from *ACS Symp. Ser.* **1980**, *142*, 193. Copyright 1980, American Chemical Society.)

spectrum; severe overlap of the anisotropy patterns from the other carbons makes impossible the distinction of the individual elements for these carbons directly from the spectrum. However, using the multiple-technique approach, Edzes arrived at the values for the CSTs in PMMA given in Table 3-3. These data are useful in studying the effect of diluents on molecular motion in this polymer. Figure 3-16 depicts the nonspinning ^{13}C spectra of PMMA polymerized in the presence of varying concentrations of chlorobenzene (CB). The MMA monomer and CB are miscible in all proportions and CB simply acts as a diluent during polymerization of methyl methacrylate but as a plasticizer for the polymer.[39,40] The plasticization effects of CB on PMMA are reflected clearly by the changes in the observed anisotropy patterns. At a concentration of 30% (w/w) CB to polymer, the low-field side of the carboxyl carbon aniso-

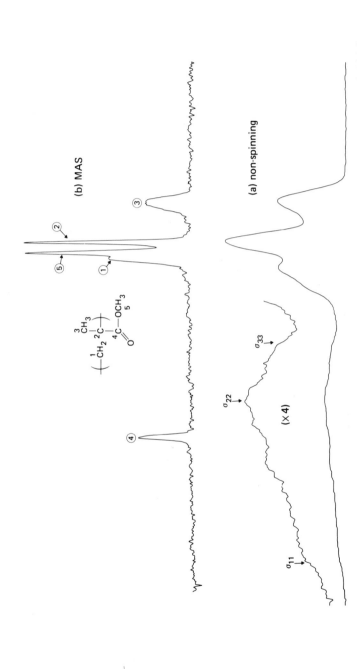

Figure 3-15. Cross-polarization spectra of glassy PMMA at 23°C (A) nonspinning (the principal elements of the carboxyl shielding tensor are marked) (B) with magic angle spinning.

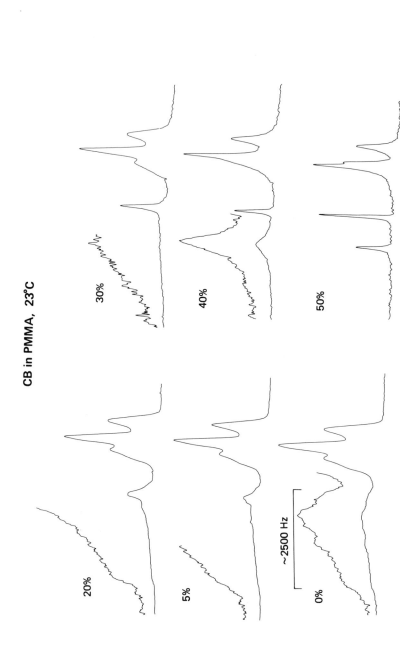

Figure 3-16. Cross-polarization ^{13}C spectra of PMMA as a function of chlorobenzene (CB) content at 23°C. Concentrations (% w/w) of CB–polymer are indicated.

TABLE 3-3. CARBON-13 CHEMICAL SHIELDING TENSOR ELEMENTS OF CARBONS IN POLY(METHYL METHACRYLATE)[a,b]

	σ_{11}	σ_{22}	σ_{33}	$\bar{\sigma}$
Carboxyl, —COO—	268 ± 2	150 ± 1	112 ± 1	177
Methoxy, —OCH$_3$	80 ± 1	63 ± 1	10 ± 2	51
Methylene, —CH$_2$—	79 ± 3	48 ± 1	29 ± 2	52
Quaternary, $-\overset{\mid}{\underset{\mid}{C}}-$	52.5 ± 0.5	43.0 ± 0.5	36.5 ± 0.5	44
Methyl, —CH$_3$	—	—	—	12–24

[a] From reference 38. Reprinted with permission of Butterworth and Co. (Publishers), Ltd.
[b] σ values are given in ppm relative to TMS. $\bar{\sigma}$ is obtained from the MAS spectrum. (Errors are given relative to $\bar{\sigma}$. The absolute accuracy of $\bar{\sigma}$ is ± 0.5 ppm.)

tropy pattern no longer displays a sharply defined σ_{11} tensor element. Correspondingly, there is a loss of some of the resolution in the downfield edge of the methoxy carbon anisotropy pattern. These results suggest the onset of motion of the ester side group at this level of plasticizer concentration. Large-amplitude motion on the time scale of the inverse of the carboxyl anisotropy pattern (ms) is apparent in the 40% (w/w) spectrum. The carboxyl anisotropy is considerably collapsed, whereas narrowing of the CST patterns in the upfield portion of the spectra suggest backbone motion has set in. A plot of the glass transition temperature (T_g) versus CB/PMMA composition is shown in Figure 3-17. At the CB concentrations of 30% (w/w) and 40% (w/w) (70% and 60% polymer) the T_g is 32°C and 18°C, respectively. Since the NMR spectra were measured at 23°C, it is clear the changes in CST spectral characteristics between these two concentrations reflect the glass transition. At 50% w/w CB/PMMA, large-scale motions rapid on the time scale of the CSTs of the polymer backbone and sidechain carbons are apparent in the collapsed anisotropy patterns.

Motional Broadening of Resonance Lines

Returning to the observation of methyl resonance broadening in the piperidine-cured epoxy resin, this phenomenon can be delineated more clearly in the context of the CPMAS ^{13}C spectra of a 90% isotactic, 70% crystalline sample of polypropylene (PP) as a function of temperature[41,42] (Figure 3-18). The methine, methylene, and methyl carbons of the repeat unit are resolved at 300 K; however, the interesting spectral feature is the progressive broadening of the methyl resonance as the temperature is lowered. At ≈ 110 K, the resonance is broadened to the point of disappearing from the spectrum. However, at temperatures below ~ 77 K, the methyl resonance narrows and reappears in the spectrum. These observations find their origin in a spin system that is subjected to both stochastic (molecular) and coherent (applied rf fields) motions. Rothwell and Waugh[43] have developed the theory for T_2 (the inverse

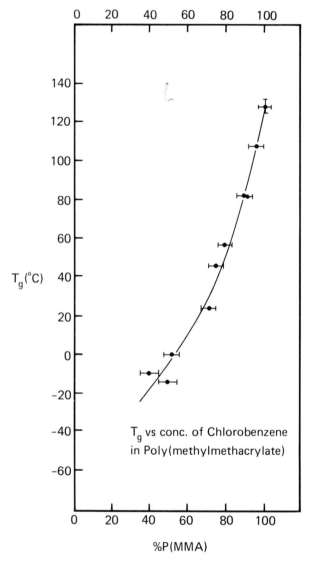

Figure 3-17. Differential scanning calorimetry (DSC)-measured T_g for PMMA–CB system as a function of wt% polymer.

of the ^{13}C linewidth) for this case. For isotropic molecular motion the appropriate expression is:

$$1/T_2 = \frac{\gamma_C^2 \gamma_H^2 \hbar^2}{5 r_{CH}^6} \left(\frac{\tau_c}{1 + \omega_2^2 \tau_c^2} \right). \tag{3-2}$$

Referring to Eq. (3-2), the profile of linewidth versus temperature shows a

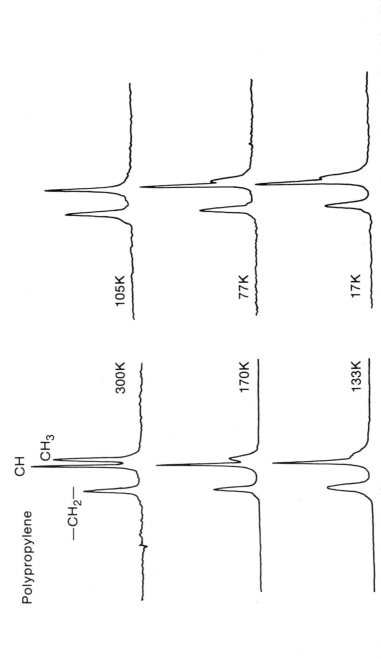

Figure 3-18. Cross-polarization MAS ^{13}C spectra of polypropylene as a function of temperature. (From reference 41. Copyright 1983 by International Business Machines Corporation; reprinted with permission.)

maximum where the correlation time for molecular motion, τ_c, is equal to the modulation period of the decoupling, $(1/\omega_2)$. For PP, this occurs around 110 K. In the "short correlation" limit $(\omega_2\tau_c \ll 1)$ (high temperature), the linewidth is reduced by the rapid motional averaging. In the case of PP, methyl motion about the C_3 axis is sufficiently fast in this regime to provide an incoherent averaging of the C–H dipolar interaction with the methyl protons and yield a narrow line. However, in the "long correlation" limit $(\omega_2\tau_c \gg 1)$ (low temperature), the linewidth is reduced by efficient decoupling of C–H dipolar interactions. The spectra of PP in Figure 3-18 are consistent with the progression of the methyl resonance through the linewidth regions as the temperature is lowered. The reappearance (narrowing) of the methyl resonance at 77 K indicates that the "long correlation" time regime has been reached.

Further proof of the progressive changes in correlation time for methyl rotation as the temperature is lowered is provided by the field dependence of the linewidth as displayed in Figure 3-19. As shown, the methyl linewidth at 160 K is independent of the applied decoupling field (ω_2), whereas at 77 K the linewidth varies approximately as the inverse square of the decoupling field. This is in accord with the expected dependence for the linewidth behavior from Eq. (3-2) for a transition from extreme narrowing to long correlation time regimes. Additionally, for a value of $\omega_2/2\pi = 57$ kHz (that of the decoupling field used to obtain the spectra in Figure 3-18), Eq. (3-2) predicts the maximum broadening to occur at $\tau_c = 2.8$ μs. Using the proton $T_{1\rho}$ relaxation data on isotactic PP to assess the temperature at which this value of τ_c obtains,[44] the maximum broadening is predicted to occur at ~ 105 K, in excellent agreement with the observations.

This broadening of the methyl resonance observed in PP is also found in polycarbonate, PMMA, and epoxy polymers. It should be a general phenomenon for polymers with rapidly reorienting side groups or main-chain carbons subject to motion. For the crystalline phase of semicrystalline systems, where the local molecular structure is relatively homogeneous, and motion characterized by a single correlation time, severe broadening should result in the "disappearance" of resonance lines from the spectra. For glassy systems, where there is more heterogeneity in the local molecular environment, and motion is characterized by a distribution of correlation times, the effect may result in significant changes in resonance lineshape as a function of temperature, because the carbons in differing environments undergo severe broadening at different temperatures. Indeed, the temperature dependence of the linewidth of the methyl carbon resonance from glassy polycarbonate (Figure 3-20) does not show broadening effects as severe as PP. The same is true of PMMA and to a lesser extent the epoxies. Thus, the phenomenon of motional broadening of some resonances in solid-state ^{13}C spectra of polymers taken as a function of temperature is understood and can be utilized (see Rothwell and Waugh[43]) to obtain information on group rotational rates and rotational barriers. Of course, cases of severe broadening limit the ability to make spectral measurements (eg, relaxation times) in such temperature intervals.

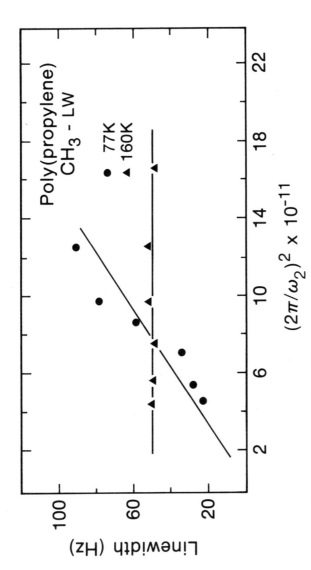

Figure 3-19. Observed linewidth of the methyl carbon of PP at two temperatures as a function of the inverse square of the decoupling field, v_1 in kHz. ●, 77 K; ▲, 160 K.

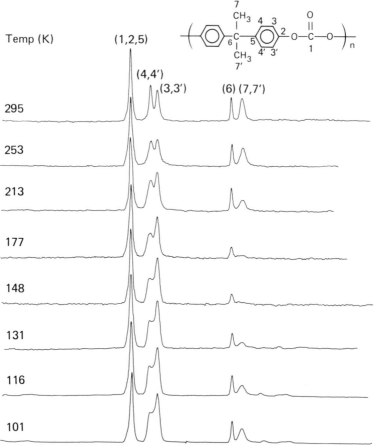

Figure 3-20. Carbon-13 CPMAS spectra of polycarbonate as a function of temperature.

Relaxation Measurements in the Solid State

Various relaxation measurements have utility for elucidating polymer dynamics in the solid state. Indeed, it was the desire to understand the origin of microscopic motions and their relationship to macroscopic behavior for glassy polymers that prompted Schaefer and Stejskal to devise CPMAS experiments.[45] These workers have carried out the pioneering studies and the overwhelming majority of all CPMAS studies on the motional characteristics of glassy amorphous polymers.

One of the principal advantages of CPMAS experiments is that the resolution allows relaxation data to be obtained on each resolved carbon type of the molecule in the solid state. If a sufficient number of unique resonances exists, the results can, in principle, be interpreted in terms of local motions (eg, methyl rotation, segmental modes in polymers). This presents a distinct advantage over the more common proton-relaxation measurements, where efficient spin diffusion[46] usually results in averaging of relaxation behavior over the ensemble of protons to yield a single relaxation time for all protons, making interpretation of the data in terms of localized motions difficult.

There are a large number of relaxation parameters of interest for the study of organic solids and polymers, and these include the ^{13}C and ^{1}H spin–lattice relaxation times (T_1^C and T_1^H), the respective spin–spin relaxation times (T_2), the nuclear Overhauser enhancement (NOE), the proton and carbon rotating-

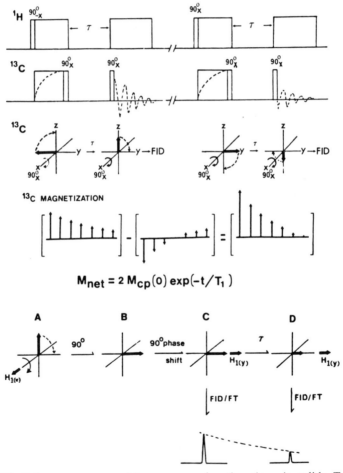

$$M_{net} = 2\,M_{cp}(0)\exp(-t/T_1)$$

Figure 3-21. Pulse sequences used to measure relaxation times in solids. Top: spin–lattice relaxation in the Zeeman frame T_1^C; bottom: spin–lattice relaxation in the rotating frame $T_{1\rho}^C$. (A) 90° rf pulse; (B) spin lock, 90° phase shift; (C) no delay; (D) after delay of τ.

JAMES R. LYERLA

frame relaxation times ($T_{1\rho}^H$ and $T_{1\rho}^C$), the C–H cross-relaxation time (T_{CH}), and the proton-relaxation time in the dipolar state (T_{1D}).[47] Not all of these parameters provide information in a direct manner; nonetheless, the inferred information is important in characterizing motional frequencies and amplitudes in solids. The initial studies of carbon relaxation in solids have empha-

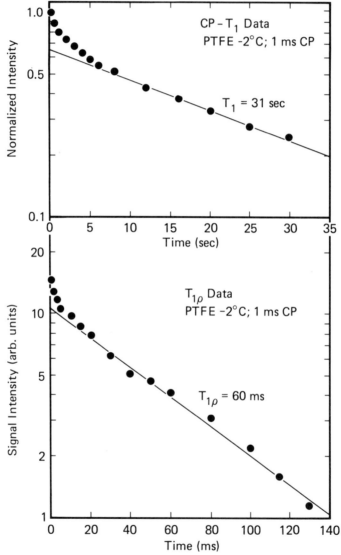

Figure 3-22. Carbon-13 NMR T_1^C and $T_{1\rho}^C$ relaxation data (intensity versus time) for PTFE at $-2°C$ and 1.4T. The sample has a crystallinity of 67%. (Reprinted with permission from *ACS Advan. Chem. Ser.* **1983**, *203*, 455. Copyright 1983, American Chemical Society.)

sized T_1 and $T_{1\rho}$ measurements, which provide information on molecular motions in the megahertz and kilohertz frequency ranges, respectively.

The pulse sequences used to measure the standard and rotating-frame relaxation times are shown schematically in Figure 3-21. The T_1 methodology[48] allows cross-polarization enhancement of the NMR signal and so differs from the usual inversion–recovery sequence used in solution measurements.[47] The $T_{1\rho}$ experiment[45] also employs CP enhancement but otherwise is similar to the corresponding liquid-state measurement (see Chapter 2). The respective time constants are derived from semilog plots of intensity versus time. In solution these plots are usually linear, indicating a single exponential relaxation process. However, in the solid state, such plots may show nonexponential behavior, particularly for glassy polymers. This is illustrated in Figure 3-22, which depicts semilog plots of typical T_1^C and $T_{1\rho}^C$ relaxation data (intensity versus time) at $-2°C$ for polytetrafluoroethylene (PTFE), a semicrystalline polymer. The decays are clearly nonexponential, indicative of multiple relaxation behavior. However, for both plots, the behavior of PTFE can be characterized by one relaxation time at long measurement times, associated with about 65–70% of the total signal intensity as judged by the y intercept. On this basis and previous ^{19}F NMR relaxation studies,[49] the relaxation component at long measurement times, which represents a single relaxation time, is ascribed to crystalline regions of the polymer (in agreement with IR analysis on the sample). The faster relaxing component of the resonance line is attributed to noncrystalline regions and is described not by a single time constant, but by a distribution of relaxation times that is in accord with the expected site heterogeneity of the glassy state.

Spin–Lattice Relaxation Studies

In approaching the limited T_1 data on glassy systems, several points on relaxation in the solid state can be illustrated by ^{13}C T_1 data on highly crystalline isotactic polypropylene.[41,42] Because of the regularity associated with crystal habit, the relaxation for each of the carbons in this polymer is governed by single exponential behavior. In addition, this is the only polymer for which relaxation has been studied by CPMAS methods over an extensive range of temperature. The data for each of the carbons in the monomer unit are presented in Figure 3-23. Over the temperature range, each carbon displays an individual relaxation time. (Of course, as discussed earlier, because of motional broadening, the methyl resonance can be followed over only a limited temperature range.) This is a result similar to that found in solution: however, unlike in solution, the backbone carbon T_1 values of PP are one to two orders of magnitude longer than the methyl carbon. Also, unlike in solution, the methine T_1 is shorter than the methylene.

To analyze these data, the rate of methyl relaxation, which appears to be dominated by methyl C_3 reorientation, is calculated based on a Bloembergen-Purcell-Pound (BPP) dipolar formalism[47] and the correlation time data from

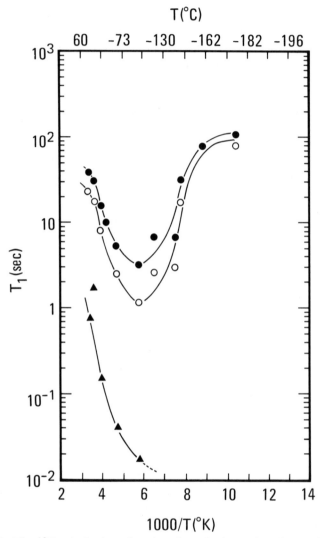

Figure 3-23. The ^{13}C spin–lattice relaxation times for isotactic polypropylene methylene ($\bullet$), methine ($\bigcirc$), and methyl ($\blacktriangle$) carbons. (From reference 41. Copyright 1983 by International Business Machines Corporation; reprinted with permission.)

proton T_1 studies.[44] This gives a value of 12 ms at $-100°C$, in good agreement with the observed value of 17 ms. Similar results are obtained at other temperatures, indicating the CH_3 T_1 mirrors methyl motion; however, the methyl motion also seems to dominate the backbone relaxation. Apparently, backbone motions are characterized by such small amplitudes and low frequencies that contributions from the direct C–H interactions to spectral density in the MHz region of the frequency spectrum are minor relative to those from the methyl group. The $1/r^6$ distance dependence of dipolar relaxation therefore accounts for both the long T_1 values of CH and CH_2 carbons relative to the CH_3 carbon and the shorter T_1 values for methine carbons relative to methylene carbons (despite there being two direct C–H interactions for the methylene carbon). The fact that the observed T_1 minimum for CH and CH_2 carbons in Figure 3-23 is close to that reported for a proton T_1 minimum (at 30 MHz)[44] in PP that was assigned to methyl reorientation provides unequivocal support for the dominance of main-chain and sidechain spin–lattice relaxation by methyl protons. The domination of the spin–lattice relaxation time for CH and CH_2 carbons in PP by methyl reorientation is clearly disappointing, because the potential for information on backbone motion from the high resolution of the CPMAS experiment is not realized. The implication is that it may not be possible to assess backbone motion in crystalline materials having rapidly reorienting side groups without resorting to deuterium substitution of these side groups. This result also has implications for relaxation data for glassy polymers.

Edzes and Veeman[50] have studied the spin–lattice relaxation of PMMA at ambient temperature as a function of tacticity and plasticizer type and content. The decay of magnetization for the various carbons showed nonexponential behavior, particularly the methoxy and quaternary carbons. They used the initial portion of the decay curves to characterize the relaxation rates, $R \; (= 1/T_1)$ by fitting the data with an expression of the form:

$$M(t) = M(0)e^{-Rt(1 + fRt)}$$

where M is the magnetization and f a "degree of nonexponentiality" factor. The relaxation rates are given in Table 3-4. Examination of the data using a dipolar relaxation formalism indicates that the methyl motion dominates the relaxation of the other carbons, except the methoxy. For the assumed C–H dipolar relaxation, the R values of the other carbons are approximately in relation to those predicted for the r_{CH}^{-6} distance between these carbons and the methyl protons. Further verification for this conclusion is found in the scaling of the various carbon relaxation rates with that of the methyl, eg, the smaller relaxation rate of the methyl in the isotactic polymer as compared to the syndiotactic polymer is also reflected in the other carbons of the isotactic polymer. For this same reason of methyl-dominated relaxation there is little effect of plasticizer on the backbone T_1s. It has been seen (Figure 3-16) that plasticizer affects PMMA backbone motion; however, the imparted backbone motions (in the Edzes and Veeman experiments) are still slow compared to the megahertz frequencies necessary for spin–lattice relaxation. Because the fast methyl motion is not strongly affected by plasticizer and because the methyl protons dominate the backbone relaxation in the undiluted polymer, there is no drastic effect of plasticizer on backbone T_1. As in PP, therefore, the methyl sidegroup dominates the backbone relaxation and little is learned about main-chain motion.

Menger et al[51] have measured T_1^C values and NOEs for crystalline and amorphous regions of semicrystalline polyoxymethylene. The data at two dif-

TABLE 3-4. Carbon-13 Spin–Lattice Relaxation Rates at 45.27 MHz of Some Solid PMMA Samples[a]

Sample	Tacticity (mm/mr/rr triad %)	Source[c]	Relaxation rates, R_1 (s^{-1})				
			α-CH$_3$	C=O	CH$_2$	—OCH$_3$[b]	—C—[b]
Commercial Perspex	4/31/65	1	19	0.10	0.10	0.16 ($f = 0.07$)	0.25 ($f = 0.07$)
Atactic PMMA	similar	1	20	0.10	0.11	0.16 ($f = 0.05$)	0.25 ($f = 0.07$)
+ 20% w/w bisdioxan[c]		1	—	0.11	0.12	0.18 ($f = 0.07$)	0.28 ($f = 0.07$)
+ 40% w/w bisdioxan[c]		1	21	0.12	0.16	0.18 ($f = 0.03$)	0.29 ($f = 0.06$)
+ 10% w/w monomer[c]		2	—	0.11	0.14	0.19 ($f = 0.06$)	0.28 ($f = 0.08$)
Isotactic PMMA	95/5/0	3	10	0.06	0.09	0.14 ($f = 0.11$)	0.17 ($f = 0.11$)
Syndiotactic PMMA	1/9/89	3	19	0.10	0.10	0.17 ($f = 0.07$)	0.28 ($f = 0.09$)
Estimated error in the relaxation rates (%)			10	15	30	10 (f:50)	10 (f:50)

[a] From reference 50. Reprinted with permission.
[b] f characterizes a nonexponential relaxation.
[c] Bisdioxan = 2,4,8,10-tetraoxaspiro-[5.5]-undecane: monomer = methyl methacrylate, MMA; PMMA sources: (1) Nujverheidsorganisatie TNO, Delft, The Netherlands, (2) Philips Research Laboratories, Eindhoven, The Netherlands, (3) Department of Polymer Chemistry, University of Groningen, The Netherlands.

TABLE 3-5. CARBON-13 RELAXATION PARAMETERS OF SOLID POLYOXYMETHYLENE AT ROOM TEMPERATURE[a]

	Amorphous part	Crystalline part
T_1 (at 45 MHz)	75 ms	15 s
NOE (at 45 MHz)	1.6	1.3
T_1 (at 15 MHz)	50 ms[b]	~1.5 s[b]
$T_{1\rho}$	17.5 ms	~0.5–3 ms

[a] Reprinted with permission from Menger, E. M.; Veeman, W. S.; DeBoer, E. *Macromolecules* **1982**, *15*, 1406. Copyright 1982, American Chemical Society.
[b] Estimated errors at least 20%.

ferent applied magnetic fields are presented in Table 3-5. Using an analysis based on relaxation controlled by a C–H dipolar mechanism, it was concluded that molecular motions with similar correlation times occur in both regions of the polymer. However, the amplitudes of motion, which possibly involve rotational oscillations of CH_2 units around the helical axis, are much more restricted in the crystalline phase. Schroter and Posern[52] have carried out preliminary T_1^C measurements on semicrystalline polyethylene. Finally, Kaplan[53] has measured ^{13}C T_1 values for the carbons of polystyrene (PS) and poly(vinyl methyl ether) (PVME) blends. The measured values show that on blending the polymers, the motions of the component homopolymers approach one another; ie, the PS chains become more flexible and the PVME chains more constrained. This suggests some, but not homogeneous, mixing of chains at the molecular level. Other than these studies, there are few examples of T_1^C studies on amorphous polymers.

Carbon Rotating-Frame Relaxation Studies

The paucity of measurements of T_1^C in amorphous glassy polymers arises in part from the long time constants involved and the difficulties in interpretation discussed above. However, in large measure it also arises from the fact that it is not the megahertz motions reflected in T_1^C determinations that are likely to be important in determining the mechanical properties of amorphous polymers below T_g.[45] Instead, it is likely to be main-chain motions in the 15–100 kHz region that are important in this connection. The ^{13}C rotating-frame relaxation time, $T_{1\rho}^C$, is sensitive to this frequency region (see Chapter 2), which is characteristic of relatively long-range cooperative motions of a polymer chain below T_g.[45] Schaefer in his pioneering work[45] has emphasized $T_{1\rho}^C$ measurements. For example, in a comprehensive paper on carbon $T_{1\rho}$ in glassy systems, Schaefer et al[45] examined seven polymers in considerable detail. The relaxation data were evaluated in terms of spin–lattice processes and interpreted in terms of distinct main- and side-chain motions in the 10–50 kHz regime. The data allowed conclusions to be drawn as to the short-range nature of certain low-frequency side-group motions (eg, the ester side-group motion in PMMA), and the long-range cooperative nature of some main-chain motions (eg, polycarbonate and polysulfone). Interestingly, a direct correlation was shown between the ratio of cross-polarization and rotating-frame relaxation times ($T_{CH}/T_{1\rho}^C$) for the main-chain carbons of a polymer with the toughness or impact strength of the polymer. Before reviewing other $T_{1\rho}^C$ results, a complication that can arise in attempting a quantitative interpretation of $T_{1\rho}^C$ as a motional parameter is discussed. Figure 3-24,[54] depicting a thermodynamic model for the carbon rotating-frame experiment, indicates the origin of the complexity.

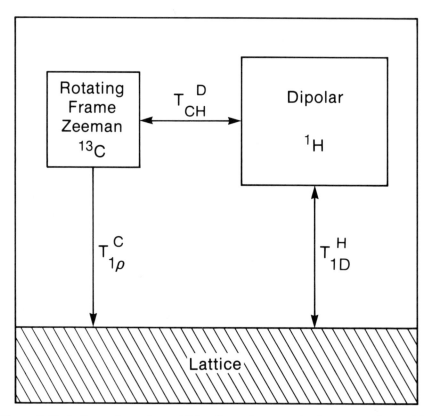

Figure 3-24. A thermodynamic model for the carbon rotating-frame experiment. The ^{13}C nuclei are coupled to the lattice by the molecular motion with $T_{1\rho}^C$ and to the dipolar reservoir by spin–spin fluctuations with the time constant T_{CH}^D. The dipolar reservoir is identified with proton–proton dipolar ordering. This is because ^{13}C–^{13}C dipolar coupling is very weak, few nuclei are involved, and the proton–carbon dipolar coupling is stirred out by the carbon irradiation. T_{1D}^H measures the dipolar spin–lattice relaxation under carbon decoupling and it is expected to differ only slightly from the conventional T_{1D}, which includes ^{13}C–^1H ordering. (From reference 54. Reprinted with permission.)

In a rotating-frame experiment, the ^{13}C nuclei are coupled to the lattice by molecular motions, the interaction being characterized by a time constant $T_{1\rho}^C$. In addition, the ^{13}C nuclei are coupled to the proton dipolar system by spin (^{13}C)–spin (^1H) fluctuations characterized by a time constant T_{CH}^D. In turn, the proton dipolar system is coupled to the lattice, the interaction being characterized by a time constant T_{1D}^H. Under magic angle spinning conditions, Garroway[55] has shown that dipolar order is destroyed rapidly so that $T_{1\rho}^C$, $T_{CH}^D \gg T_{1D}^H$. In the ^{13}C $T_{1\rho}$ experiment, the ^{13}C spin system is prepared initially in a highly nonequilibrium state and then the approach to equilibrium with the lattice monitored by decay of the initial ^{13}C magnetization. Equilibrium with the lattice can be achieved via a pathway involving $T_{1\rho}^C$ or a pathway involving T_{CH}^D or their combination depending on their relative magnitudes. Only the pathway involving $T_{1\rho}^C$ provides direct information on lattice motions; however, the experiment (in its simplest form) does not distinguish the relative contributions of the two relaxation pathways. Hence, to treat the measured rotating-frame relaxation time $T_{1\rho}^*$ as a motional parameter requires determination that the spin–spin relaxation pathway is minor, ie, $T_{CH}^D \gg T_{1\rho}^C$ or the determination of $T_{1\rho}^C$ and T_{CH}^D.

The complexity in interpretation of $T_{1\rho}^*$ measurements can be illustrated by once again resorting to the semicrystalline polymer, isotactic polypropylene. [41,42] As before, the relaxation taking place in the crystalline region of the polymer is governed largely by a single exponential process. First, the measured relaxation time is defined as $T_{1\rho}^*$ where[54,56]:

$$(T_{1\rho}^*)^{-1} = (T_{1\rho}^C)^{-1} + (T_{CH}^D)^{-1} \tag{3-3}$$

and $T_{1\rho}^C$ and T_{CH}^D are given by Eqs. (3-4) and (3-5):

$$(1/T_{1\rho}^C) = \frac{N_H \gamma_H^2 \gamma_C^2}{20 r_{CH}^6} \hbar^2 f(\tau_c) \tag{3-4}$$

where:

$$f(\tau_c) = \left\{ 2J(\omega_1) + \frac{J(\omega_H - \omega_C)}{2} + \frac{3J(\omega_C)}{2} + 3J(\omega_H) + 3J(\omega_H + \omega_C) \right\}$$

$$J(\omega_i) = 2\tau_c/(1 + \omega_i^2 \tau_c^2)$$

and:

$$(1/T_{CH}^D) = 0.5 \sin^2 \theta \, M_{CH}^{(2)} J_D(\omega_{1C}) \tag{3-5}$$

where $M_{CH}^{(2)}$ is the second moment of the carbon nucleus from the dipolar interaction with protons, θ is the off-resonance angle of the applied rf field, and $J_D(\omega) = \pi\tau_D e^{-2\pi\nu_{1C}\tau_D}$, where τ_D is the correlation time for spin fluctuations. In Eq. (3-4) it is assumed that a Lorentzian autocorrelation function is appropriate for both motional and spin spectral densities $J(\omega)$. The two pathways of relaxation have a different dependence on the applied rotating-frame field in the frequency regime where correlation times are long compared to the inverse of Zeeman field Larmor frequencies. For this case, only the $J(\omega_1)$ term is important in expression (3-4) and $(T_{1\rho}^C)^{-1} \propto \omega_1^{-2}$. On the other hand, for spin contributions T_{CH}^D has an exponential dependence on ω_1. This field dependence is one method that allows a separation of the contributions to $T_{1\rho}^*$.

A plot of $T_{1\rho}^*$ versus temperature for the three carbons in PP at a value of $v_1 = \omega_1/2\pi = 57$ kHz is given in Figure 3-25. The CH and CH$_2$ $T_{1\rho}^*$ values do not change greatly in the temperature interval ca. 170–300 K and, in contrast

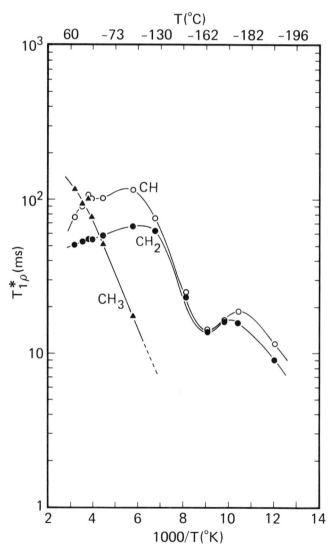

Figure 3-25. The $T_{1\rho}^*$ data for CH (○), CH$_2$ (●), and CH$_3$ (▲) carbons in polypropylene. (From reference 41. Copyright 1983 by International Business Machines Corporation; reprinted with permission.)

to the T_1^C behavior over the same temperature range, the CH_2 carbon has the shorter $T_{1\rho}^*$. Below 170 K, both backbone carbon $T_{1\rho}^*$s decrease rapidly and tend toward equality at the observed minimum at ca. 110 K. The methyl $T_{1\rho}^*$ declines rapidly as the temperature is lowered from ambient; however, because of the motional broadening, its behavior cannot be followed below ca. 150 K.

The observed temperature dependence of the $T_{1\rho}^*$ values in PP at $v_1 = 57$ kHz can be explained with the aid of the data depicted in Figure 3-26, A and B. In Figure 3-26A, the $T_{1\rho}^*$ at 299 K is plotted on semilog scale as a function of v_1. The $T_{1\rho}^*$ data for the CH and CH_2 carbons display the exponential dependence on rotating-frame field up to ca. 50 kHz, as predicted for the spin–spin (T_{CH}^D) relaxation pathway. Departures to less than exponential dependence for the data at 57 and 63 kHz indicate some motional contribution (via the $T_{1\rho}^C$ pathway in Figure 3-24) to $T_{1\rho}^*$ at these fields. Below $v_1 = 50$ kHz, the CH $T_{1\rho}^*$ is ca. 1.5–2 times longer than the CH_2 value. This is the functional dependence predicted by Eq. (3-5) when the nonbonded proton distribution about each carbon is very similar. Bonded proton contributions dominate $M_{CH}^{(2)}$, and bonded C–H dipolar interactions are identical. In this situation, τ_D is the same for both CH and CH_2 carbons and the difference in T_{CH}^D values arise from the ca. two times larger value of $M_{CH}^{(2)}$ for the CH_2 carbon. Indeed, the values of τ_D determined from the slopes of the respective lines in Figure 3-26A are within 5% (average value 28 μs).

The methyl carbon $T_{1\rho}^*$ data do not follow an exponential dependence on v_{1C} and are even independent of the rotating-frame field at > 50 kHz. This behavior is consistent with the motional pathway ($T_{1\rho}^C$) dominating $T_{1\rho}^*$ at high fields [note that in the fast motion limit ($\omega_{1C}^2 \tau_c^2 \ll 1$), a region applicable to methyl rotation at this temperature, $T_{1\rho}^C$ is not field dependent] and with increasing contributions of the spin–spin pathway at smaller rotating-frame fields.

The $T_{1\rho}^*$ data in Figure 3-26B at 120 K clearly demonstrate the importance of methyl motion on the relaxation process. At a v_1 of 63 kHz the CH has a shorter $T_{1\rho}^*$ than the CH_2, reversed from the situation at 33 kHz. The 63-kHz result is parallel to the T_1^C data, where the observed shorter values for the CH carbon relative to the CH_2 were demonstrated to arise from methyl motion dominating the relaxation. Further proof of the contribution of methyl motion to $T_{1\rho}^*$ of CH and CH_2 carbons at large v_1 values is provided by the observed minimum at 110 K in the $T_{1\rho}^*$ data (Figure 3-25). This minimum is in good agreement with a proton $T_{1\rho}$ minimum assigned to CH_3 motion.[44] Thus, the $T_{1\rho}^*$ data at $v_1 = 57$ kHz can be summarized as follows: above ca. 170 K the backbone carbon's relaxation is dominated by the spin–spin relaxation pathway, whereas below this temperature a spin–lattice pathway originating from CH_3 motion contributes strongly to $T_{1\rho}^*$. (Note that again the CH and CH_2 carbon relaxation is mirroring CH_3 motion and not backbone motion.) The methyl relaxation time is dominated by motional processes over the temperature range in which observation is possible.

The results on PP provide an indication of some of the difficulty that can be

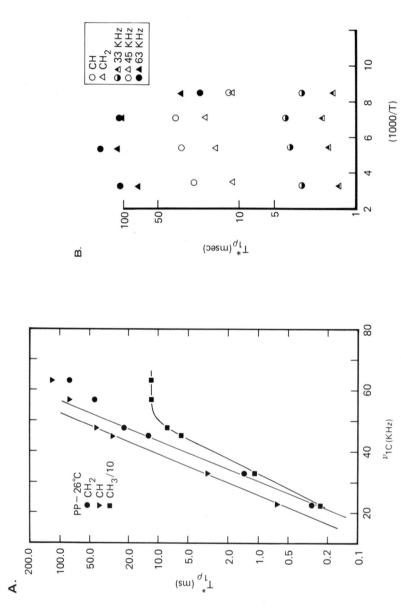

Figure 3-26. (A) The $T_{1\rho}^*$ data at 26°C for the carbons in PP as a function of the rotating-frame field. ●, CH$_2$; ▼, CH; ■, CH$_3$/10. (B) The $T_{1\rho}^*$ for methine (○) and methylene (△) carbons in polypropylene as a function of rotating-frame field and temperature; 33 kHz, half-shaded symbols; 45 kHz, open symbols; 63 kHz, filled symbols. (From reference 41. Copyright 1983 by International Business Machines Corporation; reprinted with permission.)

encountered in interpreting $T^*_{1\rho}$ data. Fortunately, the literature reveals that whereas the spin–spin process plays a dominant role in the rotating-frame relaxation of crystalline polymers, it is not a dominant feature for glassy polymers.[57] Indeed, Schaefer et al[58] have devised an elaborate protocol to distinguish the degree to which motional processes contribute to $T^*_{1\rho}$. For polymers, such as polystyrene, polycarbonate, and poly(ethylene terephthalate), at ambient temperature, motional processes are found to be dominant. This allows $T^*_{1\rho}$ to be interpreted as reflecting lattice motions and when combined with other data, such as T^C_1 and dipolar and chemical-shift side-band patterns, provides detailed insight into the glassy state. Recently Schaefer et al have reported such exhaustive studies on PS[59] and PC.[60] Before these results are reviewed, the determination and utility of C–H dipolar side-band patterns for assessing amplitudes of motion in solids are discussed in more detail.

Earlier in this chapter in the section on NMR Structural Studies of the Glassy State, it was demonstrated how measurement of $^{13}C-^{13}C$ dipolar couplings provided information on polymer structure. Here a technique to measure C–H heteronuclear dipolar coupling is discussed and the utility of the experiment to study motion in glassy polymers is illustrated. As in the homonuclear case, the $^1H-^{13}C$ spin pair must be isolated (in this case from many-body proton dipolar coupling). Schaefer et al[61] devised the pulse sequence shown in Chapter 2, Figure 2-17, to accomplish this necessity. The experiment, termed dipolar rotational spin echo, is a two-dimensional NMR technique and the dipolar patterns generated provide a measure of the amplitudes of molecular motion in a polymer. The key aspects of the experiment are the Waugh-Huber-Haeberlen (WAHUHA) pulse sequences[62] (see Chapter 9, Principles) used to suppress $^1H-^1H$ coupling and the rotation-synchronized sampling. The evolution of carbon magnetization from chemical-shift effects is refocused after two rotation periods by a 180° pulse applied after the first rotation period. The $^{13}C-^1H$ dipolar modulation is followed by varying the number of phase-shifted WAHUHA multiple-pulse sequences during the time period t_1. As an example, several chemical-shift spectra as a function of the number of WAHUHA pulses and the associated side-band patterns are shown in Figure 3-27 for poly(styrene-co-sulfone).[59] The aliphatic CH pair is a representation of a Pake doublet broken up into spinning side bands. The shape for the aromatic-carbon line is similar except for center-band intensity arising from the nonprotonated carbon. Spectra such as these are used to infer information on low-frequency motions by comparison of the observed side-band intensities with appropriate patterns observed for rigid crystalline solids. Variations are ascribed to motionally averaged powder patterns. Furthermore, it is possible to compute predicted patterns for specific motional models, allowing greater insight into the type and amplitude of motion. This is illustrated in Table 3-6, where the observed side-band intensities for the carbons of several polystyrenes are compared to those from models of motion based on 180° ring flips and 180° ring flips mixed with small-amplitude oscillations near the bottoms

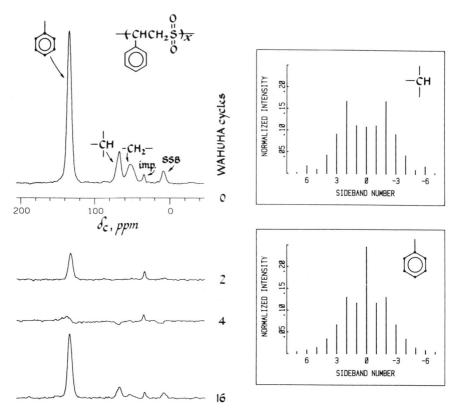

Figure 3-27. Dipolar rotational spin-echo 15.1-MHz ^{13}C NMR spectra of poly(styrene-co-sulfone) at room temperature as a function of the number of WAHUHA cycles used during ^{1}H$-^{13}$C dipolar evolution (left). Experimental dipolar side-band patterns for the two CH pairs of the polymer under magic angle spinning at 1.894 kHz are shown at the right of the figure. The intense centerband in the aromatic-carbon dipolar pattern arises from contributions from the nonprotonated carbon whose isotropic chemical shift is accidentally the same as that of the protonated aromatic carbons. (Reprinted with permission from Schaefer, J.; Sefcik, M. D.; Stejskal, E. O.; McKay, R. A.; Dixon, W. T.; Cais, R. E. *Macromolecules* **1984**, *17*, 1107. Copyright 1984, American Chemical Society.)

of the double-minimum potential well. The latter model provides a reasonable overall fit. Schaefer et al have provided a detailed introduction to the technique.[61] A discussion follows of how this experiment, combined with T_1^C and $T_{1\rho}^C$ experiments, provides a detailed picture of molecular motions in glassy polymers.

Schaefer et al have utilized the MAS side-band patterns of the C–H dipolar and carbon chemical-shift tensors to determine the amplitudes of ring and main-chain motions in a variety of polystyrenes.[59] Spin–lattice relaxation measurements (T_1^C and $T_{1\rho}^C$) were employed to characterize the frequencies of

TABLE 3-6. Comparisons of Calculated and Experimental Dipolar Rotational Side-Band Intensities for Some Polystyrenes with Magic Angle Spinning at 1894 Hz[a]

Polymer	CH pair		Experiment or motional model	Side-band number/intensity					
				0	1	2	3	4	5
Poly(p-isopropylstyrene)	Aromatic	1	Observed	0.138	0.142	0.153	0.079	0.038	0.014
		2	Observed, short T_1^C component removed	0.129	0.135	0.156	0.083	0.041	0.015
		3	Calculated, C_2 rolls, $\phi(rms) = 21°$	0.134	0.153	0.178	0.063	0.028	0.009
		4	Calculated, motion on a sphere, $\phi(rms) = 19°$	0.134	0.157	0.178	0.060	0.027	0.008
Poly(styrene-co-sulfone)	Aliphatic	5	Observed	0.104	0.110	0.164	0.091	0.045	0.012
		6	Calculated, static	0.120	0.124	0.185	0.075	0.038	0.013
Atactic polystyrene,[b] short T_1^C component	Aromatic	7	Observed	0.222	0.208	0.127	0.047	0.007	0.000
		8	Calculated, C_2 180° flips	0.198	0.213	0.128	0.033	0.014	0.004
		9	Calculated, Tonelli conformational distribution[c]	0.259	0.215	0.108	0.033	0.011	0.003

[a] Reprinted with permission from Schaefer, J.; Sefcik, M. D.; Stejskal, E. O.; McKay, R. A.; Dixon, W. T.; Cais, R. E. *Macromolecules* **1984**, *17*, 1107. Copyright 1984, American Chemical Society.
[b] Melt-quenched, high molecular weight material.
[c] Tonelli, A. E. *Macromolecules* **1973**, 6, 682.

the same motions. The dominant motion found in the glassy systems is restricted phenyl rotation, the amplitude and frequency of which vary from one substituted PS to another and from one site to another site within the same PS. For atactic PS, it is found that ca. 7% of the rings engage in 180°-flip, megahertz-rate motion. This insight into the dynamics of PS is derived from a combination of data: (1) the observed T_1^C for this fraction of rings is near a T_1^C minimum (at 15 MHz); thus, the motion is in the tens of MHz regime. (2) The extent of collapse of the dipolar tensor for the protonated aromatic carbon indicates the motion is of large amplitude. (3) The unchanged CST for the nonprotonated aromatic carbon of this fraction indicates the motion must involve 180° flips because this motion leaves the CST for this carbon unchanged. The fraction of rings undergoing the 180° flips in PS is found to be unaffected by either configuration or conformational defects in the chain. Instead, the fast ring flips appear to be part of a cooperative motion that includes the main chain. Schaefer et al speculate[59] that these motions may occur at sites in the glass where packing of chains result in main-chain flexibility. The high-frequency main-chain wiggling then leads to a low barrier to phenyl-ring motion. It is further argued that the cooperative ring flips (10 MHz at 25°C) are responsible for the weak tan-δ γ-transition in the mechanical loss spectrum of PS (1 Hz at -120°C).

In addition to those sites in PS where ring flipping occurs, $T_{1\rho}^C$ data and dipolar side-band patterns indicate measurable low-frequency (kilohertz), small-amplitude ring motion. Schaefer[59] has suggested that the same cooperative kilohertz-regime motions responsible for $T_{1\rho}^C$ are also responsible for partial averaging of the aromatic C–H dipolar tensor. The $T_{1\rho}^C$ is assumed to be given by:

$$(T_{1\rho}^C)^{-1} = K^2(\sin^2 \theta)J(\omega) \qquad (3\text{-}6)$$

where K^2 is a constant (which includes powder averaging over the solid), $\sin^2 \theta$ is the average dipolar fluctuation orthogonal to the applied radiofrequency field, and $J(\omega)$ describes the spectral density associated with the ring motion at the carbon rotating-frame Larmor frequency, in this instance 37 kHz. The $J(\omega)$ for all ring-substituted polystyrenes were shown to be about the same. If it is assumed ring rotation only occurs about the ring C_2 axis, then the relative $T_{1\rho}^C$ should be a simple function of the amplitude of the ring motion as measured by θ. These amplitudes can be estimated from the relative intensities in the dipolar CH patterns. The ratio of the second to first rotational side-band intensity for a CH pair (n_2/n_1) is given in Table 3-7 for various motional displacements. Assuming that the motion that reduces n_2/n_1 (Table 3-7) is also responsible for the $T_{1\rho}$ relaxation, comparative results of θ determination for seven substituted polystyrenes are shown in Table 3-8. The product of $\sin^2 \theta$ and $\langle T_{1\rho}^C \rangle$ is roughly constant for all seven polymers. The product for the first six polymers in Table 3-8 is constant to within about 50%, even though the $\langle T_{1\rho}^C \rangle$ themselves vary by a factor of 4. Based on rms values of θ, the ring rotations generate total angular displacements of about 40° (for sub-

TABLE 3-7. Calculated Ratio of Second to First Dipolar Rotational Side-Band Intensities for a CH Pair Undergoing Molecular Motion and Magic Angle Spinning (1894 Hz)[a]

Motional model	Total azimuthal angular displacement (°)	n_2/n_1
Restricted ring rotation[b]	0	1.50
	20	1.44
	40	1.35
	60	1.16
	80	0.93
	100	0.70
	120	0.53
	140	0.41
Ring π flips	180	0.48

[a] Reprinted with permission from Schaefer, J.; Sefcik, M. D.; Stejskal, E. O.; McKay, R. A.; Dixon, W. T.; Cais, R. E. *Macromolecules* **1984**, *17*, 1107. Copyright 1984, American Chemical Society.
[b] Aromatic CH vector undergoing fast diffusional reorientation on the surface of a 60° cone.

stituents in the ortho position) to 70° (for bulky nonpolar substituents in the para position).

Because the polystyrenes are dynamically heterogeneous (even ignoring ring flippers), both $T_{1\rho}^C$s and angular displacement fluctuation parameters must reflect averages over the entire sample.[59] The observed $\langle T_{1\rho}^C \rangle$ is the weighted average of all the $T_{1\rho}^C$s present. Since the n_2/n_1 ratio has an approximately

TABLE 3-8. Protonated Aromatic $\langle T_{1\rho}^C \rangle$s of Some Polystyrenes Scaled by the Amplitude of Root-Mean-Square Angular Fluctuations Deduced from Experimental Dipolar Rotational Side-Band Intensities[a]

System	37-kHz $\langle T_{1\rho}^C \rangle$ (ms)[b]	n_2/n_1[c]	$[n_2/n_1]_0^{*}$ [d]	θ	$\langle T_{1\rho}^C \rangle \sin^2 \theta$
Poly(p-tert-butylstyrene)	6.2	1.04	1.11	23	0.89
Poly(p-isopropylstyrene)	8.8	1.08	1.15	21	1.06
Polystyrene[e]	11.2	1.20	1.25	18	1.00
Poly(p-methylstyrene)	11.7	1.20	1.25	18	1.04
Poly(α-methylstyrene)	19.0	1.30	1.30	16	1.35
Poly(styrene-co-sulfone)	22.0	1.32[f]	1.32	15	1.38
Poly(o-chlorostyrene)	37.0	1.34	1.34	14	2.02

[a] Reprinted with permission from Schaefer, J.; Sefcik, M. D.; Stejskal, E. O.; McKay, R. A.; Dixon, W. T.; Cais, R. E. *Macromolecules* **1984**, *17*, 1107. Copyright 1984, American Chemical Society.
[b] Straight-line fit to observed decay between 0.05 and 1.00 ms after the turn off of H_{1H}.
[c] Ratio of intensities of second to first dipolar rotational side bands.
[d] Ratio of intensities of second to first dipolar rotational side bands with contributions from rings undergoing megahertz-rate flips removed.
[e] Atactic, quenched, high molecular weight material.
[f] Contributions to n_1 from the nonprotonated aromatic carbon removed, assuming $n_1/n_0 = 0.13$ for that carbon, the same ratio as is observed for polystyrene.

linear dependence on total angular displacement (and is not critically model dependent), the observed root-mean-square fluctuation parameter, θ, is also a simple weighted average. In part, the correlation of Table 3-8 (between $\langle T_{1\rho}^C \rangle$ and $\sin^2 \theta$) succeeds because it ignores details of the distributions of motions. Thus, whether polystyrene has a fraction of highly mobile rings with the remainder less mobile, or whether all rings have an intermediate mobility, becomes immaterial. Both situations result in comparable average values for $\langle T_{1\rho}^C \rangle$ and θ. The correlation of Table 3-8 fails to the extent there remain large-amplitude high-frequency motions (after removal of ring flips) which reduce n_2/n_1 but do not contribute to $T_{1\rho}^C$, or 5–10° small-amplitude low-frequency motions which can make significant contributions to $\langle T_{1\rho}^C \rangle$ but have only a minor effect on n_2/n_1. The correlation also suffers from the presence of large-amplitude motions near 10 kHz which have effects on n_2/n_1 not well represented by the calculations, which assume all motion is faster than dipolar coupling. (It is suspected this limitation is not severe since large-amplitude motions tend to be of high rather than low frequency.)[59]

Both $T_{1\rho}^C$ measurements and dipolar side-band patterns indicate that the rings and main chains of polystyrenes are engaged in a variety of small-amplitude rotational reorientations between ten and several hundred kilohertz. The lowest frequency of these motions is necessarily cooperative, because small-amplitude local motion cannot also be low frequency. Thus, the collective data provide evidence to assign the broad, intense tan-δ β-transition (1 Hz at -60 to $+20°C$) to cooperative ring- and main-chain restricted oscillations.

The above discussion of ring motion in PS is not intended to be comprehensive but to provide the reader with some appreciation of the insight possible from a carefully conceived application of various solid-state NMR experiments. For further details of motion in polystyrenes the reader is referred to the elegant paper by Schaefer, who has also carried out a similar *tour de force* on polycarbonate.[60] In the PC case, the dominant motion is found to be 180° flips about the aromatic-ring C_2 axes. These flips, which cover a range of frequencies extending to > 15 MHz, are superimposed on ca. 30° ring oscillations about the same axes. Added details of the motion in PC are given in Chapter 7.

Proton Rotating-Frame Relaxation Studies

Before this section on relaxation measurements in the solid state is concluded, the utility of proton $T_{1\rho}^H$ measurements evaluated from ^{13}C CPMAS spectra is examined in elucidating the compatibility of solid polymeric blends.[63] The $T_{1\rho}^H$ values in solids usually represent an average value of the relaxation behavior over the ensemble of protons. This is the case because the strong dipolar coupling between protons gives rise to efficient spin diffusion, thereby creating a mechanism to damp out nonequilibrium magnetization in any part of the proton spin system. The rate of spin diffusion is influenced strongly by the spatial mixing of polymeric chains and, therefore, in a blended material, measures the homogeneity of mixing. In a cross-polarization experiment, following the initial rapid buildup of carbon magnetization, the carbon signal tracks the polarization in the proton reservoir and so follows its decrease via the $T_{1\rho}^H$ process (see Chapter 2). Under high-resolution conditions, the individual carbon resonances follow the protons to which they are coupled and hence are capable of resolving $T_{1\rho}^H$ differences between protons that might be difficult to detect in a direct proton experiment. If chains are intimately and homogeneously mixed, rapid spin diffusion yields a single $T_{1\rho}^H$ for all the decays followed in the CPMAS contact time experiments. If this is not the case, it may be possible to detect different $T_{1\rho}^H$ values.

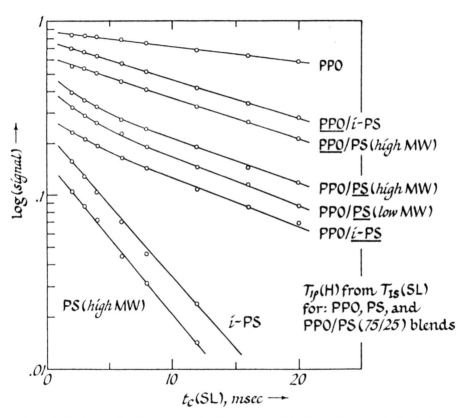

Figure 3-28. Plots of carbon magnetization generated by long-time matched spin-lock cross-polarization transfers yielding $T_{1\rho}^H$ for protons attached to main-chain carbons of PPO, PS, and a variety of 75:25 PPO–PS blends. Underline identifies the component of the blend whose main-chain proton $T_{1\rho}^H$ is reported. (Reprinted with permission from Stejskal, E. O.; Schaefer, J.: Sefcik, M. D.; McKay, R. A. *Macromolecules* **1981**, *14*, 275. Copyright 1981, American Chemical Society.)

TABLE 3-9. PROTON $T_{1\rho}^H$ OF SOME HOMOPOLYMERS AND OF POLY(PHENYLENE OXIDE)/POLYSTYRENE (25/75) SOLID BLENDS[a,b] $(T = 26°C, H_1 = 37 \text{ kHz})^c$

System	$T_{1\rho}^H$ (ms)	$\langle T_{1\rho}^H \rangle^d$ (ms)
PPO	49	49
PPO/PS-d$_8$	88	88
PPO/PS (high MW)	6.7	6.7
PPO/**PS** (high MW)	6.2	6.2
i-PS	5.3	5.3
PS (high MW)	5.6	5.6
p-Cl-PS	8.0	8.0

[a] **Boldface** identifies the component of the blend whose main-chain proton $T_{1\rho}^H$ is reported.
[b] Estimated accuracy is $\pm 5\%$. MW = molecular weight.
[c] Reprinted with permission from Stejskal, E. O.; Schaefer, J.; Sefcik, M. D.; McKay, R. A. *Macromolecules* 1981, *14*, 275. Copyright 1981, American Chemical Society.
[d] The brackets indicate a nominal initial slope.

Stejskal et al[63] have utilized CPMAS $T_{1\rho}^H$ measurements to study PS and poly(phenylene oxide) (PPO) blends. The carbon magnetization decay curves as a function of rotating-frame spin-lock (SL) time for the two homopolymers and several blends are displayed in Figure 3-28. The $T_{1\rho}^H$ of atactic and iso-tactic PS are effectively the same (5.6 and 5.3 ms, respectively) and ca. 1/10 that of PPO. The results for the blends yield intermediate $T_{1\rho}^H$ values but indicate that the main-chain carbons for the PPO and PS chains detect slight-ly different $T_{1\rho}^H$ values (see Tables 3-9 and 3-10) and that the PS decays are not single decays. There is considerable, but not complete, averaging in these

TABLE 3-10. PROTON $T_{1\rho}^H$ OF POLY(PHENYLENE OXIDE)/POLYSTYRENE (75/25) SOLID BLENDS[a,b] $(T = 26°C, H_1 = 37 \text{ kHz})^c$

Blend	$T_{1\rho}^H$ (ms)	$\langle T_{1\rho}^H \rangle^d$ (ms)
PPO/p-Cl-PS, quenched	48	48
PPO/p-Cl-PS, annealed	54	54
PPO/i-PS	20	20
PPO/PS (high MW)	18	22
PPO/PS (low MW)	15	20
PPO/**PS** (high MW)	17	6.8
PPO/ **PS** (low MW)	15	6.5
PPO/**i-PS**	15	10
PPO/p-**Cl-PS**, quenched	9	9.0
PPO/p-**Cl-PS**, annealed	9	9.0

[a] **Boldface** identifies the component of the blend whose main-chain proton $T_{1\rho}^H$ is reported. MW = molecular weight.
[b] Estimated accuracy is $\pm 5\%$.
[c] Reprinted with permission from Stejskal, E. O.; Schaefer, J.; Sefcik, M. D.; McKay, R. A. *Macromolecules* 1981, *14*, 275. Copyright 1981, American Chemical Society.
[d] The brackets indicate a nominal initial slope.

nominally homogeneous blends. Stejskal et al have attributed the initial more rapid decay (5–10% of the total) of the PS protons in the blend to small regions in the blend where PS is not dispersed fully and hence relaxes more like the homopolymer. The nonlinear decay is more pronounced for the atactic PS, suggesting that the region of poorly dispersed PS may arise from places where kinked conformations tend to fold the chain back on itself or where a few independent chains have aggregated. The greater regularity of the isotactic PS chain may tend to mitigate these effects in its blends with PPO. (Note that the nonlinearity of the decay cannot be explained by local statistical fluctuations of homogeneous mixing; see Stejskal et al[63]) The $T_{1\rho}^H$ results for PPO–p–Cl–PS blends (Tables 3-9 and 3-10) clearly show little interchain mixing for this system.

Schaefer et al[64] also have utilized the CPMAS $T_{1\rho}^H$ measurement to study the interfacial region in PS toughened by blending with a polystyrene–polybutadiene block copolymer. The results indicated that there is no organized structure near the interface of the rubber particles in the blend, eg, a structure of PS chains of the copolymer forming a dense coating around the rubber domains. Kaplan[53] has used the methodology to study blends of PS with poly(vinyl methyl ether). The reader is referred to the literature for discussion of the results. Although the studies to date are few, it is nonetheless clear the CPMAS $T_{1\rho}^H$ experiment represents another powerful NMR method for studying local polymer structure.

Summary

This chapter has demonstrated how high-resolution NMR methods for studying the structure and dynamics of solids can be utilized to provide unique information on glassy polymers. In particular, CPMAS spectra have been shown to provide information on the curing chemistry and kinetics of thermosets, mechanisms of polymerization (eg, solid-state reactions), and composition and compatibility of polymer blends. Insight into the heterogeneity of glassy systems, both motional and structural, and their interrelationships can be gained from relaxation and related solid-state experiments. Finally, the experiments offer the potential for establishing direct relationships between the molecular features of glassy polymers and their mechanical properties.

References

1. Haward, R. N. In "The Physics of Glassy Polymers", Haward, R. N., ed.; J. Wiley and Sons: New York, 1973, Chapter 1.
2. Garroway, A. N.; Ritchey, W. M.; Moniz, W. B. *Macromolecules* **1982**, *15*, 1051.
3. Grant, D. M.; Cheney, B. V. *J. Am. Chem. Soc.* **1967**, *89*, 5315.

4. VanderHart, D. L.; Earl, W. L.; Garroway, A. N. *J. Magn. Reson.* **1981**, *44*, 361.
5. Garroway, A. N.; Moniz, W. B.; Resing, H. A. *ACS Symp. Ser.* **1979**, *103*, 67.
6. Havens, J. R.; Koenig, J. L. *Appl. Spectrosc.* **1983**, *37*, 226, and references therein.
7. Fyfe, C. A.; McKinnon, M. S.; Rudin, A.; Tchir, W. J. *Macromolecules* **1983**, *16*, 1216.
8. Fyfe, C. A.; Rudin, A.; Tchir, W. J. *Macromolecules* **1980**, *13*, 1322.
9. Maciel, G. E.; Chuang, I-S.; Gallob, L. *Macromolecules* **1984**, *17*, 1081.
10. Opella, S. J.; Frey, M. H. *J. Am. Chem. Soc.* **1979**, *101*, 5854.
11. Bryson, R. L.; Hatfield, G. R.; Early, T. A.; Palmer, A. R.; Maciel, G. E. *Macromolecules* **1983**, *16*, 1669.
12. Maciel, G. E.; Chuang, I-S.; Myers, G. E. *Macromolecules* **1982**, *15*, 1218.
13. Chuang, I-S.; Maciel, G. E.; Myers, G. E. *Macromolecules* **1984**, *17*, 1087.
14. Sefcik, M. D.; Stejskal, E. O.; McKay, R. A.; Schaefer, J. *Macromolecules* **1979**, *12*, 423.
15. Schaefer, J.; Stejskal, E. O.; Sefcik, M. D.; McKay, R. A. *Phil. Trans. R. Soc. London* **1981**, *A299*, 593.
16. Havens, J. R.; Ishida, H.; Koenig, J. L. *Macromolecules* **1981**, *14*, 1327.
17. McKay, R. A.; Schaefer, J.; Stejskal, E. O.; Ludicky, R.; Matthews, C. N. *Macromolecules* **1984**, *17*, 1124.
18. Kaplan, S.; Dilks, A. *J. Polym. Sci., Polym. Chem. Ed.* **1983**, *21*, 1819.
19. Dilks, A.; Kaplan, S. *ACS Coat. Plast. Prepr.* **1982**, *47*, 212.
20. Pake, G. E. *J. Chem. Phys.* **1948**, *16*, 327.
21. Yannoni, C. S.; Kendrick, R. D. *J. Chem. Phys.* **1982**, *74*, 747.
22. Horne, D.; Kendrick, R. D.; Yannoni, C. S. *J. Magn. Reson.* **1983**, *52*, 299.
23. Yannoni, C. S.; Clarke, T. C. *Phys. Rev. Lett.* **1983**, *51*, 1191.
24. Clarke, T. C.; Kendrick, R. D.; Yannoni, C. S. *J. Physique Colloque* **1983**, *44*, C3-369.
25. Clarke, T. C.; Scott, J. C.; Street, G. B. *IBM J. Res. Develop.* **1983**, *27*, 313.
26. Gibson, H. W.; Pochan, J. M.; Kaplan, S. *J. Am. Chem. Soc.* **1981**, *103*, 4619.
27. Gibson, H. W.; Prest, W. M.; Mosher, R. A.; Kaplan, S.; Weagley, R. J. *ACS Polym. Prepr.* **1983**, *24*, 153.
28. Ito, T.; Shirakawa, H.; Ikeda, S. *J. Polym. Sci., Polym. Chem. Ed.* **1974**, *12*, 11.
29. Clarke, T. C.; Yannoni, C. S.; Katz, T. J. *J. Am. Chem. Soc.* **1983**, *105*, 7787.
30. Su, W. P.; Schrieffer, J. R.; Heeger, A. J. *Phys. Rev. Lett.* **1979**, *42*, 1698.
31. Resing, H. A.; Garroway, A. N.; Weber, D. C.; Ferraris, J.; Slotfeldt-Ellingsen, D. *Pure Appl. Chem.* **1982**, *54*, 595.
32. Williams, G.; Watts, D. C. *Trans. Faraday Soc.* **1970**, *66*, 80.
33. Inglefield, P. T.; Amici, R. M.; O'Gara, J. F.; Hung, C-C.; Jones, A. A. *Macromolecules* **1983**, *16*, 1552.
34. Spiess, H. W. *Colloid Polym. Sci.* **1983**, *261*, 193.
35. Stejskal, E. O.; Schaefer, J.; McKay, R. A. *J. Magn. Reson.* **1977**, *25*, 569.
36. Lippmaa, E.; Alla, M.; Tuherm, T. In "Magnetic Resonance and Related Phenomena", Brunner, H.; Schweitzer, D., Eds.; Colloque Ampere: Heidelberg-Geneva, **1976**, p. 113.
37. Zilm, K. W.; Conlin, R. T.; Grant, D. M.; Michl, J. *J. Am. Chem. Soc.* **1978**, *100*, 8038.
38. Edzes, H. T. *Polymer* **1983**, *24*, 1425.
39. Ouano, A. C.; Pecora, R. *Macromolecules* **1980**, *13*, 1167.
40. Ouano, A. C.; Pecora, R. *Macromolecules* **1980**, *13*, 1173.
41. Lyerla, J. R.; Yannoni, C. S. *IBM J. Res. Develop.* **1983**, *27*, 302, and references therein.
42. Fleming, W. W.; Lyerla, J. R.; Yannoni, C. S. *ACS Symp. Ser.* **1984**, *247*, 83.
43. Rothwell, W. P.; Waugh, J. S. *J. Chem. Phys.* **1981**, *74*, 2721.
44. McBrierty, V. J.; Douglass, D. C.; Falcone, D. R. *J. Chem. Soc. Faraday Trans. II* **1972**, *68*, 1051.
45. Schaefer, J.; Stejskal, E. O.; Buchdahl, R. *Macromolecules* **1975**, *8*, 291; **1977**, *10*, 384.
46. Yannoni, C. S. *Acc. Chem. Res.* **1982**, *15*, 201.
47. Lyerla, J. R. *Contemp. Top. Polym. Sci.* **1979**, *3*, 143; *Meth. Exp. Phys.* **1980**, *16A*, 241.
48. Torchia, D. A. *J. Magn. Reson.* **1978**, *30*, 613.
49. McCall, D. W.; Douglass, D. C.; Falcone, D. R. *J. Phys. Chem.* **1967**, *71*, 998, and references therein.
50. Edzes, H. T.; Veeman, W. S. *Polym. Bull.* **1981**, *5*, 255.
51. Menger, E. M.; Veeman, W. S.; DeBoer, E. *Macromolecules* **1982**, *15*, 1406.
52. Schroter, B.; Posern, A. *Makromol. Chem.* **1981**, *182*, 675.
53. Kaplan, S., *ACS Polym. Prepr.* **1984**, *25*, 356.
54. VanderHart, D. L.; Garroway, A. N. *J. Chem. Phys.* **1979**, *71*, 2773.

55. Garroway, A. N. *J. Magn. Reson.* **1979**, *34*, 283.
56. Akasaka, K.; Ganapathy, S.; McDowell, C. A.; Naito, A. *J. Chem. Phys.* **1983**, *78*, 3567.
57. Schaefer, J.; Stejskal, E. O.; Steger, T. R.; Sefcik, M. D.; McKay, R. A. *Macromolecules* **1980**, *13*, 1121.
58. Schaefer, J.; Sefcik, M. D.; Stejskal, E. O.; McKay, R. A. *Macromolecules* **1984**, *17*, 1118, and references therein.
59. Schaefer, J.; Sefcik, M. D.; Stejskal, E. O.; McKay, R. A.; Dixon, W. T.; Cais, R. E. *Macromolecules* **1984**, *17*, 1107.
60. Schaefer, J.; Stejskal, E. O.; McKay, R. A.; Dixon, W. T. *Macromolecules* **1984**, *17*, 1479.
61. Schaefer, J.; McKay, R. A.; Stejskal, E. O.; Dixon, W. T. *J. Magn. Reson.* **1983**, *52*, 123.
62. Haeberlen, U. In "High Resolution NMR in Solids: Selective Averaging", Adv. Magn. Reson. Ser., Waugh, J. S., Ed., Suppl. 1; Academic Press: New York, 1976.
63. Stejskal, E. O.; Schaefer, J.; Sefcik, M. D.; McKay, R. A. *Macromolecules* **1981**, *14*, 275.
64. Schaefer, J.; Sefcik, M. D.; Stejskal, E. O.; McKay, R. A. *Macromolecules* **1981**, *14*, 188.

4

CARBON-13 NMR OF SOLID AMORPHOUS POLYMERS ABOVE T$_g$

Richard A. Komoroski

B F GOODRICH RESEARCH AND DEVELOPMENT CENTER
9921 BRECKSVILLE ROAD,
BRECKSVILLE, OH 44141

Introduction

The development of NMR techniques to obtain high-resolution ^{13}C NMR spectra of rigid solids understandably has led to much activity on glassy and crystalline polymers in recent years[1,2] (see Chapters 3 and 5). However, since the advent of ^{13}C Fourier-transform (FT) NMR techniques for solutions in the late 1960s, it has been possible to obtain high-quality, high-resolution spectra of solid polymers well above their glass transition temperatures (T$_g$). Work in this area was in progress for several years before high-resolution work appeared on rigid polymers. Both the amorphous regions of semi-crystalline polymers and totally amorphous polymers have been studied. The former category is addressed in Chapter 5, and the latter category is the subject of this chapter.

At this point it is appropriate to define more clearly the class of polymers that is of present concern. All polymers above T$_g$ are of interest here, although most of the discussion relates to polymers well above their T$_g$, on the order of 50–100°C. This somewhat artificial distinction derives from the molecular mobility requirements for obtaining usable high-resolution ^{13}C NMR spectra. The terms "rubber," "elastomer," and "polymer well above T$_g$" are used here

Komoroski (ed): High-Resolution NMR Spectroscopy of Synthetic Polymers in Bulk

somewhat interchangeably. Use of the first two terms is not meant to imply that the polymers in question have been cured (crosslinked). Crosslinked systems are dealt with specifically later on in this chapter (see Crosslinks). This chapter also is concerned with those portions of multiphase, noncrystalline systems that are above their T_g.

Since the observation of high-resolution ^{13}C spectra of bulk cis-polybutadiene and cis- and trans-polyisoprene by Duch and Grant,[3] a steady stream of reports has appeared on the use of standard scalar-decoupled ^{13}C NMR for solid polymers above T_g. It is the purpose here to describe some of the characteristics of this approach and what can be learned of a fundamental nature. Where appropriate, comparisons are made to studies using techniques designed for rigid solids.

Observation of ^{13}C Spectra above T_g

Comparison to ^{1}H NMR

Results for solid rubbers were reported relatively early in the history of high-resolution proton NMR.[4] Resonances from individual proton types are barely resolved for cis-polybutadiene at 56.4 MHz and at room temperature the linewidths are about 250 Hz.[5] Although the ^{13}C NMR spectra of bulk elastomers do not exhibit the resolution characteristic of the solution state, resonances from individual carbon types are resolved easily at low field in most cases, with ^{13}C linewidths typically in the range of 5–30 Hz.

There are several reasons that proton spectra of solid polymers well above T_g are relatively featureless. The small chemical-shift range (10 ppm = 600–2000 Hz at typical magnetic fields) and the presence of H–H scalar couplings are two important and oft-quoted disadvantages of ^{1}H NMR relative to ^{13}C NMR. Of course, carbons typically are scalar decoupled from protons in the ^{13}C NMR experiment. Contributions to the linewidth from dipolar relaxation (ie, the "natural" linewidth) are reduced for ^{13}C relative to ^{1}H by a factor of about 16 because of the inverse dependence on the square of the gyromagnetic ratio.[6] This assumes that the same motional processes are modulating both the ^{13}C and ^{1}H linewidths. At about 50°C or more above T_g, the backbone segmental motions in amorphous high polymers produce sufficiently narrow ^{13}C lines. Nonprotonated carbons have relatively narrow lines even for ^{13}C because of the reduced interaction with hydrogen nuclei in the sample. Additional broadening of the proton spectrum relative to the ^{13}C spectrum may arise from the presence of slower inter- and intramolecular motions that modulate the H–H dipolar interaction but do not affect the C–H dipolar interaction. Schaefer has invoked slow interchain motions resulting from chain entanglements to explain why the proton linewidths for cis-polyisoprene are significantly broader than the ^{13}C linewidths.[7]

It should be pointed out that the application of a multiple-pulse sequence to elastomers yields substantial line narrowing,[5,8] reducing the ^{1}H linewidths to approximately 40 Hz for cis-polybutadiene. More is said in Chapter 9 concerning the use of such multiple-pulse techniques for producing high-resolution NMR spectra of isotopically abundant nuclei in polymers.

Experimental Techniques

Standard One-Pulse Experiment. The experimental requirements for obtaining high-resolution ^{13}C NMR spectra of elastomers are those of standard ^{13}C FT NMR for the most part[9,10] and are not repeated here. Several differences should be noted, however. These differences make the acquisition of the ^{13}C spectra of bulk elastomers more trouble free than those of liquids.

The ^{13}C T_1s of protonated carbons in rubbery polymers are usually very short and are close to the minimum value expected when spin relaxation is dominated by a C–H dipolar mechanism.[11] This is about 110 ms for a CH carbon at 67.9 MHz and is lower at lower observation frequencies (see Carbon-13 Spin-Relaxation Parameters below). Nonprotonated carbons have T_1s correspondingly reduced from values typical of small organic molecules, on the order of 0.5–1.5 s. Hence the very large T_1 range often seen for the ^{13}C NMR spectra of liquids is not present. Repetitive pulsing can be rapid, with little or no attenuation of peaks caused by their long T_1s.

The linewidths observed for bulk, rubbery polymers that are totally amorphous and unfilled are usually in the range of 5–30 Hz about 100°C above T$_g$ and at 15–25 MHz. For certain polymers or at higher magnetic fields, linewidths can be as large as 200 Hz. Hence the instrumental resolution requirements are very modest from the viewpoint of high-resolution NMR. Broadening of 1–2 Hz or more caused by magnetic field inhomogeneity or incomplete scalar decoupling can be tolerated. The larger linewidths relative to the solution state tend to offset the increased sensitivity available from rapid pulsing and high concentration in the bulk state.

Reduced but substantial nuclear Overhauser enhancements (NOE) are observed for the ^{13}C spectra of bulk elastomers obtained using standard, ungated proton decoupling[7,11–14] (see Carbon-13 Spin-Relaxation Parameters below). Generally all backbone carbons have the same NOE, facilitating quantitative analysis.

Cross-Polarization Magic Angle Spinning (CPMAS). Cross polarization (CP) is an inefficient process for elastomers because the static dipolar interaction necessary for efficient cross polarization has been eliminated or greatly reduced by chain mobility—just the factor that permits observation of the spectra by standard techniques. However, C–H cross polarization can be carried out for elastomers[15] and even for liquids.[16] In the latter case the C–H scalar coupling provides the pathway for transfer of polarization from hydrogen to carbon.

Figure 4-1 shows solid-state ^{13}C magic angle spinning (MAS) NMR spectra

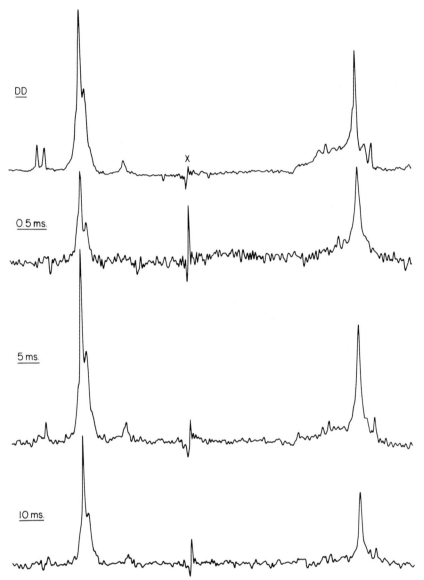

Figure 4-1. Solid-state ^{13}C NMR spectra of an emulsion SBR. The top spectrum was obtained using the single-pulse experiment with DD MAS. The others were obtained using cross polarization with contact time indicated. All spectra were accumulated with the same number of scans but are plotted out at different vertical gains. The centerband artifact is marked "X." (Reprinted with permission of the copyright owner, Rubber Division, ACS, Inc., from reference 15.)

of an emulsion styrene–butadiene rubber (SBR) containing about 23 wt% styrene.[15] The top spectrum was obtained using the standard free-induction decay (FID) from the one-pulse experiment, whereas the bottom three were obtained using cross polarization (CP) with three different contact times. All the spectra of Figure 4-1 were acquired with the same number of scans but are plotted at different vertical gains. The CP spectrum is a strong function of the contact time, with 5 ms being the best value among the three shown. This is somewhat longer than the 1–2 ms typical for glassy polymers.[1] The CP spectra have reduced signal-to-noise ratios relative to the normal spectrum, and the C-1 phenyl peak at 146 ppm is barely visible, even in the 5-ms spectrum. The spectra in Figure 4-1 demonstrate that CP generally is not recommended for obtaining ^{13}C spectra of polymers well above T_g. It is informative to compare these spectra to those of the semicrystalline poly(hydroxybenzoic acid) shown in Figure 2-11.

The ability to obtain a cross-polarized spectrum suggests that there may be at least a small, residual component of the static dipolar interaction remaining. Cross polarization also may be occurring via a reduced dipolar broadening modulated by molecular motion. Because the proton and carbon linewidths for elastomers are much reduced relative to glassy polymers, the cross-polarization process for elastomers probably requires a close match and high stability of the carbon and proton radiofrequency (rf) fields. The cross polariz-ation spectra of elastomers also are expected to be highly sensitive to the rate and anisotropy of the segmental motions, for these determine the amount of residual dipolar interaction, whether static or not. Different components or carbons may cross polarize very differently because of differences in mobility. Polymer chain segments near fillers or crosslink sites may polarize much more rapidly than those of the bulk material, possibly allowing them to be observed separately. Of course, as the temperature is lowered, cross polarization becomes more efficient. No studies have yet appeared to define the tem-perature at which cross polarization becomes a more efficient and reliable process for obtaining spectra than the standard one-pulse experiment. (See Appendix.) It is expected that this temperature is near or slightly above T_g.

Magic angle spinning and high-power decoupling can be used with some success on elastomers. Although the gains are not as dramatic as for rigid polymers, they may be sufficient to justify their use, particularly for blends or heterogeneous or filled polymers. More will be said on this topic in the sec-tions on Contributions to the Linewidth and Presence of Filler.

Carbon-13 Spin-Relaxation Parameters

For bulk polymers above T_g and in solution, the magnetic resonance par-ameters that primarily relate to molecular dynamics are the spin-relaxation times T_1 and T_2. Spin-lattice processes are sensitive to motions at or near the nuclear Larmor frequencies, which are typically 5–500 MHz. Spin-spin processes are sensitive to both low-frequency motions (10^2–10^3 Hz) and motions near the Larmor frequency. There are several mechanisms that can contribute to the spin relaxation of ^{13}C nuclei when they are present in natural abundance.[17] The ^{13}C dipolar relaxation mechanism is the most useful to analyze molecular rotational motions in rubbery polymers as well as small molecules. Other mechanisms can be ignored safely because their contributions to relaxation are usually small.

The observation of standard ^{13}C spectra is carried out routinely under conditions of complete scalar proton decoupling. Each proton-coupled multiplet collapses into a single, narrow resonance. In addition, the accompanying saturation of the proton energy levels results in a non-Boltzmann distribution of the ^{13}C spins in their own energy levels. This phenomenon, called the nuclear Overhauser enhancement (NOE), depends on the extent to which C–H dipolar relaxation is the operative relaxation mechanism.[17] It depends on motions near the Larmor frequency, but in a manner different from that of T_1. For a carbon that is fully relaxed by the C–H dipolar mechanism, up to a threefold enhancement of intensity can be obtained.

An important additional benefit accompanying proton saturation is the simplified mathematical treatment that results. It can be shown that the ^{13}C T_1, T_2, and NOE are given by[18]:

$$1/T_1 = \tfrac{1}{10}\gamma_C^2\gamma_H^2\hbar^2 \sum_i r_i^{-6}[f(\omega_H - \omega_C) + 3f(\omega_C) + 6f(\omega_H + \omega_C)] \qquad (4\text{-}1)$$

$$1/T_2 = 1/(2T_1) + \tfrac{1}{20}\gamma_C^2\gamma_H^2\hbar^2 \sum_i r_i^{-6}[4f(0) + 6f(\omega_H)] \qquad (4\text{-}2)$$

$$NOE = 1 + (\gamma_H/\gamma_C) \frac{[6f(\omega_H + \omega_C) - f(\omega_H - \omega_C)]}{[f(\omega_H - \omega_C) + 3f(\omega_C) + 6f(\omega_H + \omega_C)]} \qquad (4\text{-}3)$$

Here γ_C and γ_H are the carbon and proton gyromagnetic ratios, respectively; ω_C and ω_H are the corresponding resonance frequencies; and r_i is the distance between the ^{13}C nucleus of interest and the ith relaxing proton. The spectral density functions $f(\omega_i)$ are the Fourier transforms of the autocorrelation functions of second-order spherical harmonics. They describe the power available at angular frequency ω_i from the fluctuating interaction. The form of the $f(\omega_i)$ depends on the model used to describe the molecular motion. A number of models have been used.

Here it need only be realized that T_1, T_2, and the NOE are indicators of the high-frequency motions characteristic of liquids and polymers well above T_g. For the purposes of this discussion it is sufficient to relate the spin-relaxation parameters to molecular motion via the simplest model available, that of isotropic rotational diffusion. In this case the $f(\omega)$ have the form:

$$f(\omega) = \tau_R/(1 + \omega^2\tau_R^2) \qquad (4\text{-}4)$$

where τ_R is the rotational correlation time, the time it takes for the rotation to traverse 1 radian. Here isotropic means that all spatial angles are available equally to the CH vector.

Figure 4-2 shows the behavior of T_1, T_2, and the NOE as a function of correlation time for a C–H vector rotating isotropically in a field of 63.43 kG. There are several major features characteristic of the curves in Figure 4-2. A given value of T_1 corresponds to two values of the correlation time, except at the minimum. These correspond to the so-called "fast" and "slow" solutions for T_1, depending on the value of τ_R. Both T_2 and the NOE decrease monotonically with increasing τ_R. The region where $T_1 = T_2$ is called the region of extreme narrowing, and for protonated carbons:

$$1/T_1 = 1/T_2 = N\gamma_C^2\gamma_H^2\hbar^2 r_{CH}^{-6}\tau_R \qquad (4\text{-}5)$$

and NOE = 3. Here N is the number of attached hydrogens and r_{CH} is the C—H bond length. Implicit in Eq. (4-5) is the assumption that only directly bonded hydrogens contribute to the relaxation behavior of the carbon in question. This assumption is usually a good one for elastomers and small molecules in view of the r^{-6} dependence of the dipolar interaction. In rigid systems, more distant protons can dominate the dipolar relaxation of particular carbons because the dipolar interaction to neighboring protons is modulated only by very inefficient motions or is essentially static (see Chapter 3). Hence for all motional considerations being equal, T_1 depends inversely on the number of attached hydrogens, whereas the NOE does not depend on this factor. Nonprotonated carbons have T_1s about an order of magnitude longer than for protonated carbons. Nonprotonated carbon T_1s are not interpreted as readily in terms of molecular motion because in this case both the spectral densities and the C–H distances vary, and other relaxation mechanisms are more likely to contribute.

With increasing resonance frequency, the range of correlation times that satisfies the extreme narrowing condition is reduced. In addition, the minimum point on the T_1 curve occurs at a longer T_1. In extreme narrowing, all relaxation parameters are independent of frequency. Outside of extreme narrowing, T_1 decreases with decreasing ω_C.

Typical organic molecules in nonviscous solvents usually have correlation times for overall rotation in the extreme narrowing region.[18] Most naturally occurring macromolecules, such as the globular proteins, as well as highly structured (ie, hydrogen-bonded) liquids at low temperatures, often have correlation times for overall rotation in the intermediate and slow regions of the curves in Figure 4-2. For polymers, overall molecular reorientation should be a factor only in dilute solution or perhaps for very low molecular weight chains in the undiluted state. It should not be a concern for high molecular weight chains in bulk. The predominant motional feature is segmental mobility, either of the backbone or of the side chains. In order to observe the influence of chain segmental mobility on the ^{13}C spin-relaxation parameters, it is necessary that the correlation times governing these motions be comparable to or less than that governing overall molecular rotation. This is always the case

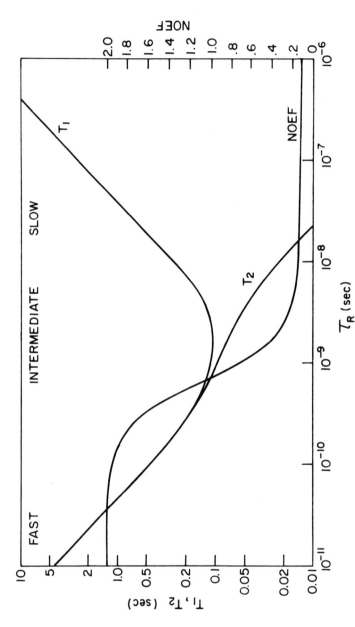

Figure 4-2. Plots of T_1, T_2, and NOEF versus τ_R for an isolated CH vector rotating isotropically in a field of 63.43 kG. The bond length, r_{CH}, is 1.09 Å. The T_1 and T_2 versus τ_R plots are log–log, whereas the NOEF versus τ_R plot is semilogarithmic. (Reprinted with permission from reference 11.)

for high molecular weight polymers in concentrated solution or in bulk at most temperatures. Such segmental motions also must occur at rates comparable to or greater than the Larmor frequency if their influence on T_1 is to be observed.

It is well known that the spin-relaxation parameters of high molecular weight polymers in solution and in bulk do not adhere to the simple isotropic model outlined above. Usually, $T_1 \gg T_2$, even when the T_1 indicates that the correlation time for segmental motion is in the extreme narrowing region. Reduced NOEs are observed also. There can be a number of reasons for the above phenomena. Often, more complicated models for segmental motion involving several or a distribution of correlation times, or librational motions, are invoked.[14,19–21] An important feature of these distributions is the inclusion of longer correlation times to account for the reduced NOEs, short T_2s, and broad linewidths. Other factors, such as a residual static dipolar interaction or nonmotional influences, can contribute to the linewidth. Because there are many possible contributions to the linewidth and T_2 is dependent on slow, long-range motional modes (see Chapter 7), the quantitative interpretation of linewidths in terms of localized motion is difficult.

The discussion that follows is little concerned with detailed modeling of the segmental motions. Instead, it relies on the spin-relaxation parameters as general probes of polymer segmental motion. Of more immediate interest is the behavior of the relaxation parameters with changes in molecular weight, temperature, and chemical structure. Whereas the spin-relaxation parameters can be viewed as another probe of polymer motion, along with mechanical and dielectric relaxation, the NMR parameters have the advantage that they can probe each resolvable site in the chain segment. Moreover, unlike such probes as electron spin resonance, the NMR probe does not perturb the chain motion or environment by introduction of a foreign group.

Contributions to the Linewidth

The factors that broaden ^{13}C NMR lines of solid polymers have been discussed previously in some detail.[5,8,22–24] The discussion here is concerned only with polymers well above T_g. The major line-broadening mechanisms for ^{13}C can be separated into those that produce a distribution of resonance frequencies and those that arise from relaxation processes. The first category includes (i) inhomogeneity in the static magnetic field, (ii) variations in bulk magnetic susceptibility within the sample, (iii) a distribution of isotropic chemical shifts, and (iv) a residual chemical shift anisotropy resulting from incomplete motional narrowing. In the second category are (a) the so-called natural linewidth, usually determined by modulation of the dipolar interaction by molecular motion as in liquids, and (b) residual static dipolar interaction from incomplete motional narrowing. These latter two sources correspond, respectively, to cases 2 and 3 discussed in Chapter 2.

Static field inhomogeneity is usually negligible for bulk polymer studies using high-resolution magnets. Contributions from the other mechanisms producing a distribution of resonance frequencies depend linearly on the field strength. These remaining mechanisms display different sensitivities to MAS. Residual CSA is eliminated by MAS, whereas a distribution of chemical shifts is not. The portion of magnetic susceptibility contributions that display a $3 \cos^2 \theta - 1$ dependence are eliminated by MAS; the remainder are not.[22,25]

Interpretation of linewidths in terms of molecular motion is severely hampered by contributions from several factors. Often the nonmotional factors can dominate. It is important to assess the various contributions because they limit the ultimate resolution attainable by application of various line-narrowing techniques. For example, polyisobutylene (PIB) is a polymer whose linewidths are dominated by the natural linewidth, mainly because the backbone segmental motion is very slow and so the linewidth is large relative to other rubbery polymers with the same T_g. However, the linewidths are field dependent,[24,26] indicating other contributions. Table 4-1 shows the results for PIB at 45°C at three fields for samples at several molecular weights (see Effect of Molecular Weight). The 67.9-MHz data are not for samples identical to those used at the other frequencies and hence these data can be compared only semiquantitatively to those at the lower frequencies.

TABLE 4-1. CARBON-13 NMR LINEWIDTHS (Hz) FOR POLYISOBUTYLENE AT 45°C[a]

Molecular weight	Configuration	Carbon								
		CH$_2$			C$_q$			CH$_3$		
		22.6	50.3	67.9	22.6	50.3	67.9	22.6	50.3	67.9
2.65×10^3	Flowed	81	69	151	19	20	39	35	51	66
4.5×10^4	Flowed	89	104	201	38	41	63	60	84	111
3.5×10^6	Plug	98	125	—	47	52	—	67	92	—
3.5×10^6	Chunks	95	119	208	49	55	86	62	100	134

[a] Estimated accuracy ±10%. Data at 22.6 and 50.3 MHz are from reference 24. Data at 67.9 MHz are from reference 26. The samples used at 67.9 MHz were not identical to those used at the lower frequencies.

The field dependence can result in large linewidths for the high-molecular-weight polymers at high frequencies, even when the field-dependent component is not dominant at low frequencies. Application of dipolar decoupling (DD) and MAS results in a narrowing of all lines by 20–30 Hz at 30°C and 22.6 MHz[24] for the PIB of the highest molecular weight. The major contribution at low field is from the natural linewidth for polyisobutylene, which appears to be an exception because of its slow backbone mobility. A 20-Hz contribution is major for many other elastomers because natural linewidths of 5–15 Hz often are observed for unfilled, uncured polymers.[7,13,14]

A comprehensive study of the contributions to the ^{1}H and ^{13}C NMR linewidths of bulk cis-polybutadiene has been reported.[5,8] Various coherent averaging techniques [DD, fast and slow MAS, multipulse decoupling (see Chapter 9)] and the field and temperature dependence were used to test possible broadening contributions. The experimental results for cis-polybutadiene and the behavior of possible line broadening mechanisms in various tests are shown in Table 4-2.[8] The major contributor to both the ^1H and ^{13}C linewidths is low in frequency ($\sim 10^3$ Hz) and inhomogeneous in nature, and is attributed to a rapid motional component that is spatially inhibited by chain constraints. This is another name for a residual static dipolar interaction, and it corresponds to case 3 in Chapter 2, Line Narrowing Schemes and Molecular Motion. Since the linewidths did not change with magnetic field, frequency-dependent broadening factors like residual chemical shift anisotropy (CSA) and magnetic susceptibility variations were essentially absent. This study is a good example of how a systematic assessment of linewidth contributions can sort out motional models and why observed linewidths can be difficult to interpret in terms of localized motions.

Results so far indicate that DD and MAS at least marginally improve the ^{13}C spectra of totally amorphous polymers above T_g. The nature of the study at hand must determine whether application of line-narrowing techniques is worthwhile relative to increasing the temperature.

For semicrystalline and filled systems, line-narrowing techniques are of more value. The resonances are broader for these systems than for the comparable totally amorphous or unfilled system.[27] For semicrystalline polymers, raising the temperature does not remove the broadening contribution from the crystallites while still below T_m.[23] Contributions to the linewidths of semicrystalline systems are discussed in more detail in Chapter 5.

It appears that the presence of crystallinity or filler can broaden the lines in the rubbery, amorphous regions by more than one mechanism. For example, for the semicrystalline trans-polybutadiene the lines narrowed from 90 to about 30 Hz with MAS, whereas DD had no effect;[24] also, the linewidths were field-dependent. This and other evidence indicates that magnetic susceptibility variations are the cause of the broadening in this case. Residual CSA was eliminated as a major contributor because the sp^2 and sp^3 carbons had essentially the same linewidth under all conditions, even though the CSAs of these two carbons differ by more than a factor of 6. It is too not surprising that line broadening for semicrystalline trans-polybutadiene arises from a different mechanism than for its totally amorphous counterpart, cis-polybutadiene.[8] On the other hand, this was not found for polyethylene,[23] for which incomplete motional narrowing was a major source of broadening.

Filled systems may be expected to behave like semicrystalline polymers in this regard. Susceptibility differences seem to be a prime candidate.[15] However, Schaefer determined that incomplete motional narrowing of a homogeneous nature was the major contributor to the linewidths for filled cis-polyisoprene.[7] This conclusion was based on a hole-burning experiment,[28] which is another way to assess contributions from the two categories of

TABLE 4-2. EXPERIMENTAL RESULTS AND PREDICTED LINE NARROWING FOR ^{1}H AND ^{13}C NMR COHERENT AVERAGING EXPERIMENTS FOR VARIOUS CHAIN DYNAMICS MODELS[a,b]

Model	^{1}H NMR				^{13}C NMR		
	REV-8	MAS	Slow MAS[c]	Decrease H_0	DD	DD MAS	MAS
Quasi-rapid isotropic motion	no	no	no	no	no	no	no
Spin diffusion	yes	yes	no	no	no	no	no
Chemical exchange							
(1) Both sites homogeneously broadened	yes	no	no	no	partial	yes	no
(2) Minor site inhomogeneously broadened	no	no	no	no	no	no	no
Anisotropic motion							
(1) Homogeneously broadened	yes	yes	no	no	yes	yes	yes
(2) Inhomogeneously broadened	yes	yes	SSB	no	yes	yes	yes
Magnetic susceptibility broadening	no	yes	SSB	yes	no	yes	yes
cis-1,4-Polybutadiene experimental	yes	yes	SSB	no	yes	yes	yes

[a] Table entries refer to whether substantial narrowing is expected for each mechanism or is experimentally observed.
[b] Reprinted with permission from English, A. D. *Macromolecules* **1985**, *18*, 178. Copyright 1985. American Chemical Society.
[c] "SSB" designates observation of spinning side bands.

broadening (homogeneous and inhomogeneous) mechanisms given earlier in this section. In a hole-burning experiment a broad line is irradiated by monochromatic rf at various points across the line. If the major source of broadening is inhomogeneous, a hole appears the width of which is the true homogeneous linewidth. If homogeneous broadening dominates, the entire line is saturated.

Effect of Molecular Weight

The only systematic study of the effect of molecular weight on ^{13}C NMR spectral parameters has been for polyisobutylene.[24,26] This polymer is an excellent one for such a study because it is totally amorphous, has a well-defined low glass transition temperature, and is available in a wide range of molecular weights. Also, the ^{13}C spectrum is not complicated by configurational isomerism as it is for vinyl polymers.

Table 4-3 gives the ^{13}C T$_1$ and NOEF (NOE $-$ 1) for bulk polyisobutylene at 67.9 MHz and 45°C as a function of molecular weight. Except for the possibility of a small variation for the CH$_2$ carbon, these two parameters are independent of molecular weight from 1350 to 3.5×10^6. The two lowest molecular weights are below the critical molecular weight for chain entanglement, M$_c$ = 17,000, ie, for one entanglement per chain.[29] The highest molecular weights are well above this critical value. Hence, it can be concluded that the relatively fast motions that determine T$_1$ and the NOEF are not influenced by chain entanglements or by the bulk viscosity. This is somewhat surprising because different carbons may have been expected to be influenced in different ways by chain entanglement, molecular packing, and other interchain interactions. The lack of dependence on chain length demonstrates that only very localized motions contribute to the T$_1$ and the NOEF. The T$_1$ and NOEF of

TABLE 4-3. CARBON-13 SPIN-RELAXATION PARAMETERS FOR BULK POLYISOBUTYLENE AT 67.9 MHz AND 45°C[a]

Molecular Weight $\times 10^{-3}$	CH$_2$		C$_q$		CH$_3$	
	T$_1$ (s)[b]	NOEF[d]	T$_1$ (s)[c]	NOEF[d]	T$_1$ (s)[c]	NOEF[e]
1.35	0.142	0.38	1.49	0.52	0.123	1.37
2.65	0.157	0.49	1.61	0.62	0.117	1.38
45.0	0.163	0.48	1.52	0.62	0.119	1.37
1000	0.176	0.40	1.57	0.86	0.123	1.24
3500	0.186	0.42	1.63	0.63	0.115	1.39

[a] From reference 26.
[b] Estimated accuracy, ± 10–15%.
[c] Estimated accuracy, ± 10%.
[d] Estimated accuracy, ± 0.2.
[e] Estimated accuracy, ± 0.1.

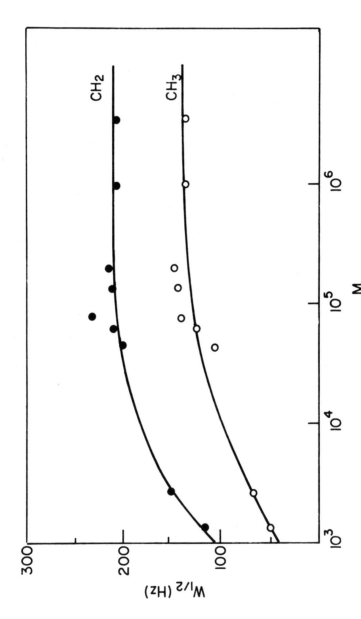

Figure 4-3. Semilogarithmic plot of linewidth versus molecular weight for the protonated carbon resonances of bulk polyisobutylene at 45°C and 67.9 MHz. (Reprinted with permission from reference 11.)

PIB both in bulk and in solution have been interpreted in more detail as a function of molecular weight.[19]

Of more interest is the behavior of the linewidths. Figure 4-3 shows the molecular weight dependence of the protonated carbons of PIB at 67.9 MHz.[26] The trend shown in Figure 4-3 has been confirmed at two lower resonance frequencies[24] (see Table 4-1). The linewidths, and hence the T_2, are invariant to molecular weight above about 3000 at 45°C. It also can be concluded that the slower modes of polymer segmental motion, which are characterized by long correlation times and contribute to T_2, do not depend on chain entanglements above some low critical value. These results are important in that they eliminate molecular weight as a direct influence on the spectra and spin-relaxation parameters, except at very low molecular weight. Of course, molecular weight can exert an indirect influence in its effect on crystallinity and morphology.

Effect of Temperature Relative to T_g

Spectral Collapse

In view of the above discussion on the effect of molecular motions on the ability to observe high-resolution ^{13}C NMR spectra of bulk polymers, it is not surprising that temperature plays a large role. Unlike the situation for non-viscous liquids or solutions, the linewidths observed for bulk rubbers are in the range where moderate changes in temperature can have a major effect on spectral resolution. The dipolar interaction, via the dipolar relaxation mechanism, controls the linewidth. This is an example of case 2 in Chapter 2 for the effect of molecular motion on the linewidth. Here the frequency of motion is much too fast for dipolar decoupling to be useful. It should be noted that, as the temperature is lowered toward T_g, a reduced component of the static dipolar interaction may contribute to the linewidth.

Results for PIB illustrate the general behavior. Figure 4-4 shows the temperature dependence of the ^{13}C spectra of bulk PIB at two extremes of molecular weight, both of which have the same T_g. With decreasing temperature the lines broaden, then merge, and the "high-resolution" spectrum collapses. For PIB of molecular weight 3.5×10^6, this collapse temperature, T_c, is in the vicinity of $-10°C$. The collapse occurs at a somewhat lower temperature for the low molecular weight polymer. Hence it is necessary to be well above (about 70°C or so) the glass transition temperature to observe a high-resolution ^{13}C spectrum.

This difference in T_g and T_c is not surprising, since NMR is a high-frequency experiment. The T_g corresponds to an onset of molecular motion, and there is a relationship between the rate of this motion and the experimental time available for its observation. Experiments involving long observation

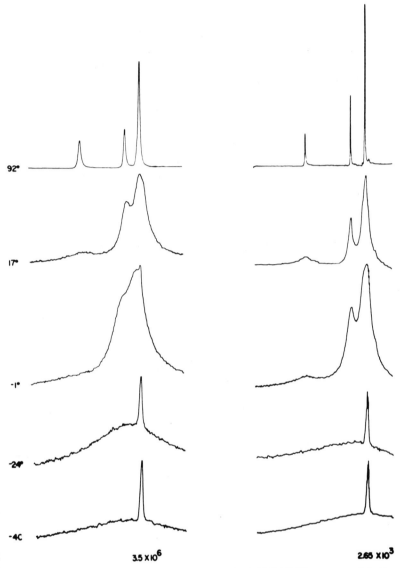

Figure 4-4. Temperature dependence of the standard ^{13}C NMR spectra of bulk poly-isobutylene at two molecular weights. A 5000-Hz region is shown. The spectra are plotted at different vertical gains. The relatively narrow resonance seen at low temperatures arises from the methyl carbon septet of acetone-d_6, which was used for field-frequency locking. (From Komoroski, R.; Mandelkern, L. *J. Polym. Sci., Polym. Symp.* **1976**, *54*, 201. Copyright 1976 John Wiley and Sons, Inc. Reprinted by permission of John Wiley and Sons, Inc.)

times are sensitive to slow motions, whereas for short observation times only rapid motions can be detected. Because molecular motion increases with increasing temperature, a low-frequency experiment (such as differential scanning calorimetry, DSC) will detect the transition at a lower temperature than the high-frequency experiment.

The temperatures at which the ^{13}C spectra collapse have been reported for a number of polymers.[30,31] The results are listed in Table 4-4. The difference between T_c and T_g ranges between about 30 and 87°C. The temperature difference shows no relationship to T_g itself. Except for PIB, the ratio of the temperature at which the spectra collapse to T_g is in the range 1.2–1.3. For PIB this ratio is 1.4, an experimentally significant difference.

TABLE 4-4. THE T_g AND ^{13}C NMR SPECTRAL COLLAPSE TEMPERATURES OF SOME POLYMERS[a]

Polymer	T_c (°C)	T_g (°C)
Isotactic poly(methyl methacrylate)	110	52
Poly(isopropyl acrylate)	47	−11
Poly(n-butyl acrylate)	0	−54
Atactic polypropylene	30	−20
trans-Polyisoprene	−28	−58
cis-Polybutadiene	−65	−102
cis-Polyisoprene	−30	−70
Polyisobutylene	−10	−70
Poly(ethylene oxide)	−38	−70

[a] From reference 30. Reprinted by permission of John Wiley and Sons. Copyright 1978, John Wiley and Sons.

The observation that the ratio of the spectral T_c to the T_g is a constant for most polymers has led to the identification of T_c with the proposed transition T_{ll} for totally amorphous polymers.[32] Considerable controversy has surrounded the existence of this transition.[33-36] However, regardless of the other evidence relating to the existence of the T_{ll} transition, the identification of T_c with T_{ll} does not appear to be justified. A number of factors can influence the temperature below which a high-resolution spectrum cannot be observed for a given nucleus. These include chemical-shift separation of the resonances in question, NMR operating frequency, the strength of the decoupling field, the degree of crystallinity or presence of filler particles, the degree of crosslinking, and the effect of motion of the polymer backbone on the linewidth (T_2 resulting from dipolar relaxation). The temperature dependence of the spectra of amorphous polymers results from the last factor, as well as from the possible introduction of a reduced but finite component of the static dipolar interaction at lower temperatures.

From a practical point of view it is difficult to determine just when T_c occurs for a given polymer. This determination depends on instrumental parameters to a large extent and can be "adjusted" artificially by varying the spectral width observed. Moreover, it is hard to explain why the ^{13}C spectrum, as opposed to spectra of other nuclei, should bear any necessary, special relationship to polymer transitions. Clearly, a different collapse temperature would be observed for the spectrum of a different nucleus, such as ^{15}N or ^{29}Si.

Recent ^{29}Si NMR work on solid silicone gums and resins confirms the inability to identify T_c with T_{ll}.[37] For blends of silicone resin with polydimethylsiloxane or poly(dimethylsiloxane-co-diphenylsiloxane) gums, T_c/T_g ratios determined by ^{29}Si NMR were from 1.04 to 1.16, measurably lower than the values of 1.2–1.3 determined for a variety of polymers by ^{13}C NMR. The lower T_c/T_g ratio for ^{29}Si relative to ^{13}C is consistent with the reduced dipolar interaction of ^{29}Si with protons in silicon-containing polymers. This reduced dipolar interaction relative to the ^{13}C case arises from the larger Si—H distances relative to the C—H bond distance.

In the same study the authors were able to observe different T_c for the different components in the resin–gum blends. From this and other evidence they concluded that the blends were incompatible mixtures of resin and gum. The fact that the T_c of the individual phases did vary with composition suggested a slight intermixing.[37]

Williams-Landel-Ferry Behavior

The behavior of the ^{13}C NMR parameters of bulk polymers above T_g has been shown to be no different than that of other techniques that measure relaxation phenomena. Specifically, the average correlation time for segmental motion, as determined by ^{13}C T_1s and NOEs, has been shown to obey the WLF equation.[38] The empirical relationship developed by Williams, Landel, and Ferry (WLF)[39] gives the ratio of the relaxation time at a temperature

TABLE 4-5. The WLF Parameters of Some Totally Amorphous
and Semicrystalline Polymers

Polymer	Literature[a]		^{13}C NMR[b]	
	C_1	C_2	C_1	C_2
Ethylene–propylene (63 mol% Pr)	13.1	40.7	15.4	67.7
cis-Polyisoprene	16.8	53.6	15.0	68.7
Polyisobutylene	16.6	104.4	14.1	91.5
Atactic polypropylene	—	—	15.2	60.7
Ethylene–butene (26 mol% Bu)	—	—	13.5	40.6
Poly(vinyl acetate)	15.6	46.8	15.1	62.0
Poly(isopropyl acrylate)	—	—	12.2	53.6
cis-Polybutadiene	11.3	60	15.1	54.3
Linear polyethylene			12.5	34.3
Poly(ethylene oxide)			12.5	26.3
Poly(trimethylene oxide)			11.8	14.7
trans-Polyisoprene			12.5	26.3

[a] Mechanical or dielectric relaxation.
[b] From Dekmezian, A.; Axelson, D. E.; Dechter, J. J.; Borah, B.; Mandelkern, L. *J. Polym. Sci., Polym. Phys. Ed.* **1985**, *23*, 367.

T to that at the transition temperature T_g for a polymer in the range T_g to $T_g + 100$ or so. The WLF equation is:

$$\log a_T = -C_1(T - T_g)/[C_2 + (T - T_g)] \tag{4-6}$$

where C_1 and C_2 are constants for a particular amorphous polymer. Initially, they were thought to be universal constants applicable to all polymers.[40] Table 4-5 contains the WLF parameters for a number of totally amorphous polymers as determined via ^{13}C NMR spin-relaxation parameters and, where

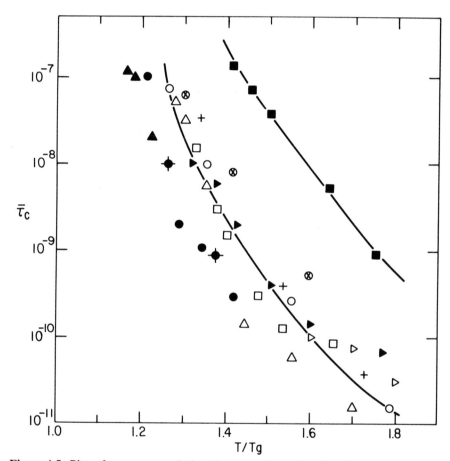

Figure 4-5. Plot of average correlation time τ_c against the reduced temperature T/T_g for a number of amorphous polymers. ○ cis-Polybutadiene; □ ethylene–butene copolymer (26 mol% Bu); + cis-polyisoprene; △ ethylene–propylene copolymer (63 mol% Pr); ✦ atactic polypropylene at 67.9 MHz; ● at 37.7 MHz; ■ polyisobutylene; ⊗ poly(isopropyl acrylate); ▲ polyvinyl acetate; ▷ ethylene–propylene copolymer (50 mol% Pr); ▶ ethylene–propylene copolymer (26 mol% Pr). (From Dekmezian, A.; Axelson, D. E.; Dechter, J. J.; Borah, B.; Mandelkern, L. *J. Polym. Sci., Polym. Phys. Ed.* **1985**, *23*, 367. Copyright 1985 John Wiley and Sons, Inc. Reprinted by permission of John Wiley and Sons, Inc.)

available, by mechanical or dielectric relaxation.[38] Good agreement is obtained for a given polymer between the constants determined by NMR and those determined by other techniques. The abnormally high value for C_2 of PIB is confirmed by NMR.

Figure 4-5 shows the dependence of the average correlation time for segmental motion versus the ratio T/T_g for the polymers in Table 4-4.[38] All the polymers exhibit approximately the same motional behavior versus the reduced temperature, except for PIB. Figure 4-5 indicates that, at the same reduced temperature, the polyisobutylene backbone motions are significantly more restricted. This confirms the conclusion made previously from ^{13}C spin-relaxation data at 45°C that the backbone mobility of PIB is about two orders of magnitude slower than for cis-polyisoprene or polybutadiene.[24,26]

Crosslinks

Low levels of crosslinking can have substantial effects on the physical properties of elastomeric materials. Little effort has been directed toward the investigation of crosslinking of polymers above T_g using ^{13}C NMR techniques. The first ^{13}C NMR study of crosslinking that did not involve solvent-swollen systems was concerned with the effect of typical low levels of crosslinking on the spectra and T_1 of butyl rubber.[41] At levels of 1.7 and 2.7 crosslinks per 1000 carbon atoms, there was no measurable effect either on the T_1 of the bulk of the carbons in butyl rubber, or on the spectra, and hence T_2. This observation is consistent with the independence of the spin-relaxation parameters with molecular weight up to 3.5×10^6.[26] Hence, typical crosslink levels do not effect the rapid segmental motions that determine T_1, or the slower modes of motion that determine T_2, for the bulk of the polymer. The crosslink sites themselves were not observable because of their very low concentration.

High levels of crosslinking do affect the ^{13}C spectra of bulk elastomers. At sufficiently high concentrations the crosslink sites themselves can be seen. Patterson et al[42] followed the peroxide curing of both natural rubber and cis-polybutadiene to high crosslinking levels. Their work illustrates the need for a multitechnique approach to such studies. Figure 4-6 shows the spectra of cis-polybutadiene as a function of the amount of curing agent. The top set of spectra are the normal FT spectra obtained with MAS. The lines broaden with increasing level of curative. At 5 parts of curative per hundred of rubber (phr) a shoulder is barely visible that may arise from carbons at the crosslink sites. The bottom set of spectra was acquired with cross polarization and magic angle spinning. New peaks attributable to crosslink sites are seen in the 30–40 ppm region. The results in Figure 4-6 indicate that, at these levels of crosslinking, the crosslink sites are less mobile than the bulk of the elastomer chains, because the crosslink sites are quite intense relative to the main peaks in the CPMAS spectra. This effect cannot be attributed to an increase in T_g, because this increases relatively little in this case. (See Appendix.)

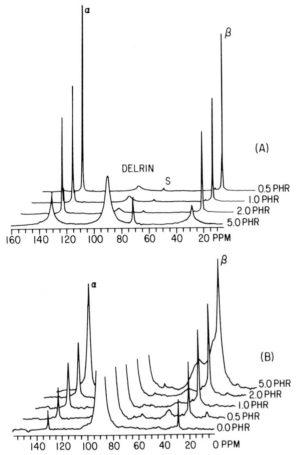

Figure 4-6. Superposed ^{13}C spectra of cis-polybutadiene cured with varying amounts of dicumyl peroxide. Samples were cured at 149°C for 2 h in a hot press with the amount of peroxide shown (in parts per hundred of rubber). (A) Spectra obtained under standard conditions with MAS. Peaks marked "s" are from the rotor. (B) Spectra obtained with CPMAS. (Reprinted with permission of the copyright owner, Rubber Division, ACS, Inc., from reference 42.)

The crosslinking of ethylene–propylene rubber (EPR) upon γ-irradiation has been studied using CPMAS NMR.[43] Small, new peaks from crosslink sites were detected in the spectrum of the polymer after γ-irradiation. It was concluded on the basis of a comparison to the spectra of model compounds that both short chain branches and network junctions were being observed. Unfortunately, the signal-to-noise ratio in the published spectra was low, calling the quantitative results into question.

The above studies illustrate the problems that can be encountered in the study of polymer crosslinking. The first is that of sensitivity. The direct observation of crosslink sites at levels typical of cured rubbers has not yet been accomplished by NMR. As of this writing, repeated attempts in our laboratory on sulfur vulcanized natural rubber have not been unambiguously successful.[44] Another problem, pertinent to industrial applications, is the assignment of peaks and the differentiation of resonances from crosslink sites from those of the degradation products of the curative, accelerator, or other additives. Resolution of the small peaks from crosslink sites from the major polymer resonances may also present a problem. Several approaches, including spectral editing techniques, are currently being pursued in the author's laboratory. (See Appendix.) More work, perhaps coupled with instrumental advances, is needed before high-resolution NMR of solids can be a useful tool for studying polymer crosslinking in common curing systems. The introduction of short crosslinks containing isotopically enriched, NMR-active atoms may be one approach for model systems.

An alternative that can provide increased spectral resolution is swelling of the polymer. Depending on the degrees of crosslinking and swelling, linewidths of cured, filled elastomers or glassy polymers can be reduced significantly and begin to approach those found for the corresponding solutions.[45,46]

Presence of Filler

Elastomers usually are used as composite materials in industrial products. Fillers, such as carbon black or reinforcing silica, as well as processing oils, tackifiers, and other polymers, are added. The compositional analysis of these materials can be complicated by their composite nature, particularly when carbon black, which is optically opaque, is present. Moreover, the interaction of elastomers with fillers and its effect on mechanical properties is a subject of both practical and theoretical importance. Carbon-13 NMR techniques for both liquids and solids undoubtedly will make contributions in these areas.

The presence of filler can broaden the lines in a high-resolution spectrum of an elastomer beyond those seen for the pure polymer. As mentioned earlier, this broadening can arise from susceptibility differences between polymer and filler or from incomplete motional narrowing caused by restriction of polymer chain mobility by the filler. By raising the temperature to around 100°C (for typical elastomers), this broadening sometimes can be mitigated sufficiently for successful use of standard FT NMR.[47] Even at elevated temperature the spectral resolution for filled materials (at typical filler levels) is substantially worse than that of the pure polymers.

Resolution can be improved significantly by employing MAS in the standard scalar-decoupling (SD) experiment.[15,48] The effects of both low-frequency incomplete motional narrowing[7] and isotropic susceptibility differences[22] are removed by MAS. The improvement is sufficient for the identification and quantification of the elastomeric components in simple, filled vulcanizates. Figure 4-7 shows the spectra of three tire-tread compounds containing differ-

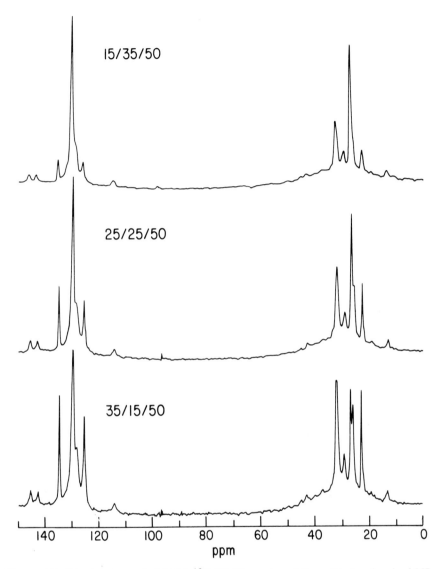

Figure 4-7. Dipolar-decoupled MAS ^{13}C NMR spectra of three filled, vulcanized NR–cis-BR–SBR blends. Weight percentages of the three components are given. (Reprinted with permission of the copyright owner, Rubber Division, ACS, Inc., from reference 15.)

ent amounts of natural rubber (NR), cis-polybutadiene (BR), and emulsion styrene–butadiene rubber (SBR).[15] Although the spectra are not of the quality of comparable solubilized or pyrolyzed materials, enough peaks are resolved for identification and a reasonable quantitation. The results in Figure 4-7 were obtained at 22.6 MHz. Because much of the broadening for filled elastomers can be removed by MAS, it cannot result from an isotropic distribution of chemical shifts. Hence, higher magnetic fields should improve the spectral resolution (and sensitivity) for such systems.

Incorporation of surface-active carbon black into various rubbers can improve such mechanical properties as strength and stiffness. Dispersion of the filler and the interaction between the polymer and filler are important for the development of suitable mechanical properties. Pulsed ^{1}H NMR has been applied to filled elastomers for some time.[49] These and other studies have demonstrated the existence of one or more types of "bound rubber" at the surface of the filler particles. It is reasonable to expect high-resolution, solid-state NMR techniques to contribute to our understanding of polymer–filler interactions.

Filled polymers can be approached as other mixed-phase polymers, and the spectra of the mobile and more rigid components separated. Schaefer[50] has reported the CP spectrum without MAS of a filled polyisoprene. The spectrum, which is from carbons near the filler surface, displays lineshapes typical of rigid systems. This is not surprising, for the CP experiment should sample preferentially carbons in immobile environments (cross polarizing from protons in rigid environments). The ability to observe only polymer bound to filler particles should be useful in expanding our knowledge about how fillers reinforce elastomers and the molecular criteria for good reinforcement. Studies of the above type must be approached with caution, because, depending on conditions, the discrimination by CP against mobile chains is not 100% effective. The recently developed electron–nuclear polarization technique,[51] in which the carbon signal is generated by CP from free radicals (directly or via protons), may be the preferred approach for observing polymer bound to carbon black.

Heterogeneous Systems and the Interface

General Aspects

Heterogeneity in polymeric systems is encountered widely. Of some importance are incompatible systems containing a glassy (or crystalline) phase and a rubbery phase, often deriving from block copolymers with blocks of "hard" and "soft" segments. Such block copolymers can function as thermoplastic elastomers in which the glassy phase (eg, polystyrene) can serve as a cross-linking medium for the rubbery phase (eg, polybutadiene). The glassy cross-links can be melted temporarily or dissolved for processing. In another case, a brittle, glassy thermoset matrix, such as epoxy resin, can be toughened substantially by incorporation of a relatively small amount of an incompatible rubber.

There are several features of these systems that are important for their ultimate physical properties. The nature of the interface between the polymeric phases is of prime importance because it is at the interface where stresses occur or are concentrated. Along with the nature of the interface is the problem of phase mixing in general, or to what degree the polymer chains of one phase can be found in the other phase, and the effect on the chain motions in the other phase? In compatible systems, to what specific molecular interaction or functional group can the compatibility be attributed?

High-resolution NMR techniques are just beginning to have an impact in this area. The rubbery portions can be observed separately using standard scalar-decoupled ^{13}C NMR or possibly CP with delayed decoupling. The glassy portions can be seen separately using CP. The entire spectrum can be obtained with the one-pulse experiment with dipolar decoupling. This battery of techniques has considerable analytical utility for multiphase systems, as recently demonstrated for butadiene in acrylonitrile–butadiene–styrene (ABS) copolymer[52] and in butadiene-grafted polypropylene.[53]

Rubber-Modified Epoxies

A small amount of rubber in the form of discrete particles in a glassy thermoset resin, such as epoxy resin, can greatly improve its crack resistance and impact strength. This improvement is achieved without significantly decreasing thermal or mechanical properties.[54] The rubber often used is a low molecular weight butadiene–acrylonitrile copolymer having carboxyl (CTBN) or other functional end groups that react with the epoxy. The chemistry has been described by Riew et al.[54] They also have determined many of the criteria necessary for effective toughening. It is known from electron microscopy that epoxy resins toughened with CTBNs consist of a continuous rigid phase and a dispersed rubber phase. However, the extent of phase mixing and the nature of the interface are not well understood. These factors are of importance for understanding the mechanism of toughening.

The precise chemical composition of the rubber particles is not known. It has been hypothesized that the rubbery particles consist of a mixture of linear copolymers of CTBN–epoxy and possibly homopolymers of epoxy resin.[54] This prediction was based on the chemistry of the process, where a linear chain extended product of epoxy and CTBN is formed early in the reaction. To shed further light on the nature of the phases in these toughened epoxies, Sayre et al determined the concentration of mobile epoxy in the dispersed phase using standard ^{13}C NMR spectroscopy.[55] Figure 4-8A shows the spectrum of a prepolymer containing 2 mol of epoxy to 1 mol of CTBN. This is a model for the spectrum expected if the rubbery regions are composed of all the CTBN and the end-capping epoxy units. Peaks from the epoxy are small but visible and are resolved from the main peaks from CTBN. Figure 4-8B shows the spectrum of a cured CTBN-modified epoxy containing 10 parts of CTBN. Arrows point to barely visible humps that arise from somewhat mobile epoxy units. These results have been confirmed by us[56] at 50.3 MHz on a number of similar samples at levels of 10–30 parts of CTBN. The authors[55] conclude that

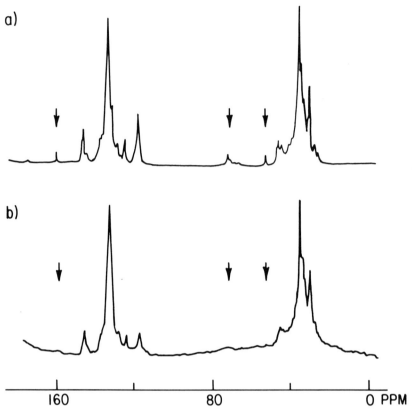

Figure 4-8. The standard ^{13}C spectra of (A) 2 mol of epoxy to 1 mol of rubber, and (B) cured rubber-modified epoxy with 10 parts of CTBN. (Reprinted with permission of Butterworth and Co., Publishers, Ltd. from reference 55.)

the peaks seen for mobile epoxy represent much less than the 16% of epoxy expected if all the end cap units are dissolved.

Unfortunately, the detection of 16% dissolved epoxy is a difficult proposition for broadened epoxy lines. It is possible that even though the epoxy units are dissolved in the rubbery portions, their signals cannot be detected under standard conditions because of broadening by residual solid-like interactions or susceptibility effects. If so, MAS may narrow these lines enough to permit their unambiguous detection. Figure 4-9 shows the SD MAS ^{13}C spectrum of a CTBN–epoxy containing 30 parts CTBN.[56] Peaks attributable to epoxy are now more clearly seen at about 155, 115, 72, and possibly 55 ppm. Additional intensity from the other, unresolved epoxy peaks appears to be occurring under the main CTBN peaks. Apparently the application of MAS narrows the signals from some of the epoxy units sufficiently for observation, but generally not to the same narrow width as the CTBN peaks. A similar result was obtained for a sample containing 10 parts CTBN. Additional decoupling power was not helpful because it began to narrow the signals from the bulk of the epoxy as well as those from the small fraction of epoxy units visible at low decoupling power.

These experiments can be interpreted as demonstrating the existence of a fraction of the epoxy molecules that is more mobile than the bulk of the rigid epoxy matrix, but not as mobile as the rubbery portion. One explanation is that this fraction represents epoxy units at the surface of the rubber particles. These surface units may be somewhat more mobile than the rigid matrix because of their proximity to the rubber. An alternative explanation is that these units have precipitated in the dispersed rubber phase as small epoxy particles. These particles may be more mobile than the bulk epoxy because of their presence in the surrounding mobile rubber. The possibility of small epoxy particles in the rubber has been put forth by Kunz et al[57] as the reason for the discrepancy they find between the volume fraction of dispersed phase as determined by transmission electron microscopy (TEM) and the amount of rubber added. The TEM gave a volume fraction nearly double that predicted from the amount of rubber added. This fact can be rationalized if there is a significant contribution to the volume of the dispersed phase from immobilized epoxy that cannot be seen by TEM.[57]

Kunz et al[57] also demonstrated the value of high-resolution ^{13}C NMR for probing the interfacial regions of multiphase polymers. Transmission electron microscopy (TEM) and energy-dispersive X-ray analysis showed relatively sharp boundaries for the rubber particles in a CTBN epoxy, whereas wider, more diffuse boundaries were observed for an epoxy modified with an amine terminated butadiene–acrylonitrile copolymer (ATBN). These results taken alone suggest greater phase mixing at the interface in the case of the ATBN–

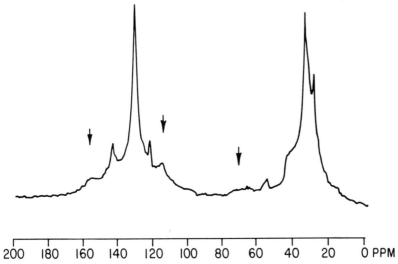

Figure 4-9. Scalar-decoupled ^{13}C NMR spectrum of a CTBN–epoxy with 30 parts of CTBN.

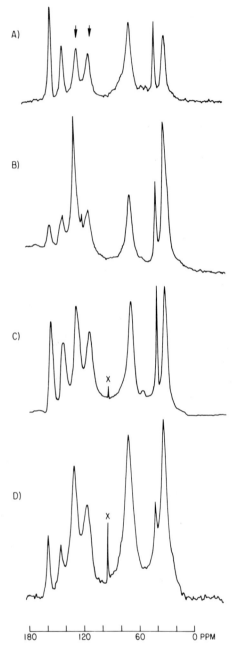

A)

B)

C)

D)

180   120   60   0 PPM

Figure 4-10. (A) Cross-polarization MAS ^{13}C NMR spectrum of a cured epoxy (Dow DER 332), contact time 1.5 ms; (B) DD MAS spectrum of a CTBN–epoxy with 30 parts CTBN, recycle time 2.2 s; (C) CPMAS spectrum of the same sample as in (B), contact time 1.5 ms; (D) CPMAS spectrum of a CTBN–epoxy containing 150 parts CTBN, contact time 0.1 ms.

epoxy. However, ^{13}C NMR (and DSC) results showed no appreciable differ-ence in the amount of mobile epoxy (as determined by standard techniques without MAS) in the dispersed rubber phases of the two systems. The diffuse appearance in the micrographs was attributed to the irregular, nonspherical character of the surface in the ATBN case.

It is also of interest to determine whether any rubber has been immobilized in the epoxy matrix. It is more difficult to answer this question by ^{13}C NMR than the reverse case because no CTBN peaks are resolved from the spectrum of the epoxy. Because rigid chains are now being sought, CPMAS NMR is the technique of choice. Figure 4-10A shows the CPMAS spectrum of a cured epoxy without rubber modifier. The arrows indicate the peaks overlapping with the major resonances of the CTBN. This overlap is borne out in Figure 4-10B, which is the DD MAS spectrum showing both rigid and mobile regions of a CTBN–epoxy containing 30 parts of CTBN.

Figure 4-10C shows the CPMAS spectrum of the same sample with a typical contact time of 1.5 ms. Changes in the relative peak intensities indicate a contribution from the CTBN portion. However, it is not clear whether this contribution arises from a weak cross polarization of the entire rubbery phase or from an efficient cross polarization of rubber at the interface or dissolved in the rigid phase. The CPMAS spectrum in Figure 4-10D is that of a sample containing 150 parts CTBN at a contact time of 0.1 ms. At this very short contact time the nonprotonated carbons of the epoxy matrix are reduced sub-stantially in intensity relative to the protonated carbons, as expected. There should be no contribution from the slowly cross-polarizing bulk of rubber. Only immobile CTBN should cross polarize to any extent. Peak intensity ratios in Figure 4-10D suggest the presence of a small fraction or rigid CTBN at the interface or in the epoxy matrix. Additional work is necessary to quan-tify these results.

Thermoplastic Polyesters

Hytrel is a thermoplastic polyester elastomer containing poly(butylene terephthalate) hard segments and poly(tetramethylene ether) soft segments in a range of compositions. Electron microscopy has shown that Hytrel polymers exist as two-phase systems. Jelinski et al[58] have used ^{13}C NMR without CP to observe the mobile domains in Hytrel polymers as a function of composition and temperature. They have also characterized molecular motion in the hard segments using CSA powder patterns and spin-relaxation parameters.[59–61]

Scalar-decoupled spectra could be observed for all Hytrel samples studied (0.80–0.96 mol fraction hard segments). Figure 4-11 shows the SD MAS and DD MAS spectra of a Hytrel containing 0.8 mol fraction hard segments. Broad signals are seen in the aromatic and carboxyl regions of the SD MAS spectrum. These signals are attributed to terephthalate groups flanking the soft segments. It was found that the number of mobile carbons exceeded by 10% the amount expected based on the known fraction of soft segments. This result

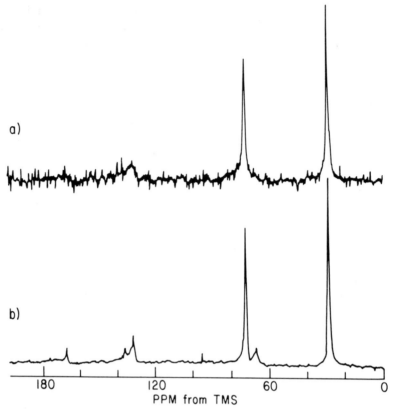

PPM from TMS

Figure 4-11. Scalar-decoupled MAS (A) and DD MAS (B) ^{13}C spectra of a solid Hytrel (0.8 mol fraction hard segments) at 50.3 MHz. Both spectra were obtained with 1-s recycle delay and were scaled to compensate for differences in the number of accumulations. (Reprinted with permission from Jelinski, L. W.; Schilling, F. C.; Bovey, F. A. *Macromolecules* **1981**, *14*, 581. Copyright 1981 American Chemical Society.)

suggests that the terephthalate groups at the domain boundaries are relatively mobile, whereas soft segments near the boundaries are not noticeably restricted.

They also found that the linewidths of the SD spectra increased with increasing mole fraction of hard segments but were essentially independent of temperature between 30 and 110°C. In this regard the Hytrel polymers behave as semicrystalline homopolymers below the melt transition (Chapter 5).

Summary

Carbon-13 NMR techniques for liquids and for solids have utility for studying amorphous polymers above T_g and the mobile portions of heterogeneous systems. A multitechnique approach can provide additional information of an analytical or fundamental nature not available using any single technique. Detailed modeling of the chain motions of elastomers is not necessary to use spin-relaxation parameters to study transitions, phase structure, and the effects of modifications, such as fillers and crosslinking.

References

1. Schaefer, J.; Stejskal, E. O. *Top.* ^{13}C *NMR Spectrosc.* **1979**, *3*, 283.
2. Fyfe, C. A. "Solid State NMR For Chemists"; C. F. C. Press: Guelph, 1984.
3. Duch, M. W.; Grant, D. M. *Macromolecules* **1970**, *3*, 165.
4. Gutowsky, H. S.; Saika, A.; Takeda, M.; Woessner, D. E. *J. Chem. Phys.* **1957**, *27*, 534.
5. English, A. D.; Dybowski, C. R. *Macromolecules* **1984**, *17*, 446.
6. Abragam, A. "The Principles of Nuclear Magnetism"; Clarendon Press: Oxford, 1961.
7. Schaefer, J. *Macromolecules* **1972**, *5*, 427.
8. English, A. D. *Macromolecules* **1985**, *18*, 178.
9. Farrar, T. C.; Becker, E. D. "Pulse and Fourier Transform NMR"; Academic Press: New York, 1971.
10. Fukushima, E.; Roeder, S. B. W. "Experimental Pulse NMR—A Nuts and Bolts Approach"; Addison-Wesley: Reading, 1981.
11. Komoroski, R. A.; Mandelkern, L. In "Applications of Polymer Spectroscopy", Brame, E. G., Ed.; Academic Press: New York, 1978, Ch. 5.
12. Schaefer, J.; Natusch, D. F. S. *Macromolecules* **1972**, *5*, 416.
13. Schaefer, J. *Top.* ^{13}C *NMR Spectrosc.* **1974**, *1*, 149.
14. Schaefer, J. *Macromolecules* **1973**, *6*, 882.
15. Komoroski, R. A. *Rubber Chem. Technol.* **1983**, *56*, 959.
16. Chingas, G. C.; Garroway, A. N.; Moniz, W. B.; Bertrand, R. D. *J. Am. Chem. Soc.* **1980**, *102*, 2526.
17. Lyerla, Jr., J. R.; Levy, G. C. *Top.* ^{13}C *NMR Spectrosc.* **1974**, *1*, 79.
18. Allerhand, A.; Doddrell, D.; Komoroski, R. *J. Chem. Phys.* **1971**, *55*, 189.
19. Jones, A. A.; Lubianez, R. P.; Hanson, M. A.; Shostak, S. L. *J. Polym. Sci., Polym. Phys. Ed.* **1978**, *16*, 1685.
20. Howarth, O. W. *J. Chem. Soc., Faraday Trans. II* **1980**, *76*, 1219.
21. Broeker, H. C.; Schöla, E. *Makromol. Chem.* **1981**, *182*, 643.
22. VanderHart, D. L.; Earl, W. L.; Garroway, A. N. *J. Magn. Reson.* **1981**, *44*, 361.
23. Dechter, J. J.; Komoroski, R. A.; Axelson, D. E.; Mandelkern, L. *J. Polym. Sci., Polym. Phys. Ed.* **1981**, *19*, 631.
24. Komoroski, R. A. *J. Polym. Sci., Polym. Phys. Ed.* **1983**, *21*, 2551.
25. Doskocilova, D.; Schneider, B. *Macromolecules* **1972**, *5*, 125.
26. Komoroski, R. A.; Mandelkern, L. *J. Polym. Sci., Polym. Symp.* **1976**, *54*, 201.
27. Komoroski, R. A.; Maxfield, J.; Mandelkern, L. *Macromolecules* **1977**, *10*, 545.
28. Bloembergen, N.; Purcell, E. M.; Pound, R. V. *Phys. Rev.* **1948**, *73*, 679.
29. Fox, T. G.; Loshaek, S. *J. Appl. Phys.* **1955**, *26*, 1080.
30. Axelson, D. E.; Mandelkern, L. *J. Polym. Sci., Polym. Phys. Ed.* **1978**, *16*, 1135.
31. Mandelkern, L. *Pure Appl. Chem.* **1982**, *54*, 611.
32. Boyer, R. F.; Heeschen, J. P.; Gillam, J. K. *J. Polym. Sci., Polym. Phys. Ed.* **1981**, *19*, 13.
33. Boyer, R. F. *J. Polym. Sci., Polym. Phys. Ed.* **1985**, *23*, 1.
34. Chen, J.; Kow, C.; Fetters, L. J.; Plazek, D. J. *J. Polym. Sci., Polym. Phys. Ed.* **1985**, *23*, 13.
35. Boyer, R. F. *J. Polym. Sci., Polym. Phys. Ed.* **1985**, *23*, 21.
36. Orbon, S. J.; Plazek, D. J. *J. Polym. Sci., Polym. Phys. Ed.* **1985**, *23*, 41.
37. Newmark, R. A.; Copley, B. C. *Macromolecules* **1984**, *17*, 1973.
38. Dekmezian, A.; Axelson, D. E.; Dechter, J. J.; Borah, B.; Mandelkern, L. *J. Polym. Sci., Polym. Phys. Ed.* **1985**, *23*, 367.
39. Williams, M.; Landel, R. J.; Ferry, J. D. *J. Am. Chem. Soc.* **1955**, *77*, 3701.
40. Ferry, J. D. "Viscoelastic Properties of Polymers", 3rd Ed.; Wiley; New York, 1981, Ch. 11.
41. Komoroski, R. A.; Mandelkern, L. *J. Polym. Sci., Polym. Lett. Ed.* **1976**, *14*, 253.
42. Patterson, D. J.; Koenig, J. L.; Shelton, J. R. *Rubber Chem. Technol.* **1983**, *56*, 971.

43. Sohma, J.; Shiotani, M.; Murakami, S.; Yoshida, T. *Radiat. Phys. Chem.* **1983**, *5*, 413.
44. Komoroski, R. A.; Savoca, J. L. unpublished observations, 1984–85.
45. Yokota, K.; Abe, A.; Hosaka, S.; Sakai, I.; Saito, H. *Macromolecules* **1978**, *11*, 95.
46. Ford, W. T.; Balakrishnan, T. *ACS Adv. Chem. Ser.* **1983**, *203*, 475.
47. Carman, C. J. *ACS Symp. Ser.* **1979**, *103*, 97.
48. Schaefer, J.; Chin, S. H.; Weissman, S. I. *Macromolecules* **1972**, *6*, 798.
49. Serizawa, H.; Ito, M.; Kanamoto, T.; Tanaka, K.; Nomura, A. *Polym. J.* **1982**, *14*, 149, and references therein.
50. Schaefer, J., in "Molecular Basis of Transition and Relaxation"; Gordon and Breach: London, 1978, p. 103.
51. Wind, R. A.; Anthonio, F. E.; Duijvestijn, M. J.; Smidt, J.; Trommel, J.; de Vette, G. M. C. *J. Magn. Reson.* **1983**, *52*, 424.
52. Jelinski, L. W.; Dumais, J. J.; Watnick, P. I.; Bass, S. V.; Shepherd, L. *J. Polym. Sci., Polym. Chem. Ed.* **1982**, *20*, 3285.
53. Barron, P. F.; Busfield, W. K.; Morley-Buchanan, T. *Polymer* **1983**, *24*, 1252.
54. Riew, C. K.; Rowe, E. H.; Siebert, A. R. *ACS Adv. Chem. Ser.* **1976**, *154*, 326.
55. Sayre, J. A.; Assink, R. A.; Lagasse, R. R. *Polymer* **1981**, *22*, 87.
56. Komoroski, R. A.; Nichols, C. E., unpublished results, 1983.
57. Kunz, S. C.; Sayre, J. A.; Assink, R. A. *Polymer* **1982**, *23*, 1897.
58. Jelinski, L. W.; Schilling, F. C.; Bovey, F. A. *Macromolecules* **1981**, *14*, 581.
59. Jelinski, L. W. *Macromolecules* **1981**, *14*, 1341.
60. Jelinski, L. W.; Dumais, J. J.; Engel, A. K. *Macromolecules* **1983**, *16*, 403.
61. Jelinski, L. W.; Dumais, J. J.; Watnick, P. I.; Engel, A. K.; Sefcik, M. D. *Macromolecules* **1983**, *16*, 409.

Appendix

Since the preparation of this chapter, additional work has appeared relating to the study of cured elastomers by solid state ^{13}C NMR.

The chemistry of sulfur vulcanization is a topic of substantial industrial importance. Much of what is known comes from vulcanization studies on low-molecular-weight model alkenes.[1] It is most desirable to examine the cured rubber directly to determine the type and concentrations of the various cross links formed. In the initial work described here,[2,3] natural rubber (NR) vulcanized with relatively high levels of sulfur is examined by standard single-pulse FTNMR at elevated temperature. The sample was cured to a level high enough to see readily the peaks due to cross link sites, but not so high as to raise T_g significantly.

Figure A4-1 shows the ^{13}C spectrum of NR cured with sulfur alone (8 parts per 100 parts NR) for six hours at 280°F.[2] A number of new peaks is seen in the spectrum of Figure A4-1. Two of these can be assigned to trans-polyisoprene units (T) resulting from isomerization during vulcanization. The other new resonances are due to the cross link sites themselves. It is not yet known if the increased linewidths of some of these peaks are due to restricted mobility or to overlap of peaks from different structural types. Using ^{13}C chemical shift substituent effects derived from model compounds, predicted chemical shifts were calculated for the major types of cross links expected in sulfur-cured NR.[3] The calculations show that each of the new peaks in the spectrum of sulfur-cured NR may arise from several types of cross links.

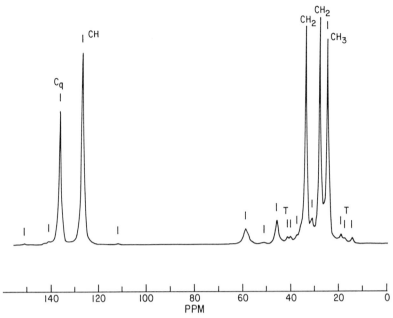

Figure A4-1. Single-pulse ^{13}C NMR spectrum of a sulfur-cured NR at 90°C. The label T indicates peaks from trans-polyisoprene units.

More discrimination of carbon types is necessary. The interpretation of ^{13}C NMR spectra of organic compounds is aided greatly by spectral editing techniques, such as the Distortionless Enhancement by Polarization Transfer (DEPT) sequence, which provides unambiguous determination of hydrogen multiplicities on carbon.[4] The value of the DEPT sequence lies in its ability to provide relatively clean subspectra, each of which arises from only one type (CH, CH$_2$, CH$_3$) of carbon. The DEPT technique has recently been applied to polymers in solution.[2,3,5,6] One point of concern in these studies was the effect of the short spin-spin relaxation times (T$_2$) of the polymer on the ability to acquire usable DEPT spectra, particularly for quantitative purposes.

Figure A4-2 shows the DEPT spectra of the cured NR. Surprisingly, it is possible to get usable DEPT spectra even on a cured rubber at about 140°C above T$_g$. Comparison of Figures A4-1 and A4-2 shows that the signal-to-noise ratio has been reduced greatly. During the DEPT sequence, which lasts on the order of 3J/2, much more of the magnetization has dephased before detection because of the short T$_2$'s of the bulk polymer. However, it is still possible to assign the various cross-link peaks as to multiplicity. Tentative assignments have been based on these results.[3]

Curran and Padwa[7] have investigated the use of CPMAS for the study of cross-linked rubbers, in particular polybutadiene. Most significant among their results is the discovery that, for an uncross-linked elastomer, most efficient cross polarization occurs not at the exact Hartmann-Hahn match, but at

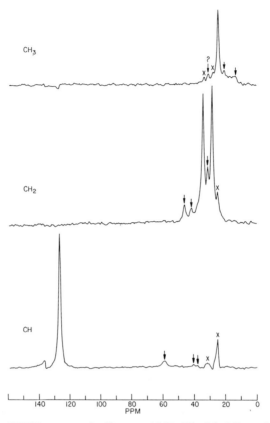

Figure A4-2. The DEPT spectra of sulfur-cured NR. The label X marks residual peaks from other subspectra. Arrows indicate peaks due only to cross-link sites.

the match plus or minus multiples of the spinning frequency. This is demonstrated for polybutadiene in Figure A4-3(A). This behavior is similar to that for adamantane, for which it was found that MAS amplitude-modulated the CH dipolar coupling, and frequency-modulated the homonuclear HH coupling.[8] Adamantane and polybutadiene are similar in that both experience substantial molecular mobility in the solid state, and hence have reduced CH and HH dipolar interactions. These results suggest that CP may be an efficient way to obtain ^{13}C spectra of bulk rubbers, if care is taken to find the frequency with the maximum CP transfer rate.

With increasing cross-link density in the polybutadiene, more efficient cross polarization is obtained at exact match. This arises from the presence of a significant static CH dipolar interaction in the cross-linked rubber. More maxima are seen in the dependence of CP transfer rate with H_{1C}, and the range of rf fields, over which efficient CP occurs, increases. In other words, the situation begins to approach that of a glassy polymer. Figure A4-3(B) shows that substantial cross-polarization is occurring both at exact match and at

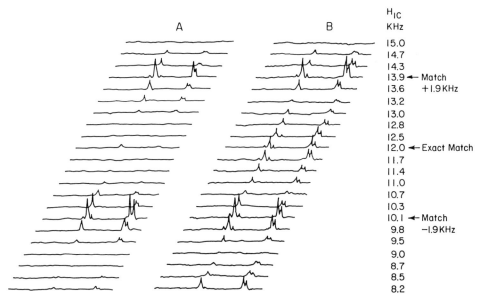

Figure A4-3. (A) Cross-polarization ^{13}C spectra of uncross-linked polybutadiene as a function of H$_{1C}$, with 1.9 kHz MAS; (B) cross-polarization ^{13}C spectra of a cross-lined polybutadiene (molecular weight between cross links of 1020) as a function of H$_{1C}$, with 1.9 kHz MAS. A contact time of 2 ms was used in both cases.

multiples of the spinning frequency for a cross-linked polybutadiene with a molecular weight of 1020 between cross links. The authors propose that the ability of MAS to modulate dipolar couplings may provide a measure of cross-link density.

References

A1. Porter, M. in "Organic Chemistry of Sulfur," S. Oae, Ed. Ch. 3, p. 71, Plenum, New York, 1977.
A2. Komoroski, R. A.; Shockcor, J. P.; Savoca, J. L. *Proceedings of the 32nd Sagamore Army Materials Research Conference*, July 22–26, 1985, Battelle Press, 1986.
A3. Komoroski, R. A.; Shockcor, J. P.; Gregg, E. C.; Savoca, J. L. *Rubber Chem. Technol.* **1986**, in press.
A4. Doddrell, D.; Pegg, D. T.; Bendall, M. R. *J. Magn. Reson.* **1982**, *48*, 323.
A5. Barron, P. F.; Hill, D. J. T.; O'Donnell, J. H.; O'Sullivan, P. W. *Macromolecules* **1984**, *17*, 1967.
A6. Newmark, R. A. *Appl. Spectrosc.* **1985**, *39*. 507.
A7. Curran, S. A.; Padwa, A. R. *Macromolecules* **1986**, *19*, in press.
A8. Stejskal, E. O.; Schaefer, J.; Waugh, J. S. *J. Magn. Reson.* **1977**, *28*, 105.

5

CARBON-13 SOLID-STATE NMR OF SEMICRYSTALLINE POLYMERS

David E. Axelson

ENERGY, MINES, AND RESOURCES CANADA
CANADA CENTER FOR MINERAL AND ENERGY TECHNOLOGY
EDMONTON COAL RESEARCH CENTER
P.O. BAG 1280, DEVON, ALBERTA, CANADA T0C 1E0

Introduction

As a result, in part, of their unique phase structure, semicrystalline polymers constitute an important and intriguing area of study within the field of polymer research. This chapter addresses the question of how solid-state nuclear magnetic resonance ^{13}C studies may help to characterize the details of semicrystalline polymer structure and morphology. Although it is not possible truly to capture the subtleties and complexities of this subject matter in a short review, a number of areas of general importance are pointed out upon which future studies may be based. It is important to recognize that the morphological features discussed relate in some manner to the physical and mechanical properties of the polymers in question. In fact, without an essentially complete understanding of the morphology of the phases it is not possible to obtain adequate, predictive structure–property correlations.

A cursory (and necessarily incomplete) introduction to the structure of semicrystalline polymers is given here before the NMR results are discussed. It must be noted that this field is in a rapid state of flux and that the details associated with the morphological features to be presented remain to be

Komoroski (ed): High-Resolution NMR Spectroscopy of Synthetic Polymers in Bulk

definitively elucidated. Needless to say, there is considerable controversy surrounding much of the theoretical and experimental work performed to date.

Given sufficient structural regularity and the proper experimental conditions (time, temperature, pressure, concentration, or stress) certain polymers crystallize from the melt, or solution, on cooling. Semicrystalline polymers crystallized from the melt usually consist of lamellar crystalline regions separated by noncrystalline regions. Within the crystalline component the polymer chain axes are transverse to the faces of the lamellae, which are of the order of 50–500 Å thick. The lateral dimensions may be much larger. The regions between the crystalline lamellae may be a few Angstroms to 200 Å or so in size. Because the lengths of the polymer chains are many times greater than the average lamellar thickness, each molecule must pass through the same or different lamellae many times. The manner in which this requirement is met has been the subject of considerable controversy and is of foremost importance for an understanding of the morphology of the crystal–amorphous interphase or transition region.[1–9]

A schematic representation of the primitive crystallite is shown in Figure 5-1.[10,11] Three distinct regions are depicted (note that this figure is only meant to define the terms to be used). There is a crystalline region, an interfacial region, and an amorphous (or interzonal) phase. The crystalline component is indicated by the vertical straight lines and comprises chain sequences in ordered or preferred conformations. (Although not depicted, the nature and distribution of imperfections or defect structures within this phase are of importance also.) The interfacial region is very diffuse and ill defined, being many monomer units thick with crowded and/or distorted segmental packing. Some chains pass through this zone, whereas others return to the crystallite. Although relatively disordered and highly irregular, the interfacial

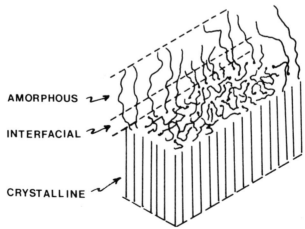

Figure 5-1. Schematic representation of primitive crystallite. (Adapted from Mandelkern, L. *J. Polym. Sci., Polym. Symp.* **1973**, *43*, 1. Copyright 1973. Reprinted by permission of John Wiley and Sons, Inc.)

structure must be considered as an entity separate from the amorphous component.

In the fully disordered, amorphous state the polymer chains assume their characteristic random configurations. The transition from the near perfect order of the crystal to the randomness of the amorphous state cannot, however, occur abruptly because the continuity of the long polymer chains imposes severe constraints on the transition.[12] Compared with the amorphous phase, therefore, there should be differences in the ordering, or orientation, as well as the overall long-range and local segmental mobility of the polymer chains involved in the transition region. Although there may be no overall noncrystalline phase orientation within the polymer, there may be local orientation of molecules with respect to the crystals.

A variety of structures within the interfacial phase may be envisioned, including loose chain segments attaching one part of a crystallite to the same crystallite, tight loops arranged similarly, chain ends terminating in the interfacial or amorphous zones, and tie molecules (both taut and loose) connecting different crystallites. The properties of the polymer reflect the relative number, type, and distribution of these chain entities.

Numerous models have been proposed for the disposition of the polymer segments in the noncrystalline component of semicrystalline polymers. The description noted in the previous paragraph is consistent with the "switchboard" or random reentry model. This model involves the notion that the majority of chains extend from the crystalline lamellae and protrude into the interphase for a distance that is well beyond the surface of the crystallite. Whereas some proportion of these chains may reenter the crystallite from which they emerged at an adjacent position, the majority of them are considered to reenter at positions more distant from adjacent sites or they connect different crystallites. As opposed to this, the other model most often invoked is the regularly folded or adjacent reentry model in which the long-chain molecule reverses its direction immediately on emerging from the face of the lamella and returns to it in adjacency to the crystalline stem of the previous passage. There is considerable controversy surrounding the relative validity of these models, although the experimental evidence is generally not consistent with the adjacent reentry-type structure.

The crystalline lamellae may organize themselves into superstructures having dimensions that are much larger than the crystals themselves.[13-15] For unoriented systems crystal growth starts from a relatively small number of nuclei and proceeds to develop radially outward through the formation of branched lamellae and fibrils. Many types of supermolecular structure may be formed depending on crystallization temperature, molecular weight distribution, branching number, type and distribution, and polymer.[13-15] For instance, these growing regions may have spherical symmetry, with growth in a particular crystallographic direction oriented along the radius. These "spherulites" grow until they impinge on each other, resulting in a volume-filling polygonal structure, or else their growth may be terminated by quenching the sample to a temperature low enough to inhibit growth. Variables in this process include the number of spherulites, their size distribution, and their perfection. Other morphologies observed ranged from rodlike to small, randomly organized lamellae. In addition, all crystallization does not necessarily occur at the growing spherulite front, but secondary crystallization may occur within the spherulite already formed. The material that crystallizes later may differ from that which crystallizes in the primary process. Molecular weight and compositional fractionation also may occur on crystallization. The supermolecular structure of the polymer greatly influences the nature and extent of deformation of the solid polymer, which is of prime interest in determining the commercial applications of these materials.

Even from this short discussion it is apparent that there are many variables to be considered in terms of a complete characterization of the polymer microstructure and supermolecular structure. Thus, one must obtain information concerning the phase structure, ie, the relative mass fractions of the crystalline, interfacial, and amorphous phases; the crystal size distribution; the distances

TABLE 5-1. APPLICATIONS OF SOLID-STATE NMR TO SEMICRYSTALLINE POLYMERS

Characterization of Transitions and Dynamic Processes
  Crystal–crystal
  Phase transitions
  Glass transition
  Secondary transitions: α, β, γ mechanisms
    Variable-temperature studies
    Comparison with dynamic mechanical analysis, dielectric relaxation
    Isotopic enrichment (^{13}C, ^{2}H)
    Physical aging
Characterization of Orientation
  Interfacial phase
  Amorphous phase
  Effect of drawing
  Effect of annealing
  Effect of processing (eg, blown film, injection molding)
Degree of Crystallinity Measurements
  Comparison with laser Raman, broad-line NMR, wide-angle x-ray scattering, heat of fusion, density
Morphology–Structure Determination
  Melt-crystallized, solution-crystallized polymers
  Phase structure analysis: amorphous, interfacial, crystalline mass fractions
  Molecular dynamics of phases
  Distribution of branches between phases
  Crystal size, crystal defects (number, type, distribution)
  Domain size
  Compatibility of blends; effect of fillers
  Interface organization
  Diffusion of gases, solvents, additives
  Solid-state reactions
  Insoluble polymers (oxidized, irradiated, crosslinked, mechanically or thermally degraded, etc)
  Analysis of molecular dynamics; correlations with physical properties

between phases; the nature of the chain orientation within each phase; the nature of defects within each phase, and the distribution of branches between phases for branched semicrystalline polymers. The type of supermolecular species formed and their associated parameters must be ascertained. More subtle considerations include a description of the frequency distribution of molecular motions within each phase as a function of temperature and the nature and extent of any interactions between phases.

As imposing as this list may seem, these factors represent but a fraction of the information required to characterize the polymer. It is important to remember that all these considerations are crucial in elucidating the relationship between polymer structure and critical physical properties, such as high- and low-temperature impact strength, tear resistance, and stress crack resistance, to name a few possibilities.

Finally, the reader should approach studies of semicrystalline polymers with the idea that these materials are highly complex and dynamic entities. As a consequence, the power of NMR studies, both liquid and solid-state, becomes evident in that it is able to elucidate not only the details of the chemical structure of a material, but also its dynamic behavior. Table 5-1 summarizes some of the important properties and areas of study related to semicrystalline polymers that have involved NMR research.

Experimental Considerations

Because of the complexity of the techniques of solid-state NMR research, as well as the complexity of the samples under investigation, it is useful and necessary to discuss some of the experimental problems. There are two aspects to this discussion; the first involves the effect of sample preparation on the NMR results, and the second the effect of the experimental parameters on the quality and nature of the spectra obtained. Proper quantitative analysis requires careful attention to all these factors.

Sample Preparation

The crystallization procedure and associated thermal history are prime factors in determining the morphological features and physical–mechanical properties of a given polymer. For this reason alone it is dubious, at best, whether a comparison of polymer NMR parameters under uncontrolled conditions has any practical significance. It is therefore imperative that as much care and caution be applied to the sample preparation procedures as some apply to setting up the NMR experiment itself. Nuclear magnetic resonance research of semicrystalline polymers actually requires a careful, multidisciplinary approach for best results to be obtained.

However, for some polymers, such as polyethylene, the requirements for optimum sample size for signal-to-noise purposes and optimum sample type for maximum control of morphological features are almost mutually exclusive. The signal-to-noise (SN) ratio would be maximized by using either (i) a rotor machined entirely from the polymer of interest or (ii) a solid cylindrical plug machined to fit exactly inside the sample volume of a magic angle rotor. Consider the latter case, which we have investigated by preparing cylindrical samples (10 mm in diameter by 10–15 mm in length) and subsequently performing numerous NMR and morphology studies. These results were compared with those obtained by preparing thin films of various thicknesses. With respect to the NMR parameters studied, Table 5-2 summarizes the effect of sample thickness on crystalline component ^{13}C spin–lattice relaxation (T_1) values. The T_1 values are constant for thicknesses of the order of 40 μm but increase for thicker samples until a T_1 value of almost 1000 s is observed for a 10-mm diameter sample.

TABLE 5-2. EFFECT OF SAMPLE THICKNESS ON CRYSTALLINE-COMPONENT ^{13}C T_1 VALUES FOR A SELECTED POLYETHYLENE SAMPLE

Sample thickness (μm)	T_1 (s)
20	95
37	93
75	135
150	175
10^4	1000

Although this wide range in T_1 values cannot be expected for most polyethylene samples (see Spin Lattice Relaxation and Polymer Morphology), it serves to illustrate the tremendous variations in T_1 values that may arise as a result of different methods of sample preparation. (In this study all samples were quench cooled from the melt into a room-temperature water bath. However, because of thickness differences the actual thermal history of the samples varied greatly.) Therefore, it is not sufficient in terms of sample preparation to consider only the method of cooling–crystallization without also considering sample size and shape.

Large, or thick, samples may suffer from internal fissures, bubbles, and other gross structural irregularities or inhomogeneities that may affect subsequent NMR measurements. Therefore, the optimum sample form for NMR solid-state studies is generally a thin film that can be cut, rolled, and packed into a rotor with relative ease. This allows the experimenter to develop the widest range in morphological structures for a given polymer and to make valid comparisons between different polymers for the same prescribed crystallization conditions. Of course, for studies of parts of commercial products, this procedure does not hold and other problems arise. For instance, the surface morphology may vary greatly from the core, or internal, morphology for many semicrystalline products. This makes the practice of cutting, or microtoming, a nontrivial task because the thermal history and morphology must be extensively characterized aside from the NMR measurements.

Having recognized that NMR relaxation parameters may be sensitive functions of thermal history, the reader should be critical of the details of any polymer sample preparation procedures, especially those that involve correlations of NMR parameters with physical–mechanical properties. Statements regarding the temperature independence of any NMR parameters in semicrystalline polymers should be viewed with particular suspicion.

Instrumental Variables

The following discussion is meant to supplement that given in Chapter 2, with the emphasis being on those factors that may cause difficulties in the quantitative cross-polarization magic angle spinning (CPMAS) analysis of semicrystalline polymers.

Cross-Polarization Rate.[16–19] The rate at which the various structural features and phases in a semicrystalline polymer cross polarize is a function of a number of variables. Generally speaking, more mobile carbon atoms take longer to cross polarize than less mobile or immobilized carbons because molecular motion attenuates the $^{13}C-^{1}H$ and $^{1}H-^{1}H$ dipolar interactions (see Chapter 4). Should the dipolar interactions effectively be eliminated as a result of near liquid-like motion, carbons thus affected may not yield a cross-polarized signal at all. The value of T_{CH}, the cross-polarization time, also depends on the nature of the lattice motions and the exactness of Hartmann-Hahn match.

In terms of carbon types T_{CH} values decrease (roughly) in the order: nonprotonated carbon > methyl carbon (rotating) > protonated aromatic (aliphatic) methine carbon > methylene carbon > methyl carbon (static). This process is further complicated in semicrystalline polymers because the various phases may be characterized (at least in part) by significant differences in degree of molecular mobility. Therefore, the carbon type and the phase(s) in which it is located must be considered simultaneously in order to understand cross-polarization dynamics fully. In addition, the amount of molecular motion and relative differences in molecular motion are also functions of the temperature at which the experiment is performed.

The net result of these considerations is that the derivation of the optimum conditions for quantitative analysis is not generally a simple matter.

Dipolar Decoupling. In terms of its effect on spectral resolution in CPMAS experiments, the details of dipolar decoupling must be considered with great care also. Unlike high-resolution solution studies, in which moderate decoupling fields (characterized by 1–5 W of power) and efficient broadband noise modulation techniques have proved very effective, CPMAS experiments may require decoupling powers that are technologically difficult to attain (~ 1000 W). Therefore, a few comments regarding the general problem with specific references to the semicrystalline polymer research area are in order.

An analysis of the effectiveness of (dipolar) decoupling as a function of the proton bandwidth to be irradiated has shown that irradiating protons $\Delta\omega$ away from resonance with a radiofrequency (rf) field ω_1^H diminishes line broadening by an amount[20,21]:

$$\Delta\langle\Delta\omega_{CH}^2\rangle^{1/2} = \langle\Delta\omega_{CH}^2\rangle^{1/2}\cos\psi \qquad (5\text{-}1)$$

where the term in brackets is the carbon–hydrogen second moment, $\tan^{-1}\psi = \omega_1^H/\Delta\omega$ and $\psi = \pi/2$ is the "magic angle" for dipolar decoupling.[22,23] For a small deviation, ε, away from the condition $\psi = \pi/2$ the residual broadening is given by:

$$\varepsilon = 0.5\,|\Delta\sigma_H|\,\omega_1^H\langle\Delta\omega_{CH}^2\rangle^{1/2} \qquad (5\text{-}2)$$

where $\Delta\sigma_H$ is the proton chemical-shift range to be irradiated. Assuming a 10-ppm proton range and a 25-kHz linewidth in the absence of decoupling, a residual broadening of only 25 Hz at 15 MHz (^{13}C observation frequency) requires an rf field of the order of 75 kHz (18 G). Similarly, considering a $B_0 = 4.7$T and $\omega_1^H = 40$ kHz, line broadening of 25–60 Hz is predicted for irradiation only 4 ppm from the mean (proton) resonance pattern.[24]

Even more significant for those interested in achieving the highest degree of resolution in the solid state is the fact that on-resonance irradiation does not narrow a line completely. Haeberlen[20] has shown that the residual broadening, ε', is on the order of:

$$\varepsilon' = 1/134(\Delta\omega/\omega_1^H)^2\langle\Delta\omega_{CH}^2\rangle^{1/2} \qquad (5\text{-}3)$$

Heteronuclear spin decoupling processes also may be affected by H–H spin interactions that may reduce or enhance the effectiveness of the decoupling.[25,26]

Figure 5-2 illustrates the possible severity of off-resonance decoupling for the crystalline component of polyethylene. Having isolated the crystalline component resonance by a selective relaxation experiment, the position of the decoupling frequency was moved from on resonance to about 2000 Hz off resonance (10 ppm). A similar number of pulses was acquired in each case. Changes in linewidth (and SN) are quite significant when the decoupler is only 2.5 ppm from resonance (with $\omega_1^H = 50$ kHz).

There may also be interference between the decoupling field and the molecular motion within the sample. Rothwell and Waugh[27] have shown that molecular motion (characterized by a frequency, ω) even though it is incoherent, may reduce the efficiency of the coherent proton rf decoupling field, particularly when the two modulations occur on the same time scale. For instance, when $\omega_1^H = 50$–100 kHz, the linewidth (in samples such as hexamethylbenzene, hexa-

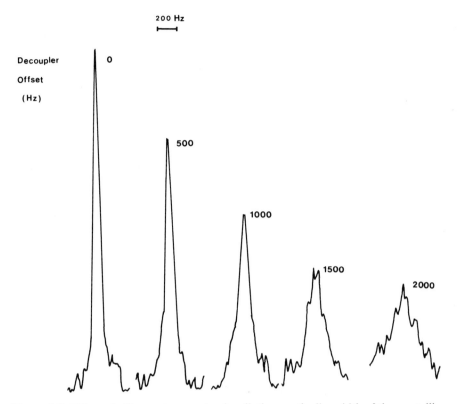

Figure 5-2. Effect of off-resonance proton irradiation on the linewidth of the crystalline component of polyethylene. The crystalline component was isolated by performing a T_1 experiment with cross polarization and using a delay between pulses of $5T_1$ (amorphous) = 1.5 s. Proton offset frequencies from resonance are as follows: (A) 0 Hz, (B) 500 Hz, (C) 1000 Hz, (D) 1500 Hz, (E) 2000 Hz. The linewidth in (A) is 100 Hz (at half-height).

methylethane, and adamantane) is narrow for correlation times (τ) less than 10^{-6} to 10^{-7} s but quickly increases by an order of magnitude or more when $\tau = 10^{-5}$ s and then decreases just as dramatically for τ longer than 10^{-4} s. In this case the linewidth depends on the sixth power of the carbon–proton distance, because $T_2^{-1} \propto r^{-6}$. Thus, protonated carbons broaden more significantly than quaternary carbons in the same sample even when the correlation times are the same.

The most striking example of this phenomenon in semicrystalline polymer studies was reported by Fleming et al[28] for isotactic polypropylene (90% isotactic, 70% crystalline). They performed a variable-temperature study of the linewidth and relaxation time (T_1, $T_{1\rho}$) behavior of this sample. At 160 K the methyl linewidth was independent of the size of the decoupling field, whereas at 77 K it varied as the inverse square of the decoupling field. They noted that this was the expected dependence for the transition between the extreme narrowing and slow motion correlation time regimes. The broadening arises when the reorientation rate of the methyl group about the C_3 symmetry axis becomes comparable to the frequency of the decoupling field. A plot of line-width versus temperature exhibits a maximum where the correlation time for molecular motion equals the frequency of the proton decoupling field. The results of this study are described in more detail in Chapter 3.

Similarly, Suwelack and colleagues[29] have observed a dependence between linewidth and magic angle spinning speed in the region between the weak and strong collision limits. When $\omega\tau = 1$ the incoherent averaging resulting from random molecular motion interferes destructively with the coherent averaging from MAS, giving rise to excess line broadening as compared with the situations $\omega\tau \ll 1$ and $\omega\tau > 1$.

The important point to remember from this discussion is that the broad distribution of molecular motions in polymers may subvert the line-narrowing effects of magic angle spinning and/or dipolar decoupling (see Chapter 2). These phenomena have important consequences for quantitative analysis because these effects may be selective not only for certain structures but also for certain phases in semicrystalline polymers (particularly in variable-temperature studies).

Recycle Delay. The recycle delay in quantitative cross-polarization experiments is a function of the proton spin–lattice relaxation time, T_1^H. Sullivan and Maciel[30] have determined the optimal pulse delay time from the point of view of maximizing signal to noise with the result that a recycle delay of 1.25 T_1^H is desirable. This conclusion is generally valid as long as the protons exhibit a single T_1^H relaxation time (as a result of rapid spin diffusion maintaining a uniform spin temperature). The relative intensities of the various carbon resonances are independent of the recycle delay under these circumstances. However, if a single T_1^H is not adequate to characterize a sample, then the recycle delay derived by this method is not appropriate. For materials with weakly coupled protons or highly inhomogeneous structures or phases, multiple T_1^H behavior may be exhibited through different resolved carbon resonances (ie, spin diffusion does not maintain a uniform proton spin temperature). The pulse repetition rate can have an effect on relative carbon intensities and a pulse delay of 4–5 T_1^H for the slowest relaxing protons in the sample should be used.

Characterization of Line-Broadening Mechanisms

As a major part of the problem of optimizing resolution and signal to noise in CPMAS studies involves determining the mechanism(s) of line broadening, Table 5-3 is presented to summarize some of the relevant literature. As a detailed discussion of the subject is beyond the scope of this chapter the reader is referred to the numerous references for a survey of the situation. Discussions of a number of line-broadening mechanisms appear in Chapters 2–4.

TABLE 5-3. Characterization of Line-Broadening Mechanisms

Method	Comments
Selective saturation[31-33]	Determines whether lines are homogeneously or inhomogeneously broadened
	Can give rough estimate of natural linewidth
2D NMR[34, 35] (spin exchange experiment)	Determines whether lines are homogeneously or inhomogeneously broadened
	Determines presence of chemical exchange or spin diffusion
	Allows calculation of natural linewidth
Variable-frequency studies[24, 36]	Determines whether lines are homogeneously or inhomogeneously broadened and to what extent (see Chapter 4)
	Useful in deciding whether high-field spectra will be of added benefit (resolution may decrease or remain unchanged with increasing field strength); a distribution of isotropic chemical shifts will cause an increase in linewidth with an increase in static field
	Magnetic field or bulk susceptibility studies
DISPA[37-40]	Used for Lorentzian lineshapes to elucidate a wide variety of line-broadening mechanisms
Sample shape[24, 41-43]	Change in sample shape or nature (eg, powder, single crystal, cylindrical plug, machined rotor) affects bulk magnetic susceptibility effects; round samples are most desirable from this point of view
Variable temperature[44-48]	Useful in characterizing relaxation-related line-broadening mechanisms, such as incomplete motional narrowing, motional modulation of CH coupling, chemical-shift anisotropy, and quadrupolar interactions
Decoupling field variations[24, 25, 36]	Off-resonance proton irradiation and incomplete decoupling mechanisms
	Qualitative analysis of frequency distribution of dipolar interactions by varying field
Magic angle spinning[24, 49-57, 109]	Qualitative analysis of frequency distribution of dipolar interactions, although this effect is probably overwhelmed by the removal of line broadening caused by chemical-shift anisotropy
	Removes susceptibility effects to large extent
	Rotor instability and missetting of angle leads to line broadening
Relaxation rate measurements[24, 59]	Allow calculation of homogeneous linewidth and modeling of molecular dynamics
Nuclear Overhauser enhancement	Used in conjunction with T_1 and T_2 to model molecular motions
Isotopic enrichment[60]	Replace ^1H with ^2H to characterize decrease in dipolar interactions because ^2H has a smaller γ (see Chapter 2)
Quadrupolar interactions[58, 61-64]	Interactions of spin greater than $\frac{1}{2}$ nuclei with spin-$\frac{1}{2}$ nuclei may cause line broadening (see Chapter 2)

Amorphous Phase Studies

Although amorphous-phase relaxation times have been reported in various studies,[31,36,65–78] there have been few extensive investigations of spin–lattice relaxation, linewidths and, to a lesser extent, nuclear Overhauser enhancement factors (NOEF) of semicrystalline polyethylene and related polymers.[31,36,69,70,73] These latter studies were initially confined solely to the characterization of the noncrystalline (amorphous) component by employing normal proton scalar-decoupling conditions on the solid samples. The fact that this component is so liquid-like enables these experiments to be performed with ease and insures that none of the crystalline component interferes.

The results have shown conclusively that the spin–lattice relaxation times of the noncrystalline phase, as studied under these conditions, are independent of the level of crystallinity, the different supermolecular structures or crystalline morphologies that can be developed in polyethylenes (whether bulk crystallized or solution crystallized), the branching concentration, and the molecular weight distribution. Because the T_1s are determined by short-range, high-frequency segmental motions, only local molecular environments are important.

Table 5-4 summarizes data related to linear and branched polyethylene samples of varying molecular weight, morphology, and degree of crystallinity. The average NT_1 (N is the number of directly bonded hydrogens) value was found to be 354 ms.[69] Subsequent studies have confirmed these observations.[72,73,77,78]

For cis-polyisoprene it was possible to obtain both amorphous and semicrystalline samples for the same polymer by a careful choice of crystallization and cooling conditions.[70] The results shown in Table 5-5 follow the trends previously discussed with respect to polyethylene, ie, an insensitivity of the ^{13}C T_1 values to degree of crystallinity differences.

TABLE 5-4. CARBON-13 SPIN RELAXATION PARAMETERS FOR THE AMORPHOUS COMPONENT OF BULK LINEAR POLYETHYLENE[a, b]

M_v[c]	$(1 - \lambda)_d$[d]	Morphology[e]	NT_1 (ms)	NOEF[f]
8.1×10^4	0.57	Spherulite	343	—
2.5×10^5	0.51	Spherulite	355	2.0
1.7×10^5	0.81	Spherulite	348	1.5
1.7×10^5	0.68	Spherulite	356	—
2.0×10^6	0.51	Spherulite	369	2.0
2.0×10^6	0.72	Spherulite	358	—
2.75×10^4	0.94	Rod	352	1.0

[a] From reference 69.
[b] All measurements taken at 45°C and 67.9 MHz.
[c] Viscosity-average molecular weight.
[d] Degree of crystallinity from density.
[e] Morphology as determined by small-angle laser light scattering.
[f] Nuclear Overhauser enhancement factor.

TABLE 5-5. CARBON-13 T_1 VALUES FOR CIS-POLYISOPRENE (ms)$^{a,\,b}$

| Carbon | T_1 | |
	Amorphous	Semicrystalline^c
α	2000	2400
β	400	460
γ	260	270
δ	240	270
ε	840	910

^{a} From reference 70.
^{b} All measurements taken at 0°C and 67.9 MHz.
^{c} Degree of crystallinity = 0.3.

The effect of degree of crystallinity differences, however, may be reflected in changes in the NOEF as evidenced by the data in Table 5-4. More detailed studies have not been performed, but the influence of low or near-zero frequency molecular motions may be evident in this parameter because the NOEF decreases as the degree of crystallinity increases. The effect does not appear (in this limited data set) until a degree of crystallinity of more than 0.5 is attained.

The linewidths of a variety of polyethylenes in the melt (140°C) ranged from 6 to 35 Hz at 67.9 MHz and from 0.8 to 6.8 Hz at 22.6 MHz for the same samples.[36] The ratio of the linewidths at the two fields was about a factor of four. Although the linewidths appeared to increase as a function of molecular weight, it was noted that there were experimental difficulties in preparing proper (void-free) samples at high molecular weights because of increasing melt viscosity.[36]

After crystallization an increase in the linewidths by factors of about 10 to 50 was measured at ambient temperature at 67.9 MHz. Crystallization therefore has a major effect on the linewidth. From the data given in Table 5-6 it was noted that the linewidth was influenced primarily by the supermolecular

TABLE 5-6. CARBON-13 NMR LINEWIDTHS FOR POLYETHYLENES AT 67.9 MHz AND 30°C^a

Polymer	Total branching (per 1000 carbon atoms)	Density	$(1 - \lambda)_d$	Morphology	NMR linewidth (Hz)
Branched					
BPE-8	7.0	0.9251	0.53	Type b/c spherulites	600 ± 30
-9	12.8	0.9158	0.47	Type b spherulites	550 ± 30
-1	17.0	0.9180	0.48	Type b/c spherulites	525 ± 25
-3	24.4	0.9250	0.53	Type b/c spherulites	490 ± 25
Linear					
LPE-4		0.9568	0.74	Type a spherulites	610 ± 30
-5		0.9535	0.72	Type a spherulites	525 ± 25
-6		0.9430	0.65	Type h random lamellae	410 ± 20
-7		0.9288	0.55	Type h random lamellae	420 ± 20

^{a} From reference 36. Reprinted by permission of John Wiley and Sons. Copyright 1981, John Wiley and Sons.

structure rather than the level of crystallinity. (The morphology description is that given by Maxfield and Mandelkern.[13,14]) The ranges and differences in the linewidths are considered real (and reproducible within the limits noted). The linewidths vary from 400 to 600 Hz as the morphology varies from random lamellae to well-developed spherulites.[36] For the same degree of crystallinity, different morphologies can be compared. The conclusion can be drawn that the broader resonances are associated with the more highly organized crystalline superstructures. As in the melt studies, the linewidths were found to be field dependent, although the percentage differences were not as great for the bulk-crystallized samples. Regardless of the field strength, the differences in linewidths ascribed to morphological changes are maintained (see Table 5-7).

TABLE 5-7. MAGNETIC FIELD DEPENDENCE OF ^{13}C LINEWIDTHS FOR POLYETHYLENE IN THE SEMI-CRYSTALLINE STATE[a]

Polymer	Morphology	Temperature (°C)	Linewidths (Hz)		Ratio
			67.9 MHz	22.6 MHz	
BPE-1	a/c	34	400	300	1.3
BPE-3	c/h	34	425	280	1.5
LPE-4	a	42	430	340	1.3
LPE-5	a	42	400	375	1.1
LPE-6	h	42	290	230	1.3

[a] From reference 36. Reprinted by permission of John Wiley and Sons. Copyright 1981, John Wiley and Sons.

Scalar decoupling and magic angle spinning (at 75.5 MHz) reduce the linewidth to 100–150 Hz, which remains invariant for spinning speeds above 1.8 kHz. Contributions to the linewidth from low-frequency molecular motions should largely be removed. Increasing the dipolar decoupling field strength under static conditions also reduces the linewidth significantly [although the chemical-shift anisotropy (CSA) remains]. Regardless of morphological form, MAS (≥ 1 kHz) and dipolar decoupling result in linewidths of 100–150 Hz. This apparently irreducible residual contribution to the amorphous linewidth in these semicrystalline polymers may be caused by chemical-shift dispersion and/or anisotropic bulk magnetic susceptibility effects.[36]

The changes of the linewidth with temperature at 67.9 MHz for two linear and two low-density (branched) polyethylenes are plotted in Figure 5-3 from room temperature through their respective melting points.[36] There is a major difference in the temperature dependence of the linear and branched polymers. The linewidths for the branched polymers decrease smoothly with increasing temperature over the whole temperature range. However, the linewidths for the linear polymers reach a plateau from about 60–90°C. At about 100°C the two curves merge and are the same in the melting range and in the melt. Hence, the difference in linewidth caused by the differences in morphology of the two samples is maintained until relatively high temperatures are reached, ie, until the fusion range. Whereas branched polyethylenes and copolymers

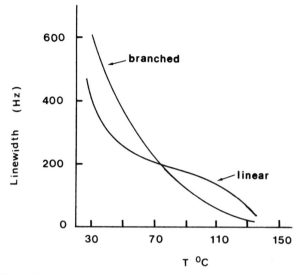

Figure 5-3. Effect of temperature on linewidths of bulk linear and branched polyethylene.

melt over a broad temperature range, the melting of linear homopolymers is relatively sharp. For example, for branched samples the onset of melting can be observed in the region from $-20°C$ to $0°C$; the fusion process continues up to the melting temperatures. On the other hand, the onset of fusion for linear polymers occurs at much higher temperatures.

For the branched samples, line broadening caused by microscopic heterogeneities (crystallites) is continuous with temperature over the range studied because the crystallization process is continuous (as is the contribution from incomplete motional narrowing). For the linear samples an essentially fixed amount of crystallinity is developed below about 110°C. The linewidth contribution from this source therefore remains constant with decreasing temperatures. However, as the temperature is decreased the broadening caused by increased restriction in the angular range of the CH vectors becomes enhanced. The combination of these two effects is postulated to lead to a plateau region and the differences between the linear and branched samples.

The linewidth difference between linear polyethylene samples, which is a reflection of the two different crystalline morphologies, is maintained from room temperature to about 100°C. This temperature range represents the contribution to the linewidth from incomplete motional narrowing. These results strongly suggest that this mechanism is responsible for the linewidth dependence on morphology. Thus, it may be that the morphology dependence of the ^{13}C linewidth is caused by the fact that the crystalline morphologies influence the angular motion of the chain CH vectors in different ways.[36]

Figure 5-4 is a plot of NT_1 versus temperature for a linear polyethylene sample. The corresponding average correlation time versus temperature plot

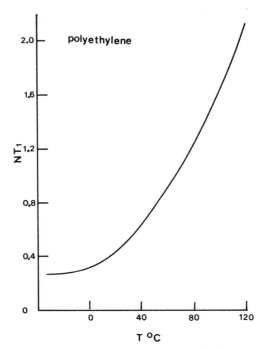

Figure 5-4. The quantity NT_1 versus temperature for linear polyethylene.

for these data is shown in Figure 5-5. The linewidth of the resonance becomes too broad to be observed below $-30°C$ to $-40°C$ for this sample (using scalar decoupling only to select the amorphous phase for observation). The fact that there is no obvious discontinuity in T_1 as the measurements are made through the melting point is illustrated in Figure 5-6 for two related polymers, poly(trimethylene oxide) and poly(ethylene oxide). In addition, the T_1s at each temperature are independent of molecular weight.

Whereas the amorphous-phase NMR parameters appear to be well behaved and somewhat insensitive to changes in polymer structure, there are specific examples where a characterization of this phase may shed new light on long-standing problems. Specifically, consider the nature of the amorphous-phase chain orientation and conformation in solution-grown crystals.[79-84]

Polymers also may crystallize from dilute solution, forming thin platelets, typically with lateral dimensions of several microns and thicknesses of the order of 100–200 Å. The chain axes are preferentially oriented parallel to the short dimension and the direction of the crystallographic a and b axes is preserved throughout the crystal so that such structures have become known as solution crystals or single crystals. These polymers are of particular interest given the constraints that must be imposed on the interfacial and amorphous-phase chain structure. Studies of such dilute solution crystals led to the hypothesis that they must possess a regularly folded structure. However,

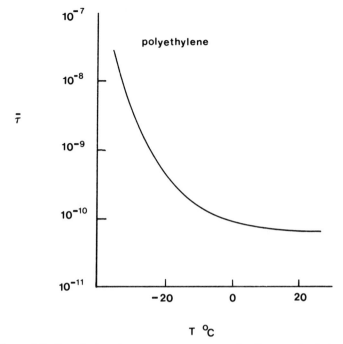

Figure 5-5. Correlation time versus temperature for linear polyethylene.

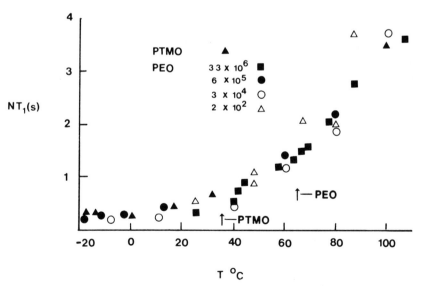

Figure 5-6. The quantity NT_1 versus temperature for poly(trimethylene oxide) (PTMO) and poly(ethylene oxide) (PEO). Arrows denote the peak melting temperatures for these polymers. (Reprinted with permission from Axelson, D. E.; Mandelkern, L. *ACS Symp. Ser.* **1979**, *10*, 181. Copyright 1979 American Chemical Society.)

several properties of these crystals have been shown to require on the order of 10–20% disorder, including density, wide-angle and low-angle X-ray analyses, selective oxidation, glass formation, dynamic mechanical behavior, and infrared and Raman spectroscopy.[85]

Because previous work noted above had indicated that the correlation time for the segmental motion in the amorphous phase of bulk-cystallized polyethylene was the same as that of the pure melt under comparable conditions, it was of interest to probe the behavior of solution-crystallized polyethylene with its distinct morphological features. It was found that the T_1 value of the amorphous phase of a dilute solution-crystallized polyethylene homopolymer was the same as the average found for the bulk-crystallized polymers.[73] This equality remained for experiments performed between 30°C and 65°C. This finding is contrary to the hypothesis that the noncrystalline region consists of a highly ordered and motionally restricted chain structure. The experimental results indicate that the fast segmental motions that determine T_1 must be the same in the noncrystalline regions of both types of samples and therefore reflect the same type of chain structure. The larger linewidths observed for the solution crystal spectra reflect differences in T_2 that may arise from incomplete motional narrowing of the dipolar interactions. This last result is consistent with the linewidth found for a bulk-crystallized sample of very high crystallinity.[69] This sample, which was 94% crystalline, had a linewidth about twice as large as for a typical 50%-crystalline sample. Because the very small amount of residual amorphous material must be near a crystallite surface, its linewidth also reflects increased motional restriction.

Kitamaru et al[72] reported dipolar-decoupled, magic angle spinning (DD MAS) experimental results obtained on bulk- and solution-crystallized (fractionated and unfractionated) linear polyethylene. Chemical shifts attributable to the crystalline and noncrystalline components were resolved into Lorentzian lines. The integrated intensities so obtained were in agreement with the crystalline fraction derived from density measurements. (As discussed in more detail in the section on Phase Structure Analysis, both of these experiments suffer from the possible deficiency that the interfacial component is not clearly resolved. Although the calculations are consistent, therefore, they do not necessarily reflect the true degree of crystallinity as measured by less ambiguous methods.) Regardless of the molecular weight distribution and thermal history, the noncrystalline component T_1 was virtually equivalent for all samples. These results support the conclusions drawn by Dechter and Mandelkern.[73]

Contrary to claims made in a study by Bovey et al,[71] others, as noted above, had reported ^{13}C studies of solution-grown single crystals. In this latter report,[71] the results of a study on solution-grown 1,4-trans-polybutadiene (TPBD) are discussed. Table 5-8 summarizes the T_1 and NOEF data for these crystals and again illustrates that the relaxation results are similar to those obtained for pure amorphous polybutadiene,[74] indicating that the motions of the fold surface chains of these single crystals are similar to those in the bulk

TABLE 5-8. SOLID-STATE ^{13}C T_1 AND NOEF DATA
FOR THE MOBILE PHASE OF TPBD SOLUTION-GROWN
CRYSTALS[a]

Carbon	T_1 (s) at:			NOEF
	+21°C	−10°C	−30°C	
—CH=	0.40	0.36	0.42	0.95
—CH$_2$—	0.18	0.16	0.24	1.19

[a] Reprinted with permission from Schilling, F. C.; Bovey, F. A.; Tonelli, A. E.; Tseng, S.; Woodward, A. E. *Macromolecules* **1984**, *17*, 728. Copyright 1984, American Chemical Society.

amorphous polymer. As was the case for polyethylene, the chemical shifts of the carbons in the folds relative to the crystalline chemical shifts are the same as bulk amorphous TPBD (Table 5-9; also see Chapter 6).

TABLE 5-9. AMORPHOUS CARBON CHEMICAL SHIFTS
RELATIVE TO CRYSTALLINE CARBON POSITION IN TPBD[a]

Carbon	Solution crystal[b]	Bulk amorphous
—CH=	−1.2	−1.2
—CH$_2$—	−2.3	−2.4

[a] Reprinted with permission from Schilling, F. C.; Bovey, F. A.; Tonelli, A. E.; Tseng, S.; Woodward, A. E. *Macromolecules* **1984**, *17*, 728. Copyright 1984, American Chemical Society.
[b] Negative sign indicates upfield direction and increased shielding.

It therefore appears that amorphous-phase studies do occupy an important part of semicrystalline polymer NMR research. Of interest to many should be the fact that such studies do not require high-power decoupling, magic angle spinning, and cross polarization for adequate spectra to be obtained. A normal high-resolution spectrometer suffices.

Separation of Phases by NMR

Chemical-Shift Effects

The separation of phases by NMR in semicrystalline polymers may be possible, without resorting to more elaborate measures, as a natural consequence of the difference in chemical shifts associated with the crystalline and noncrystalline chain conformations.[86–106] Crystal packing, hydrogen bonding, and susceptibility effects also may contribute to chemical-shift changes but are not amenable to a priori calculations of shift differences.

The chain conformation in polymers may be specified as trans (anti), (+)-gauche, or (−)-gauche (assuming a rotational isomeric state model) (see Figure

6-1). It is the positions of the γ carbons, ie, those carbons three bonds away from the carbon being observed, that determine the local chain conformation. If the carbon conformation converts from trans to gauche the γ effect predicts an upfield shift from the observed carbon shift. Although the magnitude of the effect may be highly variable, it is roughly -3.5 ppm for a single γ-carbon change and essentially additive if more than one carbon is involved. Minor differences or changes in the dihedral angle may alter the magnitude of the effect significantly.

This phenomenon can be illustrated using polyethylene NMR chemical-shift data[86, 87] and the equation:

$$\Delta\delta_{NCC-CC} = 2P_g \Delta\delta_{g-t} \tag{5-4}$$

where $\Delta\delta_{NCC-CC}$ is the observed chemical-shift difference between the noncrystalline component (NCC) resonance and the crystalline component (CC) resonance, P_g is the probability of gauche conformations, and $\Delta\delta_{g-t}$ is the magnitude of the appropriate γ effect. A $\Delta\delta_{NCC-CC}$ value of -2.36 ppm has been reported for polyethylene, leading to a value of about 0.33 for the fraction of gauche conformers in the amorphous component, in substantial agreement with previously derived values.[88–90] The effect of the crystalline-phase structure of polyethylene on the chemical shift is discussed in Chapter 6.

Polyoxymethylene (POM) has a hexagonal structure and the conformation is that of a 9_5 helix. The helical structure leads to the upfield shift of the crystalline carbons relative to the noncrystalline,[91] as compared with the all-trans crystalline structure of polyethylene.

The assignment of the crystalline polymer polyphenylacetylene to a cis-cisoid helix has been deduced from a study of the amorphous and crystalline forms by Sanford et al.[92]

A most interesting use of CPMAS in structural characterization is that reported by Bunn et al[93] with respect to syndiotactic polypropylene. The chemical-shift data for both syndiotactic and isotactic polypropylene are given in Table 5-10. The crystalline structure of syndiotactic polypropylene has the appearance of a figure eight when viewed down the helix axis (see Figure 6-2). There are two equally probable, distinct sites for methylene carbons, one lying on the axis of the helix and the other on the periphery of the helix. There are only single sites for the methine and methyl carbons. The shift difference between the methylene carbon sites of 8.7 ppm was considered to arise from the γ effect. For the low-field methylene carbon, both γ carbons are in the

TABLE 5-10. POLYPROPYLENE CHEMICAL SHIFT DATA (ppm)[a]

	Methyl		Methylene		Methine	
	Solution	Solid	Solution	Solid	Solution	Solid
Isotactic	21.58	22.5	45.54	44.5	27.86	26.5
Syndiotactic	19.90	21.0	46.14	39.6, 48.3	27.45	26.8

[a] From reference 93. Reprinted by permission of the copyright holder, the Royal Society of Chemistry.

trans conformation, whereas for the high-field ones, both γ carbons are in the gauche conformation. (Isotactic polypropylene crystallizes in a simple helical form and has only one resonance for each carbon type.)

The effect of thermal history on isotactic polypropylene has been investigated by Bunn et al and illustrates the potential of solid-state studies for characterizing thermal effects in polymers.[94] These results reinforce the comments made earlier concerning the importance of controlling thermal history in semicrystalline polymer research. Chemical-shift differences between the α-crystalline, α-quenched, and β-crystalline forms were observed. (The detection of these small shift differences, however, required that the decoupling field strength be of the order of 60 kHz or greater. Careful attention to magic angle adjustment and shimming were required also.)

For these polymers, features of interest in Figure 5-7 and Table 5-11 include: (i) the observation of well-resolved splittings of the CH_3 and CH_2 carbons in the annealed α form, (ii) the reduced resolution of splittings and the changes in intensity distribution for these regions in the quenched α form, and (iii) the symmetrical lineshapes for these resonances in the β form. These effects were rationalized in terms of the different ways in which the helices were packed in the crystalline regions of the α and β forms of this polymer. It was noted in summary that both intramolecular and intermolecular effects, as well as morphologies, gave rise to subtle chemical-shift differences in this series. The polypropylenes are discussed in more detail in Chapter 6.

TABLE 5-11. Effect of Thermal History on Chemical Shifts of Isotactic Polypropylene[a]

	Chemical shift (ppm)		
Sample	Methyl	Methylene	Methine
α crystalline	22.1, 22.6	45.2, 44.2	26.8
α quenched	22.1	44.2	26.8
β crystalline	22.9	45.0	27.1

[a] From reference 94, by permission. Copyright 1982, Butterworth and Company, Ltd.

The sensitivity of ^{13}C solid-state chemical shifts to subtle changes in structure are also illustrated well by the following example. Isotactic polybutene-1 has a 3_1 ... (gt) (gt) (gt) ... helical crystalline conformation at room temperature.[95] The chains have dihedral angles of 60° for gauche and 180° for trans conformations and are packed in a trigonal unit cell (form I). At 90°C and above it adopts a tetragonal form (form II) in which the chain conformation is an 11_3 helix with alternating "gauche" and "trans" angles of 77° and 163°, respectively. (At room temperature form II spontaneously transforms into form I in a few days.) A third polymorph (form III) may be formed by crystallization from a variety of solvents, by evaporation of $CHCl_3$ solutions, or by freeze drying of benzene solutions. This form is orthorhombic and has a

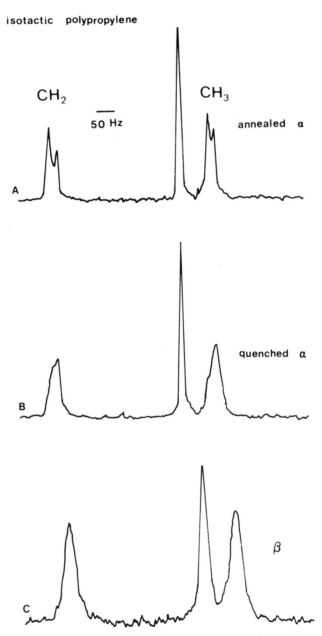

Figure 5-7. The CPMAS spectra of various forms of isotactic polypropylene (A) annealed α form, (B) quenched α form, (C) β form. (Reproduced from Bunn, A.; Cudby, M. E. A.; Harris, R. K.; Packer, K. J.; Say, B. J. *Polymer* **1982**, *23*, 694, by permission of the publishers, Butterworth and Co., Publishers, Ltd.)

4_1 helical conformation with "gauche" and "trans" angles of 83° and 159°. Although stable at room temperature, it can be transformed to form II at 94–96°C. Figure 6-4 depicts the conformations of the three forms of polybutene-1. Table 5-12 summarizes the corresponding chemical-shift differences for each form, and the spectra are shown in Figure 1-2.

TABLE 5-12. CARBON-13 SOLID-STATE CHEMICAL SHIFTS OF POLYBUTENE-1 [a,b]

	Carbon			
State	CH_3	Sidechain CH_2	α-CH	β-CH_2
Form I	1.55	15.46	20.68	27.19
Form II	0.75	17	22	28
Form III	3.05	16.96	25.21	29.78
	2.25			
Amorphous	0.00	16.40	23.46	28.32

[a] Reprinted by permission from Belfiore, L. A.; Schilling, F. C.; Tonelli, A. E.; Lovinger, A. J.; Bovey, F. A. *Macromolecules* **1984**, *17*, 2561. Copyright 1984, American Chemical Society.

[b] Referred to the CH_3 resonance of the amorphous form as zero.

The chemical-shift effects were interpreted in terms of variations in the gauche angles and concomitant γ-shielding parameter variations as a function of dihedral angle, because all three helices had a ... (gt) (gt) ... conformation. Harris and associates[96] have reported similar, although less detailed, observations regarding polybutene-1.

Gronski and colleagues[89] have reported the conformational analysis of an amorphous polymer, threodiisotactic poly(1,2-dimethyltetramethylene), in the glassy state by CPMAS variable-temperature experiments. At 220 K different conformational diads were resolved and assigned. The shift increment associated with a trans/anti–gauche conformational change was determined. In other studies[97,98] Möller and Cantow analyzed the rotational conformations of semicrystalline erythrodiisotactic poly(1,2-dimethyltetramethylene) [—CH(CH$_3$)—CH(CH$_3$)—CH$_2$—CH$_2$—]. They assigned the crystalline chain conformation from a study of the amorphous and crystalline component chemical shifts (see Table 5-13 and Chapter 6).

TABLE 5-13. CARBON-13 CHEMICAL SHIFTS FOR ERYTHRODIISOTACTIC POLY(1,2-DIMETHYLTETRAMETHYLENE)[a]

		Solid	
Carbon	Solution	Crystalline	Amorphous
CH	37.96	41.15	38.0
		40.80	
CH_2	31.11	36.30	32.74
		27.85	
CH_3	16.75	20.78	16.88
		12.74	

[a] From reference 97.

Finally, Sacchi and associates[99] illustrated how inferences could be made regarding the similarity of chain conformations in the solid state of polymers by careful analysis of chemical shifts[100] and known crystal structures by X-ray analysis. The approach was applied to the characterization of racemic and optically active isotactic poly(3-methyl-1-pentene). The crystal structure of the latter is a 4_1 helix, whereas the structure for the former remains to be determined. The NMR results indicate a great similarity in structure (Table 5-14; also see Chapter 6).

TABLE 5-14. CARBON-13 CHEMICAL SHIFTS OF ISOTACTIC POLY-(3-METHYL-1-PENTENE)[a]

		Solid	
Carbon	Solution[b]	Poly(S)-3M1P[e]	Poly(RS)-3M1P[c, e]
CH_3	12.8	14.1	13.7
CH_3'	13.8		(17.9)
CH_2	28.6	29.0	28.9 (22.8)
$CH_2{}^d$	32.3	32.5	32.4
CH	36.3	36.7	36.7
CH^d	36.7	38.0	(40.6)

[a] Reprinted with permission from Sacchi, M. C.; Locatelli, P.; Zetta, L.; Zambelli, A. *Macromolecules* **1984**, *17*, 483. Copyright 1984, American Chemical Society.
[b] Both polymers.
[c] Numbers in parentheses are shifts of smaller resonances absent in poly(S)-3M1P.
[d] Backbone carbons.
[e] The designation S refers to the optically active monomer, and RS to the racemic monomer.

High-resolution CPMAS spectra of solid isotactic, atactic, and syndiotactic poly(vinyl alcohol) (PVA) have been reported.[76] These samples all exhibit splittings of the methine carbon resonance into three peaks that (as in solution spectra) relate to the tacticity of the sample. However, the CPMAS results are not quantitative because of the presence of intermolecular hydrogen bond effects that distort the relative intensities. Intramolecular hydrogen bonds produce a large downfield shift of certain methine carbon resonances (Figure 5-8, Table 5-15). The methine carbon resonance splitting cannot be attributed to differences in conformation because the crystalline component of PVA takes the same zigzag planar conformation regardless of the tacticity. This is the only polymer for which resonances from different tactic structures have been resolved clearly.

Figure 5-9 is a spectrum of isotactic polystyrene at two different crystallinities and the difference spectrum.[105] The upper spectrum is a highly crystalline sample obtained by annealing for 3 h at 180°C, whereas the middle spectrum is the original unannealed sample of lower crystallinity. The third spectrum was obtained by subtracting one-half of Figure 5-9B from Figure 5-9A to deconvolute in effect the upper spectra to obtain a spectrum of only

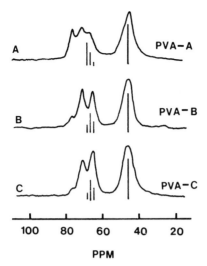

Figure 5-8. Carbon-13 CPMAS spectra of poly(vinyl alcohol) (PVA) at room temperature; 15-G decoupling, 0.85-ms contact time, 2.4-kHz spinning, 2-s recycle delay, 15 MHz (A) PVA-A (fairly isotactic) (B) PVA-B (atactic) (C) PVA-C (slightly syndiotactic) (vertical lines denote corresponding solution chemical shifts). (Reprinted with permission from Terao, T.; Maeda, S.; Saika, A. *Macromolecules* **1983**, *16*, 1535. Copyright 1983 American Chemical Society.)

the crystalline component. The increase in resolution with an increase in crystallinity was attributed to more uniform packing and chain conformation. The crystal structure of isotactic polystyrene shows that no two of the protonated ring carbons are magnetically equivalent, so the observation of five resonances is reasonable. The noncrystalline resonances are usually broader and often asymmetrical because of a distribution of conformations and/or chemical-shift dispersion.[105] The resonance position of the crystalline methylene carbons are difficult to determine in Figures 5-9A and B, but in the difference spectrum the resonance appears as a downfield shoulder on the methine resonance.

TABLE 5-15. CARBON-13 CHEMICAL SHIFTS OF POLY(VINYL ALCOHOL)[a]

	Solution				Solid			
	CH				CH			
	mm	mr	rr	CH$_2$	Peak 1	Peak 2	Peak 3	CH$_2$
PVA-A[b]	68.3	66.6	64.8	45.3	76.0	70.7	66.0	44.8
PVA-B[c]	68.4	66.8	64.8	45.5	77.4	71.2	65.3	46.5
PVA-C[d]	68.3	66.6	64.9	45.7	76.7	71.0	65.2	46.5

[a] Reprinted with permission from Terao, T.; Maeda, S.; Saika, A. *Macromolecules* **1983**, *16*, 1535. Copyright 1983, American Chemical Society.
[b] Isotactic.
[c] Atactic.
[d] Slightly syndiotactic.

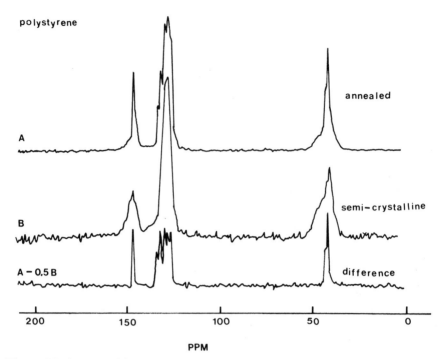

polystyrene

annealed

A

semi-crystalline

B

A − 0.5 B

difference

200 150 100 50 0

PPM

Figure 5-9. Spectra of isotactic polystyrene as a function of crystallinity (A) sample annealed for 3 h at 180°C; (B) same material but with no annealing, both spectra plotted with the same total area; (C) difference spectrum, (A) − 0.5 × (B), to effectively remove signal attributable to noncrystalline carbons giving very nearly the spectrum of purely crystalline polystyrene. (Reprinted from Earl, W. L.; VanderHart, D. L. *J. Magn. Reson.* **1982**, *48*, 35, by permission of Academic Press, Inc.)

Figure 5-10A is a rapidly quenched (from the melt) poly(ethylene terephthalate) (PET) sample with a degree of crystallinity of ≈0.[105] Figure 5-10B is the spectrum of the sample after annealing at high temperature to increase the degree of cystallinity. As the crystallinity increases, resolution improves (with the loss of the broad features attributed to noncrystalline regions).

The characterization of annealed polyethylene by solid-state NMR by VanderHart[106] illustrates the tremendous potential for understanding polymer behavior and structure–property relationships. The changes in orientation and segmental mobility in the noncrystalline component of a drawn linear polyethylene sample were monitored as a function of annealing temperature. Previous studies have determined the principal values of the polyethylene chemical-shift tensor[107,108] [$\sigma_{xx} = 51.4$ ppm (direction parallel to intramethylene H–H vector), $\sigma_{yy} = 38.9$ ppm (direction parallel to H–C–H angle bisector), and $\sigma_{zz} = 12.9$ ppm (direction parallel to chain axis)]. Inspection of a cold-drawn sample (nominal draw ratio 15) with dipolar decoupling only (no MAS) and with a pulse sequence that suppressed the crystalline component signal

PET

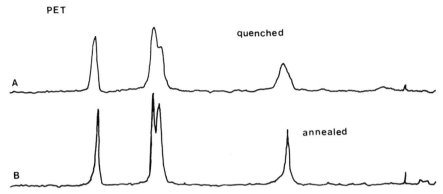

Figure 5-10. CPMAS spectra of poly(ethylene terephthalate) (PET): (A) a sample quenched rapidly from the melt resulting in no detectable crystallinity by X-ray analysis; (B) extruded PET annealed under vacuum at 500 K for 8 h and cooled to room temperature in 10 h, producing a high crystallinity. (Reprinted from Earl, W. L.; VanderHart, D. L. *J. Magn. Reson.* **1982,** *48,* 35, by permission of Academic Press, Inc.)

was undertaken. The spectrum (Figure 5-11) showed that a substantial fraction of the noncrystalline component was oriented within the 15° of the draw direction (ie, resonates in the 12–15 ppm range). Upon annealing, this oriented, mobile material decreased in intensity, with the most significant changes occurring in the 120–140°C range. In contrast, the more poorly oriented material characterized by chain resonances in the 25+ ppm range showed virtually no change in peak intensity through the whole temperature range (Figure 5-12). The author speculated that the annealing process stiffened only those chains which originally had been oriented by the drawing process.[106]

Although the emphasis in this review is on the effects of thermal history, the imposition of stress during and after crystallization can have a profound effect

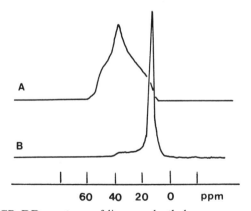

Figure 5-11. The CP–DD spectrum of linear polyethylene: upper spectrum is characteristic of an unoriented melt-crystallized sample; the lower spectrum arises from the drawn (15×) material. (Reprinted with permission from VanderHart, D. L. *Macromolecules* **1979**, *12*, 1232. Copyright 1979 American Chemical Society.)

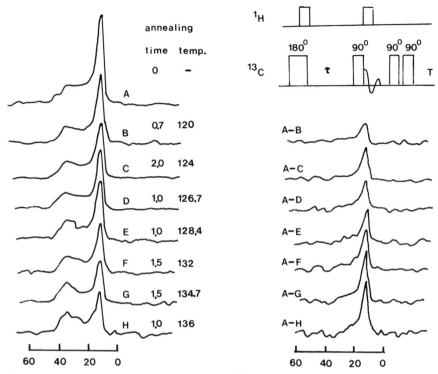

Figure 5-12. The CP–DD room-temperature ^{13}C spectra of the noncrystalline component of drawn (15 ×) linear polyethylene as a function of consecutive annealing times (hours) and temperatures (°C). Difference spectra relative to the unannealed spectrum are displayed on the right and show the principal effect of the annealing is to stiffen the originally oriented, mobile chains. The pulse sequence is shown in the upper right side of the figure. (Reprinted with permission from Vanderhart, D. L. *Macromolecules* **1979**, *12*, 1232. Copyright 1979 American Chemical Society.)

on polymer structure. However, little has been reported for ^{13}C CPMAS studies in this area. Poly(butylene terephthalate) (PBT) and related model compounds have been studied by Havens and Koenig[102] and Grenier-Loustalot and Bocelli[102a] with the objective of understanding the nature of the changes in the crystalline structure of PBT on uniaxial extension (ie, is it a predominantly reversible crystal–crystal transition?). The β form of PBT is produced by stretching to about 10% elongation and holding it under stress. As a result the butylene units are reported to have the all-trans conformation.[103] Conversely, in the absence of stress, the β modification relaxes to the α form, which is reported to have, approximately, the gauche-trans-gauche conformation.[104] One of the questions to be answered in a situation such as this is whether the structural differences manifest themselves as differences in the ^{13}C chemical shifts.

Havens and Koenig[102] had concluded, on the basis of model compounds, that the inner methylene carbons of the β form (ttt) should occur upfield of the α form (gtg). This is contrary to the expectation based on the γ-gauche effect (see Chapter 6), in which case the all-trans form should occur at lowest field. However, an X-ray diffraction study by Grenier-Loustalot and Bocelli[102a] showed that the model used by the previous authors[102] for the β form actually had the gtt conformation, and that the peak splitting in this model arose from this conformational factor. The latter authors[102a] determined the solid state ^{13}C spectra of model compounds known to be α and β by X-ray diffraction (Figure 5-13). They found that the inner methylenes occur at 24.2 ppm for the α model, and 27.8 ppm for the β model. Quenched PBT consists of both amorphous and crystalline components so that a range of conformations of the tetramethylene unit is to be expected. For this reason, the chemical shifts of the centers of the polymer resonances are anticipated to occur between those of the α and β model compounds. This is indeed observed, with the maximum of the broad line for the inner CH_2 s occurring at about 26.2 ppm.

The use of high-resolution, solid-state NMR to study oriented polymers is described in detail in Chapter 8.

Relaxation Studies

Any heterogeneous sample that exhibits nonexponential or multicomponent relaxation is a possible candidate for resolution enhancement by one of many "partial polarization" or selective saturation experiments. The type of relaxation process need not be explicitly defined here, because for the most part, one may take advantage of any relaxation process differences. This section briefly outlines a number (but not all) of the possibilities to give the reader an idea of the processes involved and, to some degree, the extent to which the experiments in question are applicable. Generalizations cannot be made here as to the best experiments to perform.

The advantages in terms of increased resolution that derive from the use of these experimental tricks are illustrated in Figure 5-14, which shows the ^{13}C spectra of trans-polybutadiene (40% crystalline).[109] Figure 5-14A is the dipolar-decoupled magic angle spinning spectrum taken with a (short) recycle delay of 0.5 s to enhance the relatively rapidly relaxing carbons in the amorphous phase and attenuate the signal from the crystalline component. Figure 5-14B is the dipolar-decoupled spectrum obtained with a long (60-s) recycle delay. Resonances from methylene carbons in the amorphous and crystalline regions are resolved clearly. The crystalline component resonances that comprise the main contribution to Figure 5-14C were obtained by introducing cross polarization (1-ms contact time) along with dipolar decoupling and magic angle spinning.

Resolution Enhancement Based on Differences in T_{CH}. Consider the most elementary steps of the cross-polarization process. As noted earlier, the rate at which various carbons polarize, $1/T_{CH}$, changes dramatically as a function of the environment of the carbon and the nature and extent of molecular motion. Quantitative analysis of certain samples therefore may be difficult, if not

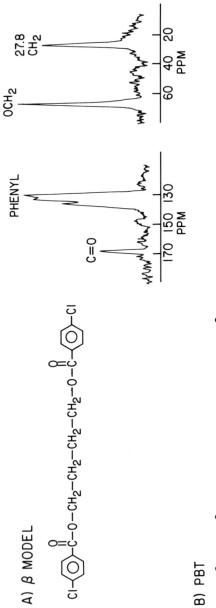

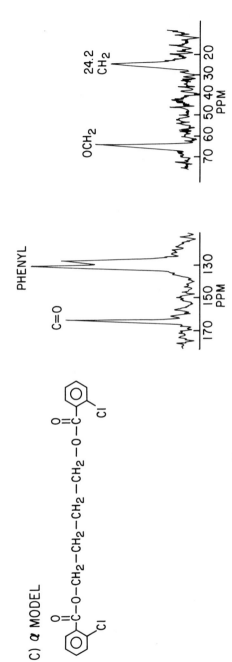

Figure 5-13. (A) Structure and CPMAS ^{13}C spectrum of a model of the β form of PBT; (B) structure of PBT; (C) structure and CPMAS ^{13}C spectrum of a model of the α form of PBT. (Adapted from reference 102a. Reprinted with permission of Pergamon Press. Copyright 1984, Pergamon Press.)

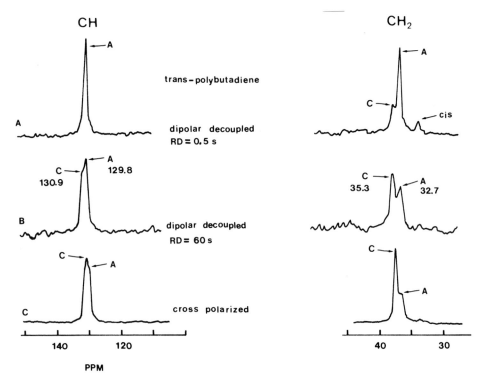

Figure 5-14. The MAS ^{13}C NMR spectra of trans-polybutadiene under several conditions. Peaks from the amorphous component are labeled A; peaks from the crystalline component are labeled C. A small fraction of cis units is visible at about 27 ppm. The exact chemical shifts are given in (B). (A) Dipolar decoupling, pulse delay = 0.5 s; (B) dipolar decoupling, pulse delay = 60 s; (C) 1-ms cross polarization. (Reproduced from Komoroski, R. A. *J. Polym. Sci., Polym. Phys. Ed.* **1983**, *21*, 2551. Copyright 1983. Reprinted by permission of John Wiley and Sons, Inc.)

impossible, because of the wide range of cross-polarization rates within a single sample. Careful attention has to be paid to the choice of experimental delay times in order to optimize the accuracy of the relative intensity measurements.

Conversely, in certain circumstances, these differences in $1/T_{CH}$ values may be exploited in order to enhance resolution. Partial or total suppression of resonances may result in a lessening of signal overlap by careful choice of the contact time.

Resolution Enhancement Based on Differences in $T_{1\rho}^H$. There appears to be a wide variation in the proton rotating-frame spin–lattice relaxation times in polymers[101] (Tables 5-16 through 5-18). Again, resolution enhancement may be effectively achieved by selection of a contact time that allows a particular component to decay to an insignificant intensity relative to another component (or components). This process is illustrated by the plot in Figure 5-15. These data were obtained for a simple polymeric sample that had undergone

TABLE 5-16. SELECTED POLYMER RELAXATION DATA: POLYETHYLENE[a]

| Relaxation time | Sample | | Comments |
	Linear	Branched	
T_1^C (s)	474	68	Crystalline component
	22–34	—	Crystalline component
	0.165	0.165	Noncrystalline component
	2	2–2.4	Noncrystalline component
$T_{1\rho}^H$ (ms)	33	5.9	Crystalline
	12	3.7	Noncrystalline
T_{CH} (ms)	0.1	0.1	Crystalline
	0.3	0.3	Noncrystalline

[a] From references 77 and 78.

TABLE 5-17. SELECTED POLYMER RELAXATION DATA[a]

Sample	Relaxation time		Comments
Polyethyleneterephthalate (PET)	$T_{1\rho}^H$	23 ms (35%)	35% crystalline, thin film,
		4.5 ms (65%)	biexponential decay
	T_1^H	1.68 s	Single component
Amorphous PET	$T_{1\rho}^H$	4.4 ms (76%)	Melt cast on cool surface
		880 ms (24%)	
	T_1^H	1.08 s	
Nordel/550	$T_{1\rho}^H$	5.2 ms (70%)	Ethylene–propylene copolymer
		1.4 ms (30%)	
	T_1^H	108 ms	

[a] From reference 101.

TABLE 5-18. RELAXATION DATA FOR α-ISOTACTIC POLYPROPYLENE[a]

Relaxation parameter	Relaxation time (ms)	Comments
$T_{1\rho}^H$	108 (48%)	Annealed sample,
	15 (25%)	three-component decay
	0.83 (27%)	
T_1^H	700	Single component
$T_{1\rho}^H$	31.7 (44.5%)	Quenched sample,
	6.4 (23%)	three-component decay
	0.67 (32.5%)	

[a] From reference 94.

two different thermal histories. (The T_{CH} values were essentially unchanged by the different treatments for these samples.) The slowly cooled sample exhibited a much longer $T_{1\rho}^H$ than the quench-cooled sample. Assuming that both components were present in the same sample (as could be the case for an annealed polymer) the selection of a contact time of about 10 ms or greater would insure that only the component from the slowly cooled fraction would be observed experimentally. This spectrum (after rescaling for the intensity lost via relaxation) could then be subtracted from the spectrum obtained at a shorter contact time (less than 1 ms) to obtain the spectrum of the other phase.

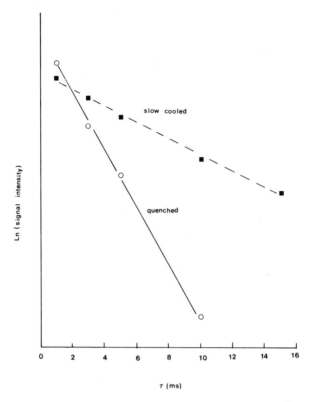

Figure 5-15. Plot of ln peak height versus variable delay, τ, in a $T_{1\rho}^{H}$ experiment.

In this manner, structural features characterized by even minor chemical-shift differences between components may be distinguished.

An NMR relaxation study of semicrystalline polyethylene oxide (PEO) samples of varying molecular weights has also produced intriguing results.[67] Acquisition of CPMAS spectra of low molecular weight PEO is very difficult because of the very short proton rotating-frame relaxation times, which results in a maximum in the signal versus contact time curve at 55–80 μs. Transfer of magnetization from protons to carbons is very inefficient. The crystalline- and noncrystalline-phase chemical shifts are virtually superimposed. Fortunately the noncrystalline component is characterized by a relatively long $T_{1\rho}^{H}$ as compared with the crystalline component (about a factor of 22 longer).[67] This can be used to discriminate between the major components. For long contact times all the crystalline component has decayed away and only the narrower, noncrystalline component remains. The linewidth of the narrow component is about 40 Hz and 300 Hz for the broad component. These line-widths are the reverse of the situation for polyethylene and most other semi-crystalline polymers discussed in this review where the crystalline component gives the narrow(er) signal.[67]

Zumbulyadis[110] has shown how more sophisticated pulse sequences may be used to isolate phases and domains in heterogeneous materials. The experiments rely on the existence of observable differences in relaxation of the protons in the different domains. The protons are first prepared in a well-defined nonequilibrium state, either in the laboratory frame or in an appropriate rotating frame (ie, the frame in which they exhibit different relaxation rates in the different domains). After some predetermined time the protons are brought into thermal contact with the ^{13}C reservoir to allow cross polarization to selectively enhance ^{13}C magnetization from specific regions of the sample.

For instance, if the protons in the various domains have distinctly different rotating-frame relaxation times, then carbons can be excited selectively by allowing the magnetization from one of the domains to decay significantly before the carbons are cross polarized. The pulse sequence consists of an Ostroff-Waugh[111] pulse sequence, during which a train of dipolar echoes is formed that decays with a time constant related to $T_{1\rho}^H$.[110] This burst of pulses is sustained until the magnetization from the faster relaxing domains is attenuated substantially. Again, only the carbons from the more slowly relaxing domains are selectively enhanced during subsequent cross polarization.

Successful application of the experiment depends on two criteria. First, for the domains to be unambiguously distinguishable, their proton relaxation times should differ by at least a factor of two. Second, the time scale of the experiment should be shorter than the time required for spin diffusion to average the magnetization from the different domains.[110]

An estimate of the domain size that can be probed by these methods[112,113] is given by $(DT)^{1/2}$, where D is the spin-diffusion coefficient and T is the time allowed for the proton magnetization to evolve prior to cross polarization. Typically, D is 2×10^{-11} to 10^{-10} cm^2/s. Selectivity based on $T_{1\rho}^H$ can probe domains on the order of 20–300 Å, assuming relaxation times in the millisecond range. For T_1-selective experiments (T $\approx$ 0.5–1 s) this translates to a characteristic domain size (eg, radius, lamellar thickness) of 300–1000 Å and above. Selectivity based on proton spin–lattice relaxation differences would therefore be most useful in the study of coarse mechanical mixtures.[110]

Finally, the delayed-contact pulse sequence may be used.[94] During a pre-contact delay, spin-locking is maintained in the ^1H channel. Cross polarization then proceeds as in the usual experiment. During the delay, those carbons, regions, or domains that are characterized by short values of $T_{1\rho}^H$ lose magnetization preferentially, so that by adjusting the delay time these species may be discriminated against. This experiment is entirely similar to the normal cross-polarization experiment when a very long contact time is used.

Selectivity Based on ^1H Spin–Lattice Relaxation. Consider a heterogeneous sample with various domains characterized by different spin–lattice relaxation times.[110] Within any domain all protons have the same proton T_1 (as a consequence of efficient spin diffusion). The domains are assumed large enough that interdomain spin diffusion can be ignored.

In this experiment a 180° proton pulse is followed by a delay chosen such that the proton magnetization from one of the domains is nulled while the proton magnetization of another domain is still inverted or has substantially recovered. A 90° proton pulse at the end of the delay brings this magnetization into the xy plane. Spin locking and cross polarization at this point will enhance the carbons selectively in the domain with a finite proton magnetization. Thus, an inversion–recovery relaxation experiment is followed by cross polarization.

This experiment is illustrated by Figure 5-16. The sample in question was a rotor made of polyoxymethylene (POM) that contained a small amount of adamantane (denoted by A in Figure 5-16A). The other resonances in the spectrum are impurities and/or additives in the rotor material (marked by *) or

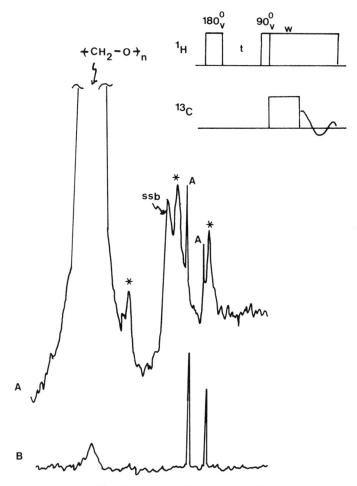

Figure 5-16. Selectivity in ^{13}C CPMAS NMR based on differences in ^{1}H spin–lattice relaxation times, T_1^H. The pulse sequence is shown in the upper right side of the figure.

spinning side bands (ssb). Because the proton T_1 of the POM is shorter than that of adamantane, the former signal may be nearly nulled (complete cancellation may not be possible if there is a distribution of proton relaxation times in this semicrystalline material), leaving behind a well-resolved adamantane spectrum (Figure 5-16B). The POM signal may be suppressed by a factor of several hundred using this technique. Other combinations of materials may be resolved similarly.

Resolution Enhancement Based on Differences in T_1^C. Another critical parameter is the carbon spin–lattice relaxation time T_1^C.[114] Figure 5-17 for polyethylene illustrates the use of relaxation time differences in a cross-polarization–^{13}C spin–lattice relaxation time (CP–T_1) experiment to enhance resolution. There are two overlapping resonances in Figure 5-17A that are not resolved under the usual simple CP experimental conditions. Each component represents a different phase characterized by different T_1^C values. In this example the downfield, crystalline-component peak has a T_1 of ≈ 100 s and the upfield, amorphous-component shoulder has a T_1 of ≈ 170 ms. A delay time of 0.096 s still allows both phases to be observed, whereas a delay of 0.9 s ($\approx 5T_1$ for the fast decay) results in the intensity of the amorphous resonance being reduced to less than 1% of its original value. One is left with a resolution-enhanced spectrum.

Another experiment, of a relatively simple nature, that can be of great use in discriminating among different carbons with a range of T_1 values is the $(180°–\tau–90°–T)_x$ pulse sequence.[106] The selectivity of the sequence depends on the T_1 values of the components and the delay times chosen. The greater the differences in the relaxation times the better the discrimination. In Figure 5-18B the usual CPMAS spectrum of semicrystalline polyethylene is presented. It was acquired with a very short contact time to minimize the contribution from the amorphous phase. The discrimination in this case, based on the differences in T_{CH}, is quite good. Figure 5-18A is the result obtained using the $(180°–\tau–90°–T)$ sequence with $\tau = T = 10$ s. The major resonance is now the low-frequency amorphous component.

No cross polarization is involved in this experiment (although magic angle spinning and dipolar decoupling are still employed in data acquisition). Magnetization builds up via T_1^C processes. Equal proton-decoupling pulses are applied to both the 180° and 90° pulses in order to minimize transient nuclear Overhauser intensity distortions from carbons with T_1 greater than the chosen delay. Two additional 90° pulses separated by 150 μs follow the observation decoupling pulse to insure ^{13}C saturation. Equal 10-s intervals were chosen in this example to reduce the intensity contributions from carbons with $T_1^C > 10$ s. Ignoring the effects of transient Overhauser enhancements, a carbon contributing unit intensity at equilibrium has an intensity $[1 - \exp(-10 \text{ s}/T_1^C)]^2$ in this experiment. Therefore, those carbons with T_1^C less than 2 s are seen at full intensity, whereas those carbons with T_1^C greater than 100 s have less than 1% of their full intensity.

Finally, a simple 90° pulse (Bloch decay experiment) followed by a delay, t,

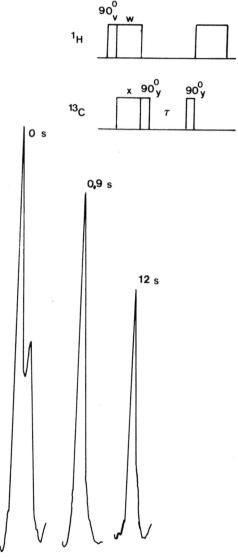

Figure 5-17. Effect of increasing delay time on resolution in a CP–T_1 pulse sequence of semicrystalline polyethylene (A) $\tau = 0.096$ s, (B) $\tau = 0.9$ s, (C) $\tau = 12$ s. The pulse sequence is shown at the top.

can be used to saturate those species that have relatively long T_1^C values,[94] just as one does in the usual progressive saturation experiment in high-resolution solution NMR relaxation experiments. This experiment is illustrated in Figure 5-19. The CPMAS spectrum of polydiacetylene exhibits two resonances for each carbon type.[115] In Figure 5-19A the simultaneous application of dipolar decoupling and magic angle spinning insures that both the crystalline and amorphous phases are observed (given the proper contact time). Only the amorphous carbons are observed in Figure 5-19B because a short (2-s) recycle time was combined with a simple Bloch decay-type experiment. In this instance, conditions do not allow the rigid, slowly relaxing crystalline component to be isolated or observed directly.

Ernst and colleagues[116] have reported a number of selective saturation pulse experiments in solid-state NMR, with applications including selective

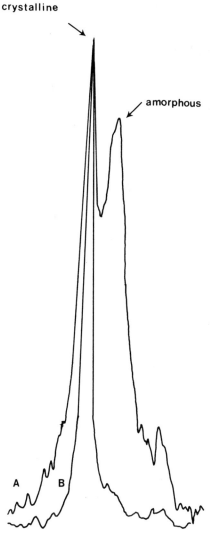

Figure 5-18. Effect of selective relaxation pulse sequence on polyethylene: $(180°-\tau-90°-T-)_x$. (A) Selective relaxation pulse sequence; (B) CP spectrum with short contact time.

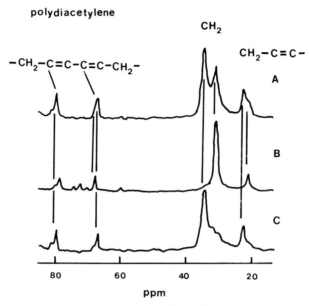

Figure 5-19. Carbon-13 CPMAS spectrum of polydiacetylene showing separation of phases: (A) CPMAS spectrum with contributions from both crystalline and amorphous regions; (B) single-pulse excitation spectrum with a recycle delay of 2 s; high-power decoupling, magic angle spinning were used. This spectrum represents the mobile, amorphous fraction. (C) Difference spectrum (A − B) to isolate the spectrum of the crystalline component. (Reprinted with permission from Havens, J. R.; Koenig, J. L. *Appl. Spectrosc.* **1983**, *37*, 226, Figure 18.)

preparation of a system in a nonequilibrium state for investigation of slow dynamic processes (chemical exchange, spin diffusion, cross relaxation) and simplification of spectra by selective excitation or selective saturation. For example, a single side band may be perturbed selectively in order to affect a family of spinning side bands. Ernst et al assumed the case of two or more overlapping side-band families whose amplitude distributions were quite different. Further, one side band is dominated by the contribution from a single family. In this case, a long unsynchronized pulse train with the carrier set either on the center band or on a side band of the resonance desired, largely eliminates the entire side-band family by saturation without affecting other resonances. Saturation of the longitudinal magnetization of a side-band family by selective irradiation of an arbitrary side band works much better the stronger the side band selected. This technique may prove to be a valuable alternative to total suppression of spinning side bands (TOSS) when the latter experiment fails.[55–57]

Dipolar Dephasing. The ^{13}C–^{1}H heteronuclear dipolar dephasing[117–119] techniques may allow qualitative or quantitative discrimination among different chemical species, domains, or phases. Following the proton–carbon contact time an extra delay, T_{dd}, is inserted before data acquisition and dipolar decoupling occurs. During this delay period the signals decay with a characteristic time constant, T_2, which depends on spin diffusion, the strength of the heteronuclear dipolar interaction, molecular motion, and magic angle spinning speeds. Therefore, if a quantitative evaluation of the number of spins associated with a particular species or phase is to be made, the decay constant for the spins in question must be determined.

In a dephasing experiment one chooses a dephasing time, T_{dd}, so that static heteronuclear dipolar interactions eliminate certain spin types from the observed spectrum. The resulting simplified spectrum may be extrapolated to $T_{dd} = 0$ in order to obtain the magnitude of the spin information that was lost by the dephasing delay. Short decays resulting from nearest-neighbor dipolar interactions can be approximated by second-order or Gaussian decays, whereas longer decays, for which the dipolar interactions are much weaker, are well fitted to first-order or Lorentzian decays. Experimental data may have to be resolved into two-component decays with Lorentzian, Gaussian–Gaussian, and Gaussian–Lorentzian components. The "best fit" overall may then be chosen.

The two-component Gaussian–Lorentzian decay of a single band $I(T_{dd})$ can be expressed as a function of the dipolar dephasing time, T_{dd}:

$$I(T_{dd}) = I_{0g} \exp\left[-0.5(T_{dd}/T_{2g})^2\right] + I_{01} \exp\left(-T_{dd}/T_{21}\right) \qquad (5\text{-}5)$$

where I_{0g} and I_{01} are the initial areas (at $T_{dd} = 0$) of the bands associated with the two decays and T_{2g} and T_{21} are the decay constants for the two decays, respectively. The total integrated intensity of the band with no dipolar dephasing $I(T_{dd} = 0) = I_{0g} + I_{01}$.

For this two-component decay it is possible to choose some value of T_{dd} that is greater than about $4T_{2g}$ so that the short Gaussian component of the band has more or less vanished and only the longer Lorentzian component remains. The initial area of the band associated with the Lorentzian component, I_{01}, can then be calculated by extrapolation of the residual area of this band back to $T_{dd} = 0$ if T_{21} is accurately known:

$$I_{01} = I(T_{dd}) \exp\left(T_{dd}/T_{21}\right) \qquad (5\text{-}6)$$

The initial area of the Gaussian component of this band follows from the equation:

$$I_{0g} = I(T_{dd} = 0) - I_{01} \qquad (5\text{-}7)$$

Suppose that one phase is characterized by a T_2 of 15 μs and the other by 192 μs. For a delay of 60 μs the faster decaying component is eliminated entirely. The resulting spectrum may then be multiplied by $[\exp(60/192)]$ to correct for lost intensity so that one effectively has extrapolated to $T_{dd} = 0$. This corrected spectrum may then be subtracted from the original spectrum to obtain the difference, which is the phase characterized by the fast decay only.

The resonances from the crystalline and noncrystalline components in polyoxymethylene may be separated by taking advantage of the differences in relaxation rates associated with the phases.[91,120,121] Although there are chemical-shift differences arising from conformational differences, they are not sufficiently large to allow clear separation on this basis alone.

Dipolar dephasing was employed to suppress the crystalline component signal (Figure 5-20). The difference in the chemical shifts of the two components is 1 ppm.[91] As the dipolar interaction is a function of molecular motion, in part, the $^{13}C-^{1}H$ dipole–dipole interaction is weaker for the more liquid-like noncrystalline component than the rigid crystalline component. The crystalline component signals therefore dephase more quickly because of tighter coupling to the protons. This is manifested in the large differences in the decay rates measured for the two sets of resonances: 17 μs for the crystalline phase and 92 μs for the noncrystalline component.[91]

Figure 5-21 is a dramatic example of the ability of dipolar dephasing to discriminate between comonomer resonances in a copolymer. The proprietary polymer in question is characterized by comonomers with very different

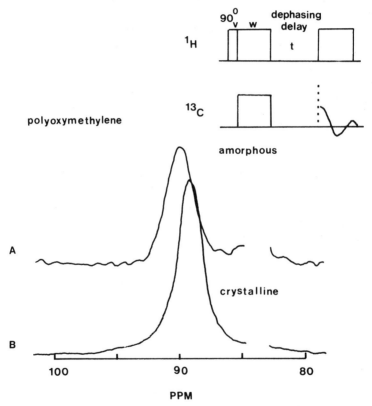

Figure 5-20. Dipolar–dephased spectra of polyoxymethylene (POM): (A) dipolar-dephased spectrum; (B) normal CPMAS spectrum. The pulse sequence used is shown at the top. (Reprinted from Cholli, A. L.; Ritchey, W. M.; Koenig, J. L. *Spectrosc. Lett.* **1983**, *16(1)*, 21, by courtesy of Marcel Dekker, Inc.)

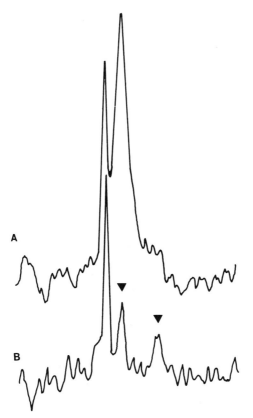

Figure 5-21. Dipolar-dephased spectrum of a proprietary copolymer: (A) normal spectrum, (B) dephased spectrum, 60-μs delay.

polyethylene

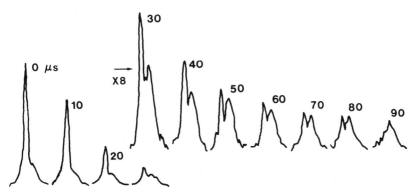

Figure 5-22. Dipolar-dephased spectra of polyethylene. The delay time is shown for each spectrum. (Reprinted from Schroter, B.; Posern, A. *Makromol. Chem.* **1981**, *182*, 675, by permission of the publishers, Huthig and Wepf Verlag, Basel.)

molecular mobilities and carbon types. In Figure 5-21A the spectrum obtained with short dephasing times (less than 20 μs) exhibits two peaks, the high-field one being quite broad. At dephasing delays of about 60 μs the presence of two previously obscured resonances (confirmed by solution NMR measurements) is revealed clearly.

As shown in Figure 5-22 for polyethylene the ability to discriminate between phases or structures may not be so dramatic.[77] However, for polyethylene

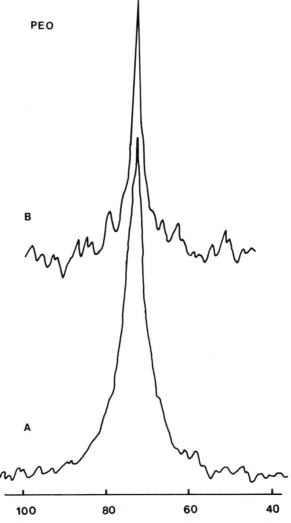

Figure 5-23. Dipolar-dephased spectrum of poly(ethylene oxide): (A) "normal" spectrum showing overlapping crystalline and noncrystalline components; (B) dephasing delay = 60 μs, which reveals the narrow noncrystalline component. (Figure courtesy of Prof. J. J. Dechter, Dept. of Chemistry, University of Alabama, Tuscaloosa, Alabama.)

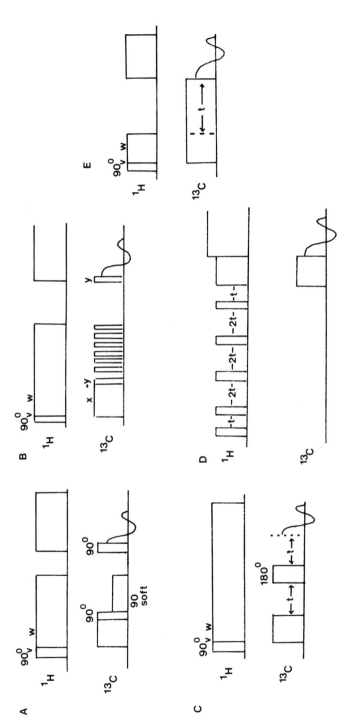

Figure 5-24. Pulse sequences that may be used to enhance resolution. (A) Selective saturation pulse sequence: the cross-polarization step is followed in the ^{13}C channel by a 90° hard pulse to put the transverse magnetization along the z direction; a 90° soft (long) pulse is applied to select a small frequency range to be saturated; the decoupler is turned off to remove the net transverse magnetization of the selected spins; finally the remaining magnetization is placed in the transverse plane by a nonselective hard pulse. (B) Selective suppression of spinning side bands[116]: the multipulse train in the ^{13}C channel is a Delays Alternating with Nutations for Tailored Excitation (DANTE) sequence. (C) Carbon-13 spin-echo T_2 sequence: the delay t must be an integral number of rotor revolutions. (D) Selective suppression based on differences in $T_{1\rho}^H$.[110] (E) Carbon-13 $T_{1\rho}$ pulse sequence with cross polarization.

oxide there is very little difference in chemical shifts between phases so that the dephasing experiment is required to separate the components (Figure 5-23).[67]

As a general guide, the decay constants for various types of carbons in rigid systems can be assigned to a certain range of values: nonprotonated $sp^2 = 75$–218 μs, protonated $sp^2 = 15$–32 μs, methine and methylenes = 12–29 μs, and methyls = 50–121 μs. However, as shown above, the phase in which the carbon in question is located is a prime determinant of the decay constant actually observed.

Figure 5-24 summarizes various pulse sequences that can be used to enhance resolution where differential relaxation rates exist in a sample. Other techniques that have not been discussed explicitly in this chapter include selective saturation (Figure 5-24A), the spin-echo pulse sequence (Figure 5-24C), and the ^{13}C rotating-frame spin–lattice relaxation experiment (Figure 5-24E). It can be seen, therefore, that there is a multitude of possibilities for separating resonances and structures on the basis of their relaxation characteristics. In fact, regardless of the depressingly broad and featureless resonances that sometimes are found in the normal CP experiment, the experimental tools available to the researcher are formidable.

Phase Structure Analysis

Given a complex decay curve, how can quantitative information be derived from the data? What pitfalls may be anticipated? This section considers these questions in terms of the analysis of polyethylene phase structure. Figure 5-25 illustrates the form of the relaxation decay when bi- or multiexponential behavior is observed in a CP–T_1 experiment (although the analysis is general in nature). This section discusses the extraction of quantitative data from these curves.

For the situation chosen there are actually three components in the decay, which are fortunately well separated in terms of their relaxation rates (a factor of 10–100 between each). The slowest decay in Figure 5-25A may be fit by any suitable least-squares procedure. The slope of the curve is given by $-1/T_1^C(3)$, to denote that it is the third component. From the least-squares fit one obtains the intercept at time $t = 0$, as well as the slope. The value of the intercept represents the initial magnetization from component 3. The total magnetization at time $t = 0$ is also known. Therefore the mass fraction of the third component is the ratio of these parameters times 100 to yield relative percentages. In a CP experiment this assumes that the initial integrated intensity is representative of the phases present, whatever differences may exist in the various relaxation rates. This problem must be addressed by the experimenter for the polymer(s) under investigation.

The next step is to subtract the calculated intensities for all delay times from the observed intensities, all the way back to $t = 0$. The result of this calcu-

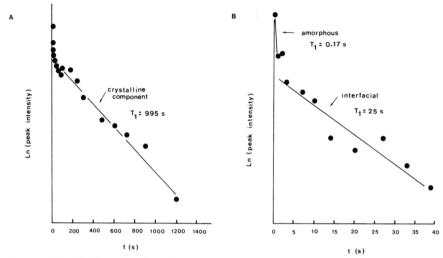

Figure 5-25. (A) Plot of ln signal intensity versus delay time, τ, in $CP-T_1$ experiment of a linear polyethylene. Solid line indicates least-squares fit to long decay portion of curve. (B) Plot of ln signal intensity versus delay time, τ, in $CP-T_1$ experiment of linear polyethylene. Data obtained by calculating intensities from least-squares fit in (A) and subtracting resultant intensities from observed values. Solid line indicates best fit to slow decay (attributed to interfacial component).

lation is shown in Figure 5-25B. There are now two distinct components remaining, a very short decay and a relatively long decay. The longer decay is again analyzed by a least-squares fit to obtain the slope and intercept, yielding the relaxation time and equilibrium magnetization for this component (component 2). Depending on the quality of the data and the number of data points, the process is repeated until all fractions have been accounted for. (In the present example the relaxation time of the fastest decaying component was already known, which obviated the need to acquire a series of data points in this time span). It is important in these experiments to choose the delay times and the total number of points that one desires to characterize the curve carefully. Preliminary, semiquantitative analyses are sometimes required for samples that are not well characterized.

This type of analysis is particularly useful in situations in which there may be no observable chemical shift difference among the components or phases. There may be no possibility of curve fitting or self-deconvolution procedures being helpful. As long as the relaxation rate differences among the phases are large enough, then quantitative results may still be obtainable.

Table 5-19 illustrates the agreement between laser Raman phase structure analysis and the NMR $CP-T_1$ decay curve analysis on the same samples of polyethylene. The degree of error associated with both methods remains to be determined at present, but these initial results are encouraging. Many obstacles remain. For instance, the major question with respect to the NMR results is whether natural abundance $^{13}C-^{13}C$ spin diffusion affects the calcu-

TABLE 5-19. PHASE STRUCTURE OF SELECTED POLYETHYLENE
SAMPLES BY NMR CP–T$_1$ AND LASER RAMAN MEASUREMENTS:
A COMPARISON

		Mass fraction	
Sample	Phase	Laser Raman	NMR CP–T$_1$
A	$\alpha_c{}^a$	0.54	0.46
	$\alpha_a{}^b$	0.31	0.27
	$\alpha_b{}^c$	0.15	0.19
B	α_c	0.65	0.74
	α_a	0.34	0.21
	α_b	0.01	0.05
C	α_c	0.84	0.91
	α_a	0.16	0.05
	α_b	0.0	0.04

^{a} Crystalline component mass fraction.
^{b} Amorphous phase mass fraction.
^{c} Interfacial phase mass fraction.

lations.[122] This subject is discussed in more detail later in this section. First, consider just one of the components determined from an analysis of the type just described, the crystalline component.

One of the most common parameters measured and reported for semicrystalline polymers is the degree of cystallinity. Numerous methods of characterization exist, including heat of fusion, density, wide-angle X-ray diffraction, NMR, and laser Raman spectroscopy. The results obtained on the same sample by these diverse methods may not be consistent, however. The major problem arises for those methods that assume a two-phase model and/or that the mass fractions of the crystalline and noncrystalline phases are additive, eg, ΔH_f and $(1 - \lambda)_d$:

$$\text{Degree of crystallinity by heat of fusion} = \frac{\Delta H_f \text{ (measured)}}{\Delta H_f \text{ (pure crystal)}}$$

$$\text{Degree of crystallinity by density} = \frac{\rho_c}{\rho} \frac{\rho - \rho_a}{\rho_c - \rho_a}$$

where ρ = measured density, ρ_c = density of the crystalline component, and ρ_a = density of the amorphous phase. It is clear that the possibility of contributions from the interfacial region must be considered.

These problems are illustrated in Figure 5-26A,[122] a plot of degree of crystallinity as derived from density, $(1 - \lambda)_d$, versus that derived from analysis of the relaxation decay curve of a CP–T$_1^C$ experiment, α_{NMR}. The $(1 - \lambda)_d$ values are consistently higher on average than the α_{NMR} over the entire range of crystallinities investigated (0.3–0.9). A similar plot with crystallinity derived from a complete analysis of the laser Raman internal mode vibrations, α_c, is shown in Figure 5-26B. Within a reasonable experimental error there is a one-to-one agreement between the two methods in this case. Despite the apparent differences among these methods, they are actually consistent. It has been determined previously that, for linear and branched polyethylenes and ethylene copolymers, $(1 - \lambda)_d$ is generally greater than α_c. It has been shown quantitatively that the reason for this disparity is that the density method contains a contribution from the interfacial region. This contribution is accounted for separately by both the NMR relaxation time and laser Raman experiments and does not appear to interfere with the determination of the mass fraction of the crystalline component.

The good agreement between the techniques is more remarkable considering the fundamentally different physical methods used and the phenomena involved. The Raman analysis determines the relative proportion of the orthorhombic crystalline ordered, all-trans, chain structure. In the NMR method adopted here it is the differences in the chain segmental mobility between the phases that distinguishes them. Although extremely different molecular properties are being determined, a quantitatively consistent measure of the fraction of the crystalline component is obtained.

Conversely, where discrepancies between these techniques arise, useful information is still obtained. Certainly, a qualitative agreement is to be expected between chain ordering and the associated molecular mobility. It is not as clear that this correlation should be quantitative. (As will be shown, there are several factors that exacerbate the problem of obtaining quantitative NMR analyses.) Given that the results may not be in good agreement, it can be argued that the numbers derived are some measure of an important property of the polymer, either chain ordering or conformations or molecular dynamics, and that these numbers are in some manner related to the polymer phase structure. For the purposes of future structure–property correlations it is suggested that the NMR results are complementary with other methods, at worst.

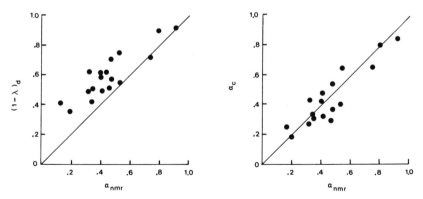

Figure 5-26. (A) Plot of degree of crystallinity as derived from density versus degree of crystallinity as derived from CP–T_1, NMR relaxation curve. Data points represent a wide range of polyethylene samples. (B) Plot of degree of crystallinity as derived from laser Raman internal mode analysis versus NMR CP–T_1 relaxation curve derived value. (Reprinted from Axelson, D. E.; Mandelkern, L.; Popli, R.; Mathieu, P. *J. Polym. Sci., Polym. Phys. Ed.* **1983**, *21*, 2319. Copyright 1983. Reprinted by permission of John Wiley and Sons, Inc.)

Although polyethylene has been the most thoroughly studied polymer in this respect to date, it appears that other polymers may be studied similarly. Carbon-13 spin–lattice relaxation studies of atactic PVA at room temperature have shown that there are three components with relaxation times of about 80 s, 6 s, and a few hundred milliseconds.[76] These were attributed to the crystalline, "medium," and amorphous components in order of decreasing T_1. A more complete analysis remains to be performed.

The generality of this type of analysis remains to be determined, because there are various problems that may arise and that affect the quantitative nature of the experiment. This is made obvious in the work of Fleming et al,[28] who reported the results of a variable temperature CPMAS study of isotactic polypropylene. Exponential decays for all carbons at each temperature were observed. Methine and methylene carbons exhibited T_1 minima at 163 K, with T_1 for these carbons being one to two orders of magnitude longer than the methyl T_1.

Of interest was the observation that the methine carbon had a shorter T_1 value than the methylene carbon over the entire temperature range, whereas the situation was reversed for the $T_{1\rho}$ data. The authors discussed these results in terms of the dominance of the relaxation behavior of the methine carbon by dipolar interactions with the protons of the attached methyl group. Because of this interaction it is not possible properly to characterize the backbone motion of the polymer independently.[28] Furthermore, the warning was offered that it might not always be possible to observe backbone motion in crystalline materials with rapidly rotating sidegroups unless they were deuterated. A more detailed discussion of this work appears in Chapter 3.

Confirmation of this problem has been demonstrated in another semicrystalline polymer. Semicrystalline and amorphous poly(methyl methacrylate) have been studied by Veeman and Menger, the results being summarized in Table 5-20.[75] There is little difference between the relaxation times in the different (isotactic, syndiotactic) samples and this effect was ascribed to the hypothesis that all T_1's are determined by high-frequency components of motion. Furthermore, it was surmised that the rapidly rotating α-CH_3 group dominated the T_1 behavior of most carbons. Thus, changes in the degree of crystallinity did not affect the various T_1 values because the α-CH_3 group rotation itself does not appear to be affected by changes in crystallinity.

TABLE 5-20. CARBON-13 T_1 DATA FOR POLY(METHYL METHACRYLATE)[a]

	α-CH_3				C=O	
Sample[b]	i	s	C_q	OCH_3	i	s
i-PMMA (amorphous)	0.053	—	5.3	5.7	12.5	—
i-PMMA (crystalline)	0.060	—	5.4	5.9	14.0	—
s-PMMA (amorphous)	—	0.048	3.4	6.0	—	6
i-s PMMA (complex)	0.053	0.034	3.6	6.1	13.3	7.2

[a] From reference 75.
[b] Ambient temperature, 45 MHz.

In what follows the question of spin diffusion is returned to and its origin and the factors that may influence it briefly summarized before the possibility of its presence in ^{13}C relaxation measurements is addressed.

Various phases, domains or sites in a polymer or heterogeneous material may provide efficient transfer of energy to the lattice because of the nature of their molecular motion.[34,35,123–125] In semicrystalline polymers, for instance, one component (noncrystalline) can exhibit molecular motions that favor relaxation (short T_1) while the other (crystalline) remains weakly coupled to the lattice (very long T_1). It is possible, through the mechanism of spin diffusion, for the component that is highly coupled to the lattice to relax resonant nuclei of other components in the spin system either totally or partially. If the process is efficient, a single T_1 is observed. Partial coupling will cause multicomponent T_1's to approach a common value. Paramagnetic impurities, lattice imperfections, mobile end groups, and side chains also can provide centers for efficient relaxation to the lattice.

The intensities of the components of T_1 and $T_{1\rho}$ are affected by spin diffusion. The longer the time scale of the relaxation the more likely that it is to be affected by spin diffusion. Therefore, proton T_1 is more likely affected than proton $T_{1\rho}$'s. In such cases the component intensities of the coupled system bear no direct relationship to the number of resonant nuclei contributing to the individual decay terms. In the total absence of any interaction between the phases in a heterogeneous sample, the T_1 (or $T_{1\rho}$) decay would be a simple superposition of the individual decays characteristic of the various phases or related morphological entities present. However, as the degree of interaction (ie, spin diffusion) becomes more significant, the relaxation decays may exhibit behavior that is more complex, resulting in an averaged value for the original decay rates. What factors affect the spin-diffusion rate? Factors that enhance spin diffusion include:

1. Large γ nuclei
2. Small internuclear distances (high density)
3. High isotopic abundance
4. Intimate mixing of the phases with fast and slow relaxation times
5. Decreasing M_2^{CH} (carbon–hydrogen second moment)

Assuming a significant contribution of spin diffusion to the crystalline component T_1's, one could propose the following qualitative argument for semicrystalline polymers. The greater the disruption of the crystalline order or crystalline–interfacial transition zone, the greater the thermal contact between the phases. This might be the case if a greater crystalline surface area were exposed or were in otherwise close proximity to the more rapidly relaxing noncrystalline component. On that basis the T_1 value of the crystalline phase within a certain distance from this interface would be reduced in magnitude by the spin-diffusion process. Given a small crystal size the entire crystalline component may be affected. However, should a substantial portion of the crystalline component be at such a great distance from the interface that the spin-diffusion contribution becomes less severe (or negligible), this portion may be characterized by a second distinguishable (crystalline) decay rate, one that is much slower compared to that for the crystalline regions affected by spin diffusion. Any thermal history or combination of structural features that leads to a morphology characterized by more intimate contact between the various phases present therefore is more conducive to the enhancement of the spin-diffusion phenomenon. The extent of this mixing determines whether single or multiple decays are observed. Thus, for the optimist, morphological information may be extracted from a study of this normally undesirable phenomenon.

One pulse sequence that can be used to determine the presence of spin diffusion is shown in Figure 5-27. It is a two-dimensional (2D) NMR experiment.[34,35] In the absence of spin diffusion the 2D spectrum appears as a series of peaks, representing the normal spectrum, along the diagonal. When spin exchange occurs, cross peaks are created that indicate the connectivity between different resonances (see Chapter 2). For linear polyethylene (Figure 5-27B) only two resonances, corresponding to the crystalline and amorphous phases, are observed. Despite the poor resolution in the one-dimensional spectrum the 2D spin exchange experiment is still possible through the use of

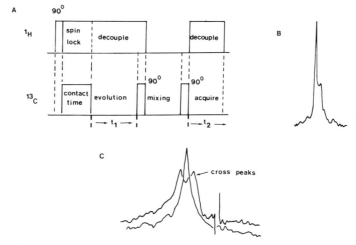

Figure 5-27. (A) Two-dimensional spin exchange experiment pulse sequence. (B) Carbon-13 CPMAS spectrum of linear polyethylene showing crystalline and amorphous resonances. (C) Selected cross sections of 2D spin-diffusion experiment: spectrometer frequency 50.3 MHz, contact time = 1 ms, recycle delay = 5 s, mixing time = 15 s, 13-G decoupling, 128 spectra × 256 data points acquired, 20 scans per spectrum; the single peak is crystalline component resonance [no amorphous component appears because T_1 (amorphous) is much shorter than mixing time]. Double peaks represent tilted cross section, showing cross peaks connecting crystalline and noncrystalline phases. (Reprinted from Axelson, D. E.; Mandelkern, L.; Popli, R.; Mathieu, P. *J. Polym. Sci., Polym. Phys. Ed.* **1983**, *21*, 2319. Copyright 1983. Reprinted by permission of John Wiley and Sons, Inc.)

small spectral widths, large data sets, high SN, and resolution enhancement techniques.

Figure 5-27C shows selected cross sections from the experiment to illustrate the results. The crystalline component resonance is shown from one cross section (the amorphous resonance does not appear because the relaxation time is very short compared to the mixing time used). The corresponding "tilted" cross section reveals that cross peaks connecting the crystalline and amorphous resonances do exist, confirming natural abundance ^{13}C spin diffusion effects in this sample. However these results do not indicate whether the extent of the spin diffusion is significant enough to cause measurable errors in the mass fractions derived from the relaxation time analysis.[126] In fact, the excellent agreement between α_{NMR} and α_c previously discussed indicates that the phenomenon did not seriously affect the quantitative analyses performed.

Aside from characterizing exchange processes between phases, the 2D experiment also allows for the study of exchange within a phase, as reported by Edzes and Bernards[127] in a study of a low-density ethylene–octene copolymer, Dowlex 2045. As noted earlier the 2D spectrum represents the correlation function between resonance frequencies ω_1 before and ω_2 after an exchange time, τ. Any exchange process that couples different resonance frequencies ω_1 and ω_2 manifests itself by off-diagonal signal intensity.

The sample under investigation was studied by 2D spin exchange CP–DD (no magic angle spinning) to insure the observation of the complete chemical-shift anisotropy powder pattern of the crystalline component. A large diagonal signal was observed with off-diagonal ridges being observed for a mixing time of 10 s. The off-diagonal pattern is specific for the change in orientation of the shielding tensor, the ridges beginning at the intersection of ω_{11} and ω_{22} and converging toward ω_{33}. This indicates that the exchange leaves the direction of σ_{33} unaltered, whereas σ_{11} and σ_{22} are interchanged. The authors concluded that the most plausible mechanism consistent with the experimental results was interchain spin exchange between ^{13}C nuclei located on neighboring molecules with mutually perpendicular orientation. (The shielding tensor is invariant for 180° jumps of the molecules about the chain axis or intrachain spin diffusion and hence these processes, if present, would be unobservable in the 2D spectrum.)

Thus slow spin exchange or molecular jumps can be measured for exchange times τ in the range $t_1 < \tau < T_1$, where t_1 is the preparation time of the 2D experiment (milliseconds) and T_1 is the spin–lattice relaxation time for the ^{13}C nuclei (5–5000 s for polyethylene crystalline phases).

Spin–Lattice Relaxation and Polymer Morphology

The study of the spin–lattice relaxation time T_1 has played a significant role in the understanding of the structure and motions of liquids for decades. The advent of solid-state NMR has made available the ability to probe the structure and motion of both crystalline and noncrystalline domains of semicrystalline polymers. Hence T_1 experiments can now be used to probe crystal structure and this section considers this topic in some detail.

There have been few reports of T_1 measurements of semicrystalline polymers and only one major study as of this writing (insofar as ^{13}C CPMAS is concerned).[122] The structural and morphological parameters that govern the T_1 of the crystalline component of polyethylene are discussed here in detail. This work should provide a solid basis for future studies in terms of understanding the polymer variables that correlate with NMR relaxation times.

There has been considerable controversy concerning the nature and extent of "defects" within the crystalline component of semicrystalline polymers.[125,128–137] Various techniques and model compounds have been employed in the hope of characterizing this phenomenon quantitatively. The contention that branches (as well as chain ends, unsaturations, and other defects) may be incorporated into the crystal lattice during crystallization is supported by considerable evidence in the literature (primarily X-ray in origin). Specifically, expansion of the unit cell, predominantly in the a-axis direction, and a concomitant decrease in the degree of crystallinity have been taken to arise from incorporation of branches within the crystalline component.

Furthermore, the extent to which the branches are accommodated was reported to depend in part on the branch length and the specific crystallization conditions. Baker and Mandelkern,[137] employing isothermal crystallization, studied ethylene–propylene and ethylene–pentene copolymers and found that methyl branches expanded the crystal lattice, whereas larger branches did not. As the crystallization conditions moved away from the isothermal regime, propyl branches increasingly contributed to unit cell expansion as more were incorporated. Longer branches, up to ten carbons in length, have been reported to be incorporated in the crystal lattice when rapid crystallization conditions are employed.[128]

The general problem remains one of (i) characterizing the distribution of chain defects between the crystalline component and the amorphous component and (ii) characterizing the degree and nature of the disorder introduced into the lattice by the defects.

The crystalline defect areas have been described variously as "amorphous" or "liquid-like."[130–132] In terms of NMR studies, this implies that there is a significant difference in the degree of segmental, vibrational, or rotational motion of the chain segments in the area of the defect as opposed to the more distant, less disrupted crystalline material. This motional difference may be amenable to characterization by T_1 measurements if the disruptive influence of the defects affect the high-frequency components of chain motion characterized by the T_1 parameter.

Qualitatively, the presence of a branch defect incorporated in the crystalline component may be considered to effectively destroy the regularity of the lattice structure within a few intermolecular distances of the defect. At a given distance each atom tends to be displaced away from the inclusion with the effect falling off roughly as the inverse square distance and resulting in a net expansion of the lattice.[134] Not only is the number of branches important, but also the branching distribution; this effect was noted by Swan,[134] who reported that "block" (ethylene–propylene) and "uniformly random" copolymers behaved differently in terms of their relative effectiveness in expanding the unit cell as the comonomer content was increased. The suggestion was made that the "blocked" sample branches were excluded preferentially from the crystal lattice or, if incorporated, were less effective in expanding the unit cell.

The effect of branching distribution on ^{13}C T_1 values has been analyzed.[125] Four samples were chosen for study, three of which could be termed "homogeneous" with respect to their branching distribution and one that could be termed "heterogeneous." Heterogeneous copolymers may be defined as those in which the copolymer molecules do not have the same ethylene–comonomer ratio.[138] Within this classification there are various distinctions that may be made and the details may be found in the references.[138] Homogeneous copolymers may be defined as those in which the molecules have the same ethylene–comonomer ratio. An empirical homogeneity index (HI) has been defined wherein a value of 0 is assigned to a heterogeneous copolymer and a value of 1.0 to a homogeneous copolymer.[138] Most samples exhibit intermediate values on this scale. It is important to note that "heterogeneity" is a relative rather than an absolute term and the HI scale is therefore to be viewed as qualitative in nature, although it remains quite useful for correlations of the type to be discussed here. There are numerous characterization methods required to place the samples in one category or the other, including low-angle laser light scattering–size exclusion chromatography, preparative fractionation, differential scanning calorimetry, and NMR. Despite the imperfections inherent in quantifying these features, it is important to note that some relevant method of branching distribution analysis is necessary to understand the behavior of copolymers. This includes not only the NMR analyses, but also the mechanical properties of these polymers.

Consider samples A and B of Table 5-21,[125] which exhibit similar degrees of crystallinity and overall branching values, but quite different degrees of branching uniformity, the former being a "typical" heterogeneous copolymer and the latter a homogeneous copolymer. Sample A has larger T_1 values, both for the quench-cooled and slowly cooled crystallization conditions as compared with sample B. The slowly cooled T_1 value of sample A is almost seven times longer than that observed for sample B. Comparison of samples C and D indicate that differences in microstructure appear more sensitive in terms of T_1 values when the samples are cooled slowly. For high degrees of branching in the homogeneous case, the effect of thermal history is not particularly significant (samples B and C, Table 5-21). These results were discussed in terms of the effect of branch number and distribution on the ability of the polymers to crystallize with the long, branch-free sequences forming the lamellae. The relative propensity for the different samples to crystallize in defect-free structures

TABLE 5-21. CRYSTALLINE COMPONENT SPIN–LATTICE RELAXATION TIMES[a] FOR SOME POLYETHYL-
ENE COPOLYMERS[b]

| Sample | Quench cooled | | Slowly cooled | | Branching[c] per 1000 carbon atoms | M_n[d] | M_w[d] | HI[e] |
	Density	T_1	Density	T_1				
A	0.9169	160	0.9261	302	11.7	20	77	0.0
B	0.9070	47	0.9186	45	12.6	44	94	0.8
C	0.9151	74	0.9277	89	7.0	40	93	0.9
D	0.9135	79	0.9270	239	5.3	57	107	1.0

[a] In seconds; all samples studied as thin films.
[b] From reference 125. Reprinted by permission of John Wiley and Sons. Copyright 1982, John Wiley and Sons.
[c] As determined by ^{13}C FT NMR solution measurements.
[d] In units of 1000; M_n, number-average molecular weight; M_w, weight-average molecular weight.
[e] Homogeneity index, an empirical parameter characterizing comonomer distribution, where a zero value represents a heterogeneous copolymer and a value of one represents a homogeneous copolymer; see reference 125 for details.

was discussed. This work led to a more elaborate characterization of a large number of quite different polyethylenes.

Aside from the NMR T_1's obtained, the parameters investigated in this more extensive study[122] also included lamellar thickness, supermolecular structure, level of crystallinity, structure and extent of the interfacial regions, molecular weight distribution, and branching number, type, and distribution. As noted earlier, correlation of trends in NMR parameters with polymer structure requires an extensive and complete data set. Many studies suffer from (sometimes unrecognized) deficiencies in this respect. Although some data are difficult and/or expensive to obtain, the limitations of the experiments and subsequent deductions should at least be noted.

The best available crystalline component T_1 data are plotted against crystal size, L_R (as determined by laser Raman longitudinal acoustic mode spectroscopy) in Figure 5-28; the data point for the longest T_1 (4500 s) has been omitted, although the plot is still linear over the long T_1 range. Also plotted in the figures are the data points for two selectively oxidized samples. Four sets of data, the two selectively oxidized samples and the two low molecular weight linear polyethylene samples, do not lie on the curve delineated by the main body of the results. These four samples have the common feature that the crystallite thickness is comparable to the extended chain length. The significance of this set of results is discussed shortly.

Some salient features of the data obtained should be noted. For L_R about 100 Å and greater, there is a relatively smooth monotonic increase of the crystalline T_1 with crystallite thickness. The curve appears to increase without limits. Even longer T_1's have been measured.[68] For crystallite thicknesses below about 100 Å, T_1 is very insensitive to the size and is in the range of 40–50 s for this data set. Again, shorter crystalline component T_1's have been observed (in highly homogeneous, high-comonomer content copolymers of ethylene with long-chain n-alkenes).[68]

Examination of the data shows that the supermolecular structure of the samples does not have a direct influence on the crystalline T_1.[122] For example, although specimens comprised of random lamellae, the (h)-type morphology, are in the lower group of T_1's these values range from 40 s to 350 s. The low values are most probably a reflection of the small crystallite sizes that usually accompany this morphological feature. On the other hand, for the most highly organized superstructures, such as the (a)-type spherulites, the T_1s range from 32 s to 1090 s. Again, no direct influence of the supermolecular structure on T_1 is observed. Among the independent variables that have been identified and measured, the crystallite thickness plays a direct and major role in determining the value of the crystalline T_1. The influence of the interfacial structure in tempering this conclusion is discussed next.

There is a remarkable increase in T_1 values when the interfacial structure is effectively removed.[122] For the oxidized samples five- to sixfold increases are observed. For the low molecular weight samples the effects are similar. The rationalization of this phenomenon lies in the behavior of one of the so-called secondary transitions that have been observed in various materials. A brief digression reviews this subject.

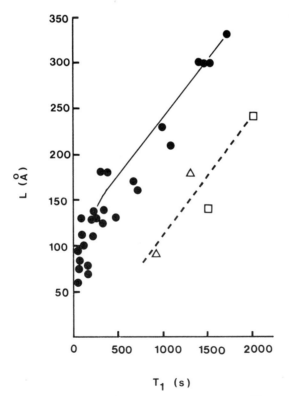

Figure 5-28. Plot of crystallite thickness L_R versus crystalline component T_1. (Reprinted from Axelson, D. E.; Mandelkern, L.; Popli, R.; Mathieu, P. *J. Polym. Sci., Polym. Phys. Ed.* **1983**, *21*, 2319. Copyright 1983. Reprinted by permission of John Wiley and Sons, Inc.)

Semicrystalline polymers exhibit a number of transitions that are detectable by a variety of techniques, such as dynamic mechanical analysis (Figure 5-29), dielectric relaxation, and heat capacity experiments.[139–143] Linear and branched polyethylenes probably constitute the polymers most widely studied and are the center of the present discussion, although the trends and principles involved extend to other types of semicrystalline polymers. These transitions have been designated as the α, β, and γ transitions in order of decreasing temperature. The γ transition is generally observed in the range of $-150°C$ to $-120°C$, the β transition in the range of $-30°C$ to $+10°C$, and the α transition between 30°C and 120°C.

Given a careful choice of model compounds and carefully controlled experimental conditions it may be possible to elucidate the origins of these transitions with more accuracy than is inherently available in non-NMR techniques (certainly all the information available must be considered in the final analysis to insure internal consistency).

The α relaxation has been attributed to motions of chain units that lie within the crystalline phase.[143–147] This conclusion resulted from a variety of studies of α-relaxation intensity and behavior as a function of degree of crystallinity and incorporation of structural changes in the main chain (chlorination, branching, copolymerization with noncrystallizable counits). The temperature at which the α relaxation is observed increases with the long period of linear polyethylene samples even for quite different derivatives of the linear structure. Crystal sizes studied have varied from a few tens of Angstroms to about 2000 Å in high-pressure crystallized polyethylene.

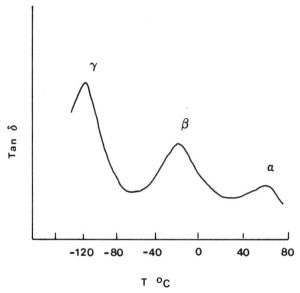

Figure 5-29. Schematic representation of a dynamic mechanical spectrum of a polymer showing three transitions, α, β, and γ.

The important suggestion that has been proposed is that the α transition is governed by the long period and thus the crystallite thickness.[146, 148] Raman longitudinal acoustic-mode experiments may be used to determine accurate crystal size distributions in polyethylenes.[149–152] A study by Mandelkern and colleagues confirmed that irrespective of the molecular weight, branching type and concentration, supermolecular structure, and level of cystallinity, T_α is governed solely by the crystallite thickness. T_α ranged from $-20°C$ to $+120°C$ corresponding to thicknesses of 60–300 Å.[143] They noted that the α-transition temperature must always be below the melting temperature T_m and that T_α and T_m have a qualitatively similar functional dependence on crystal size $(1/L)$. Hence, it was found that T_α/T_m was in the range 0.80–0.85 for the linear polyethylenes and about 0.70 for the branched polymers and copolymers. (This ratio is about 0.60–0.65 for n-hydrocarbons.)[143]

As noted in this discussion, the α transition, T_α, is thought to be a consequence of relaxations within the crystalline regions, in particular, the rotation or partial rotation of the ordered crystalline sequences. A connection with the crystalline T_1 can then be anticipated.[122] It would be expected that the larger the temperature difference between T_α and that at which the experiment is conducted, the greater would be T_1. On the other hand, if T_α were located below the temperature of the experiment the T_1 would be expected to be small and essentially independent of the crystallite size. The expected correlation between T_1 and T_α is indeed borne out by the data plotted in Figure 5-30. In this figure T_1 is plotted against T_α for common values of the crystallite thickness. The values of T_α were taken from the work of Mandelkern et al.[143] For values of T_1 greater than about 100 s, corresponding to a T_α of about 20°C, there is a definite monotonic increase of T_1 with T_α. For T_1 of the order of 100 s or less, T_α is always less than 20°C (the temperature at which the T_1 measurements were made). No significant change in T_1 is observed. Minor perturbations from the generalization that has been developed can be attributed predominantly to samples that have reduced values of (α_b/α_c), the ratio of the mass fractions of the interfacial and crystalline components. Samples for which the relative interfacial content is reduced, for the same crystallite thickness (or T_α), give much larger T_1's.

The influence of the interfacial content on T_1 has been qualitatively analyzed in terms of correlating α_b/α_c versus T_1.[122] This ratio is very close to zero for the two low molecular weight samples and the two selectively oxidized ones, ie, the samples with the apparently anomalously long relaxation times. In general, the samples with virtually no interfacial content exhibit relatively long T_1 values. More subtle effects are apparent. For example, for crystal sizes in the range of 60–80 Å, where α_b/α_c varies from 0.56 to 0.33 to 0.0 for different samples, T_1 increases by a factor greater than six. In the 175-Å range a steady decrease in α_b/α_c from 0.39 to 0.28 to 0.0 results in about a factor of five increase in T_1. Similar results are found for the larger crystallite thicknesses. Apparently, even a modest amount of interfacial content is sufficient to influence T_1 profoundly (Table 5-22).

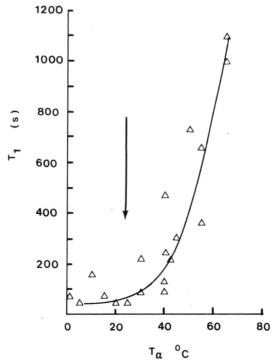

Figure 5-30. Plot of crystalline component T_1 against temperature, T_α, for the α-dynamic mechanical transition. Arrow denotes temperature at which all experiments were performed. (Reprinted from Axelson, D. E.; Mandelkern, L.; Popli, R.; Mathieu, P. *J. Polym. Sci., Polym. Phys. Ed.* **1983**, *21*, 2319. Copyright 1983. Reprinted by permission of John Wiley and Sons, Inc.)

It was found, therefore, that there is a strong coupling of the interfacial structure with the motion that takes place within the crystalline region. When these connections or junctions are severed, the relaxation within the crystalline region becomes much more retarded. The relaxations involved in the α-transition process also can be expected to be tempered by the connection with, and motion of, the disordered interfacial region. Another secondary transition that may also be related to the NMR relaxation behavior is considered next.

TABLE 5-22. COMPARISON OF THE PROPERTIES OF ORIGINAL AND SELECTIVELY OXIDIZED SAMPLES[a]

	Sample C, air cooled		Sample C, quenched 0°C	
	Original	Oxidized	Original	Oxidized
T_1 (sec)	1091	2075	220	1500
L (Å)	210	243	110	142
α_b/α_c	0.16	$\simeq 0$	0.23	$\simeq 0$

[a] From reference 122. Reprinted by permission of John Wiley and Sons. Copyright 1983, John Wiley and Sons.

The β-relaxation process has been, until recently, considerably more ambiguous in terms of its relationship to polymer structure and morphology.[143, 153–155] Beta relaxation has been observed sometimes, but not all the time, in linear polyethylene and its appearance may be a sensitive function of the sample thermal history.[143] In branched polyethylenes, including ethylene copolymers, the existence of the β relaxation is well established. The β-transition temperature for samples reported in an extensive recent study ranged from 0°C to −25°C. They were independent of crystallite thickness.[143] The intensity of the transition is related to branch content. It has been proposed that this transition results from the onset of diffusional motion at branch points. It also has been reported that because the amorphous content increases with an increase in branching content, the β transition corresponds to the glass temperature of these polyethylenes. Studies of the ^{13}C NMR relaxation parameters of the noncrystalline regions of both linear and branched polyethylenes, and the values of the correlation times that were deduced, have addressed this problem and are discussed later in this section. The results do rule out the possibility that the β transition can be identified with the glass temperature.

The fact that the β transition can be observed in linear polymers does not allow the origin of the transition to reside exclusively in the motion of the branch points. The intensity of the relaxation, however, clearly is increased in branched polymers. Carbon-13 NMR measurements show that the motions of the branch points are retarded severely at or below the β-transition temperature. The branching is to be considered not as the cause of this transition but as an enhancement.

For polyethylene, Raman internal mode analyses clearly yield the mass fractions of the crystalline, interfacial, and amorphous phases.[156] Substantial differences in these quantities arise as a function of molecular weight, chain structure, and mode of crystallization. The interfacial content of bulk-crystallized molecular weight fractions is only about 5% for molecular weights less than 2×10^5.[143] The proportion of interface increases to about 13% for higher molecular weight fractions as well as unfractionated samples. The interfacial content is relatively high for the branched polyethylenes and ethylene copolymers, regardless of the details of the molecular weight and distribution. The addition of only 0.6 mol% branches gives an interfacial content of 13%, which increases further with additional branching. Thus, the universal observation of the β transition in branched polymers and its definite existence in very high molecular weight species indicated a possible correlation between the interfacial content and the intensity of the β transition.[143] A compilation of the experimental results for the β transition in linear and branched polyethylenes is given in Table 5-23. When the interfacial content is small, less than about 5–7%, then the β transition is not observed. When the interfacial content increases above about 10% well-defined β transitions are observed. The β transition can be attributed to motion of disordered chain units, which are associated with the interfacial regions of semicrystalline polymers.[143] The

TABLE 5-23. SUMMARY OF β TRANSITION IN THE POLYETHYLENES[a]

	Observation	Interfacial content (%)
Solution crystals of linear polyethylene	Not observed	5
Solution crystals of branched polyethylene	Transition is observed	11–17
Bulk-crystallized linear polyethylene	Not observed for low molecular weights (2×10^5)	7
	Observed for high molecular weights (2×10^5)	10
Bulk-crystallized branched polyethylene	Strong relaxation is always observed	11–21

[a] From reference 143.

phase structure, NMR relaxation times, and secondary transitions of these polymers therefore are related intimately.

Aside from the types of motions described so far there are other dynamic processes in the crystalline regions to be considered. The work of Reneker and Mazur centered on the characterization of the effect of stochastic diffusion of a defect (known as a point dislocation) in crystalline polyethylene.[157] It had been suggested that some relaxation processes observed in crystalline polyethylene were a consequence of the diffusive motion of this defect along the crystalline stems in lamellae. The motion of the defect results in a 180° rotation of the part of the molecule traversed. Such restricted, discontinuous motion has been observed in certain ^{13}C data.

Specifically, it was suggested that the narrowing of ^{13}C (dipolar) satellite lines in polyethylene CPMAS spectra was the consequence of "first passage times" of a defect along a particular route.[158] The complete narrowing of the lines would require that all CH_2 groups rotated at a rate faster than a minimum rate deduced from the linewidth. To rotate all these groups in a stem, a defect must traverse the entire stem length at least once. The satellite linewidth was observed to narrow when all the CH_2 groups rotated ca. 180° at a rate of 700 times per second (at 70°C). This was interpreted to indicate that the defect passed completely through each stem more often than once every 1/700th of a second.

Finally, consider those studies that relate in some manner with the glass transition. The glass transition temperature T_g is a phenomenon characteristic of the amorphous phase.[145, 159–162] If the sample in question is totally crystalline then no glass transition is observed, whereas the greater the fraction of amorphous material the more readily the transition is observed. Roughly speaking, T_g is the temperature at which the polymer chain segments have acquired sufficient thermal energy for large-scale rotational motion or considerable torsional oscillation to occur about the majority of bonds in the main chain. Below T_g these motions are very infrequent and the majority of the chain segments have relatively fixed conformations. The transition is not necessarily sharp and a single transition temperature, usually quoted, actually represents the midpoint of the transition.

Why is the glass transition temperature important? All physical properties of amorphous and semicrystalline polymers that are dependent on segmental relaxation show a major change on heating through the glass transition. These properties include viscous flow, mechanical and dielectric relaxation, creep, crystallization, diffusion, and chemical reactivity, to name but a few. The effect of structural modifications on the glass transition, as well as other polymer transitions already discussed, may be characterized by NMR relaxation studies.

Even basic, unambiguous T_g data are lacking for some polymers despite years of study by a variety of techniques. The question of the exact value of the glass transition temperature of linear polyethylene has been the subject of great debate and experimentation over the years. Values between $-125°C$ and $-5°C$ have been reported.

Carbon-13 studies of semicrystalline polymers using scalar decoupling only have shown that well-defined high-resolution spectra may be obtained at room temperature and above (amorphous phase only).[155] It has also been shown that for both semicrystalline and amorphous polymers, as well as low molecular weight substances, with established T_g's that ^{13}C spectra collapse at temperatures well above T_g. For the purposes of one study reported by Axelson, an unfractionated high molecular weight (2×10^6) linear polyethylene was subjected to NMR analysis.[155] Spectra were obtained for temperatures down to $-35°C$ to $-40°C$ and spin–lattice relaxation times and nuclear Overhauser enhancement factors (NOEF) obtained. The T_1 value of 140 msec and NOEF of 0.15 (linewidth at half-height of 2 kHz) at $-40°C$ corresponds to a correlation time of 1.1×10^{-8} s (single correlation time model). The correlation time at the glass transition temperature is of the order of 100 s.[163–165] Thus the NMR evidence is that the correlation time at $-40°C$ is about 10 orders of magnitude smaller than that which corresponds to the true glass transition temperature. Conversely, an extremely conservative upper estimate for T_g may be made, ie, $-40°C$. Assuming a distribution of correlation times model would not affect these conclusions materially.

The temperatures at which the ^{13}C spectra collapse, T_c, for a number of polymers are shown in Table 4-4. These values depend, inter alia, on instrumental conditions, sensitivity, molecular weight, degree of crystallinity, chemical-shift separations, NMR operating frequency, motional characteristics of the backbone, glass transition temperature, instrument time, and patience.

As a result of this study an investigation of the glass transition in a set of amorphous homopolymers and copolymers was undertaken.[166,167] The T_g values of these polymers are well established and noncontroversial. This work was performed with the objective of improving the understanding of the T_c/T_g ratio.

Because frequency–temperature-dependent motion in amorphous polymers (eg, dynamic mechanical and dielectric relaxation measurements) has been analyzed successfully by the Williams-Landel-Ferry (WLF) relation,[168] it was suggested that the NMR data should also follow the WLF relations.

The relation in question is given by Equation 4-6 in Chapter 4, where a_T is the shift factor (which is actually a ratio that relates a specific viscoelastic property at a temperature T relative to the value of the property at T_g), and C_1 and C_2 are constants. Thus, the constants may be derived by plotting $T - T_g$ versus $(T - T_g)/\log \tau$, where τ is the average correlation time for overall molecular motion of the polymer in question at a given temperature. Straight lines resulted in all cases. As noted in Table 4-5, the constants C_1 and C_2 determined by ^{13}C NMR are in good agreement with the corresponding quantities determined by other methods.

The last four samples in Table 4-5 are semicrystalline polymers for which the reported glass transition temperatures may be in some dispute. The data adhere well to the WLF equation. Unfortunately, C_1 and C_2 values have not been derived for these polymers by dynamic mechanical or dielectric relaxation experiments, so a comparison with other methods is not possible as yet.

The assignment of the glass transition temperature and its identification with the β and γ transitions has been a subject of considerable controversy. The observation by high-resolution, scalar-decoupled ^{13}C NMR of a signal in linear polyethylene at temperatures down to $-40°C$ rules out the identification of T_g with the β transition for at least the linear polymer.

The same situation applies to branched polyethylenes and ethylene copolymers in that the β transition is intense, varies with the degree of crystallinity, and commonly is identified with the T_g of these species. An experimental procedure similar to that already described was employed.[154,155,166,167] The temperature at which the spectra could no longer be seen was determined as well as the correlation time as a function of temperature for the backbone methylene carbons. The low-frequency dynamic mechanical (DMA) spectra were obtained for all of the samples studied, so that an unequivocal, well-defined β transition was established independently.[154] Resolvable spectra were obtained at temperatures coincident with (or nearly so) the DMA-determined β-transition temperatures, but not at lower temperatures. Again, having determined that the correlation times were of the order of 10^{-9} s, the β transition of these branched polymers cannot be identified with the glass temperature.[154]

For the same hydrocarbon-type branched polymers there is not sufficient resolution in the chemical shifts between the backbone methylene and branch-point methine carbons to carry out relaxation-time measurements. However, for ethylene–vinyl acetate copolymers the resolution is sufficient to measure T_1 values and NOEF's independently to determine correlation times as a function of temperature. These copolymers also exhibit well-defined β transitions. It was found that resolvable ^{13}C spectra for the methylene backbone carbons could be obtained at or within a few degrees of the independently determined β transition.[166,167]

The resonance associated with the methine carbon is lost about 20°C above this transition temperature. Combining the linewidth, T_1, and NOEF data to yield correlation times showed that the molecular motion of the methine carbon is reduced significantly as the β transition is approached. The methylene carbon resonance is resolvable through the β-transition region. The results support the hypothesis that the branch point is involved in the β transition. As noted earlier, though, the more general structural basis for the β transition is that it is a property of the interfacial region.[143] This condition satisfies the requirement that the sample be partially crystalline in order to observe this transition. The noncrystallizing counits, and so the branch points, must be concentrated preferentially within the interfacial region.[166,167]

Summary

Aside from the abundance of information that can be obtained by the methods already outlined in this chapter, there are a number of areas, not covered here, that remain to be studied more extensively. For instance, the relationship between relaxation parameters and the physical properties of semicrystalline polymers has only begun to be probed.[68,169] Unlike many glassy polymers, for which $T_{1\rho}^C$ experiments are generally useful for correlations with impact strength,[170–172] this relaxation parameter usually is dominated by spin–spin interactions in semicrystalline polymers (at least in the crystalline regions). Therefore, it is not highly correlated with molecular motion within the polymer. This problem may be obviated by the use of a relaxation parameter that is correlated with molecular motion (spin–lattice processes), namely T_1^C. Although there are problems associated with the quantitative correlation of this relaxation time with polymer structure (as discussed in this review) it does provide a viable alternative to the rotating-frame relaxation-time measurements. Furthermore, the sometimes long relaxation times associated with T_1^C should be viewed not with fear but as an indication that subtle changes in the chemical or phase structure of semicrystalline polymers may manifest themselves as significant differences in the measured relaxation times. This (optimistic) viewpoint is supported to date by the data presented in this review.

One of the most exciting areas of interest for the near future is the use of variable-temperature studies of relaxation parameters in semicrystalline polymers. As resolvable resonances are obtained for a wide variety of phases and carbon types, it will be of interest to reinvestigate the structural origins of the various secondary transitions observed by non-NMR methods. These studies may be enhanced further by the judicious use of specifically deuterated and ^{13}C-enriched monomers (aside from the use of NMR-active nuclei other than ^{13}C).

It should be apparent from this brief review of semicrystalline polymer NMR that few, if any, areas of concern in the field of solid-state morphology are not amenable to study by these techniques. As an understanding of the relationship between relaxation times, chemical shifts, and the details of polymer phase structure develops, new and fruitful avenues of research will open up.

As the old controversies surrounding the phase structure of semicrystalline polymers are resolved, new and valuable structure–property relationships of importance will emerge. In fact, detailed solid-state NMR studies of polymers may force those in the field to reevaluate many of the long-standing concepts of semicrystalline polymer morphology held so dear. This alone would provide a valuable contribution to this field of research.

References

1. Geil, P. H. "Polymer Single Crystals"; Wiley-Interscience: New York, 1963.
2. Stein, R. S. In "Structure and Properties of Polymer Films", Lenz, R. W., Stein, R. S., Eds.; Plenum Press: New York, 1972.
3. Mandelkern, L. "Crystallization of Polymers"; McGraw-Hill: New York, 1964.
4. Wunderlich, B. "Macromolecular Physics", Vol. 1, "Crystal Structure, Morphology, Defects"; Wiley-Interscience: New York, 1976, Vol. 2, "Crystallization", Vol. 3, "Crystal Melting", 1980.
5. Flory, P. J., "Principles of Polymer Chemistry"; Cornell Univ. Press: Ithaca, NY, 1953.
6. "Structural Studies of Macromolecules by Spectroscopic Techniques", K. J. Ivin, Ed.; Wiley-Interscience: New York, 1976.
7. Rabek, J. "Experimental Methods in Polymer Chemistry"; Wiley-Interscience: New York, 1980.
8. Kitamaru, R.; Horii, F. *Adv. Polym. Sci.* **1978**, *26*, 137.
9. Stein, R. S. *Appl. Polym. Symp.* **1973**, *20*, 347.
10. Mandelkern, L. *Acc. Chem. Res.* **1976**, *9*, 81.
11. Mandelkern, L. *Prog. Polym. Sci.* **1970**, *2*, 165.
12. Flory, P. J.; Yoon, D. Y.; Dill, K. A. *Macromolecules* **1984**, *17*, 862.
13. Maxfield, J.; Mandelkern, L. *Macromolecules* **1977**, *10*, 1141.
14. Mandelkern, L.; Maxfield, J. *J. Polym. Sci., Polym. Phys. Ed.* **1979**, *17*, 1913.
15. Mandelkern, L.; Glotin, M.; Benson, R. A. *Macromolecules* **1981**, *14*, 22.
16. Alemany, L. B.; Grant, D. M.; Pugmire, R. J.; Zilm, K. W. *J. Am. Chem. Soc.* **1983**, *105*, 2133.
17. Pines, A.; Gibby, M. G.; Waugh, J. S. *J. Chem. Phys.* **1973**, *59*, 569.
18. Demco, D. E.; Tegenfeldt, J.; Waugh, J. S. *Phys. Rev.* **1975**, *B11*, 4133.
19. Schaefer, J.; Stejskal, E. O.; Buchdahl, R. *Macromolecules* **1977**, *10*, 384.
20. Haeberlen, U. In "High Resolution NMR in Solids. Selective Averaging"; Adv. in Magn. Reson. Ser., Suppl. 1. Waugh, J. S., Ed., Academic Press: New York, 1976.
21. Garroway, A. N.; Moniz, W. B.; Resing, H. A.; *ACS Symp. Ser.* **1979**, *103*, 67.
22. Bloch, F.; *Phys. Rev.* **1948**, *111*, 841.
23. Sarles, L. R.; Cotts, R. M. *Phys. Rev.* **1948**, *111*, 853.
24. VanderHart, D. L.; Earl, W. L.; Garroway, A. N. *J. Magn. Reson.* **1981**, *44*, 361.
25. Mehring, M.; Sinning, G. *Phys. Rev.* **1977**, *B15*, 2519.
26. Abragam, A.; Winter, J. C. *R. Acad. Sci. (Paris)* **1959**, *249*, 1633.
27. Rothwell, W. P.; Waugh, J. S. *J. Chem. Phys.* **1981**, *74*, 2721.
28. Fleming, W. W.; Lyerla, J. R.; Yannoni, C. S. *ACS Symp. Ser.* **1984**, *247*, 83.
29. Suwelack, D.; Rothwell, W. P.; Waugh, J. S. *J. Chem. Phys.* **1980**, *73*, 2559.
30. Sullivan, M. J.; Maciel, G. E. *Anal. Chem.* **1982**, *54*, 1615.
31. Axelson, D. E.; Mandelkern, L. *ACS Symp. Ser.* **1979**, *103*, 181.
32. Schaefer, J.; Chin, S. I.; Weissman, S. I. *Macromolecules* **1972**, *5*, 798.
33. Bloembergen, N.; Purcell, E. M.; Pound, R. V. *Phys. Rev.* **1948**, *74*, 679.
34. Szeverenyi, N. M.; Sullivan, M. J.; Maciel, G. E. *J. Magn. Reson.* **1982**, *47*, 462.
35. Bronniman, C. E.; Szeverenyi, N. M.; Maciel, G. E. *J. Chem. Phys.* **1983**, *79*, 3694.

36. Dechter, J. J.; Komoroski, R. A.; Axelson, D. E.; Mandelkern, L. *J. Polym. Sci., Polym. Phys. Ed.* **1981**, *19*, 631.
37. Marshall, A. G.; Roe, D. C. *Anal. Chem.* **1978**, *50*, 756.
38. Roe, D. C.; Marshall, A. G.; Smallcombe, S. H. *Anal. Chem.* **1978**, *50*, 764.
39. Marshall, A. G. *J. Phys. Chem.* **1979**, *83*, 521.
40. Marshall, A. G.; Bruce, R. E. *J. Magn. Reson.* **1980**, *39*, 47.
41. Alla, M.; Lippmaa, E. *Chem. Phys. Lett.* **1982**, *87*, 30.
42. Drain, L. E. *Proc. Phys. Soc. London* **1962**, *80*, 1380.
43. Schneider, B.; Doskocilova, D.; Pruvcova, H. "Magnetic Resonance in Chemical Biology", Nearak, J. N.; Adamic, K. J., Eds.; Marcel Dekker: New York, 1975, p. 127.
44. Macho, V.; Kendrick, R.; Yannoni, C. S. *J. Magn. Reson.* **1983**, *52*, 450.
45. Lyerla, J. R.; Yannoni, C. S.; Fyfe, C. A. *Acc. Chem. Res.* **1982**, *15*, 208.
46. Yannoni, C. S.; Reisenauer, H. P.; Maier, G. *J. Am. Chem. Soc.* **1983**, *105*, 6181.
47. Hanson, B. E.; Sullivan, M. J.; Davis, R. J. *J. Am. Chem. Soc.* **1984**, *106*, 251.
48. Lyerla, J. R.; Fyfe, C. A.; Yannoni, C. S. *J. Am. Chem. Soc.* **1979**, *101*, 1351.
49. Waugh, J. S.; Maricq, M. M.; Cantor, R. *J. Magn. Reson.* **1978**, *29*, 183.
50. Maricq, M. M.; Waugh, J. S. *J. Chem. Phys.* **1979**, *70*, 3300.
51. Andrew, E. R. *Prog. NMR Spectrosc.* **1971**, *8*, 1.
52. Herzfeld, J.; Berger, A. E. *J. Chem. Phys.* **1980**, *73*, 6021.
53. Andrew, E. R.; Farnell, L.; Firth, M.; Gladhill, T.; Roberts, I. *J. Magn. Reson.* **1969**, *1*, 27.
54. Beams, J. W. *Rev. Sci. Instr.* **1930**, *1*, 667.
55. Dixon, W. T. *J. Chem. Phys.* **1982**, *77*, 1800.
56. Dixon, W. T.; Schaefer, J.; Sefcik, M. D.; Stejskal, E. O.; McKay, R. A. *J. Magn. Reson.* **1982**, *49*, 341.
57. Dixon, W. T. *J. Magn. Reson.* **1981**, *44*, 220.
58. Komoroski, R. A. *J. Polym. Sci., Polym. Phys. Ed.* **1983**, *21*, 1569.
59. Kaufman, S.; Slichter, W. P.; Davis, D. D. *J. Polym. Sci., Polym. Phys. Ed.* **1971**, *9*, 829.
60. Schaefer, J. *Macromolecules* **1972**, *5*, 427.
61. VanderHart, D. L.; Gutowsky, H. S. *J. Chem. Phys.* **1968**, *49*, 261.
62. Hexem, J. G.; Frey, M. H.; Opella, S. J. *J. Am. Chem. Soc.* **1981**, *103*, 224.
63. Naito, A.; Ganapathy, S.; McDowell, C. A. *J. Magn. Reson.* **1982**, *48*, 367.
64. Bohm, J.; Fenzke, D.; Pfiefer, H. *J. Magn. Reson.* **1983**, *55*, 197.
65. Schaefer, J. In "Topics in Carbon-13 NMR Spectroscopy", G. C. Levy, Ed.; Wiley-Interscience: New York, 1974, Vol. 1, p. 149.
66. Komoroski, R. A.; Mandelkern, L. In "Applications of Polymer Spectroscopy", E. G. Brame, Ed.; Academic Press: New York, 1978, p. 57.
67. Dechter, J. J. *J. Polym. Sci., Polym. Lett. Ed.* **1985**, *23*, 261.
68. D. E. Axelson, unpublished observations.
69. Komoroski, R. A.; Maxfield, J.; Sakaguchi, F.; Mandelkern, L. *Macromolecules* **1977**, *10*, 550.
70. Komoroski, R. A.; Maxfield, J.; Mandelkern, L. *Macromolecules* **1977**, *10*, 545.
71. Schilling, F. C.; Bovey, F. A.; Tonelli, A. E.; Tseng, S.; Woodward, A. E. *Macromolecules* **1984**, *17*, 728.
72. Kitamaru, R.; Horii, F.; Murayama, K. *Polymer Bull.* **1982**, *7*, 583.
73. Dechter, J. J.; Mandelkern, L. *J. Polym. Sci., Polym. Phys. Ed.* **1979**, *17*, 317.
74. Jelinski, L. W.; Dumais, J. J.; Watnick, P. I.; Bass, S. V.; Shepherd, L. *J. Polym. Sci., Polym. Chem. Ed.* **1982**, *20*, 3285.
75. Veeman, W. S.; Menger, E. M. *Bull. Magn. Reson.* **1980**, *2*, 77.
76. Terao, T.; Maeda, S.; Saika, A. *Macromolecules* **1983**, *16*, 1535.
77. Schroter, B.; Posern, A. *Makromol. Chem.* **1981**, *182*, 675.
78. Schroter, B.; Posern, A. *Makromol. Chem. Rap. Commun.* **1982**, *3*, 623.
79. Fischer, E. W. *Z. Naturforsch.* **1957**, *12a*, 753.
80. Keller, A. *Philos. Mag.* **1957**, *2*, 1171.
81. Till, P. H. *J. Polym. Sci.* **1957**, *24*, 30.
82. Stamm, M.; Fischer, E. W.; Dettenmaier, M.; Convert, P. *Faraday Disc. Roy. Soc. Chem.* **1979**, *68*, 263.
83. Yoon, D. Y.; Flory, P. J. *Faraday Disc. Roy. Soc. Chem.* **1979**, *68*, 288.
84. Sadler, D. M.; Keller, A. *Macromolecules* **1977**, *10*, 1129.
85. Mandelkern, L. *Ann. Rev. Mater. Sci.* **1976**, *6*, 119.
86. Earl, W. L.; VanderHart, D. L. *Macromolecules* **1979**, *12*, 762.

87. Fyfe, C. A.; Lyerla, J. R.; Volksen, W.; Yannoni, C. S. *Macromolecules* **1979**, *12*, 764.
88. Dalling, D. K.; Grant, D. M. *J. Am. Chem. Soc.* **1974**, *96*, 1827.
89. Gronski, W.; Hasenhindl, A.; Limbach, H. H.; Möller, M.; Cantow, H.-J. *Polym. Bull.* **1981**, *6*, 93.
90. Flory, P. J. In "Statistical Mechanics of Chain Molecules"; Wiley-Interscience, New York, 1969.
91. Cholli, A. L.; Ritchey, W. M.; Koenig, J. L. *Spectrosc. Lett.* **1983**, *16*, 21.
92. Sanford, T. J.; Allendoerfer, R. D.; Kang, E. T.; Ehrlich, P.; Schaefer, J. *J. Polym. Sci., Polym. Phys. Ed.* **1981**, *19*, 1151.
93. Bunn, A.; Cudby, M. E. A. *J. Chem. Soc., Chem. Commun.* **1981**, 15.
94. Bunn, A.; Cudby, M. E. A.; Harris, R. K.; Packer, K. J.; Say, B. J. *Polymer* **1982**, *23*, 694.
95. Belfiore, L. A.; Schilling, F. C.; Tonelli, A. E.; Lovinger, A. J.; Bovey, F. A. *Macromolecules* **1984**, *17*, 2561.
96. Harris, R. K.; Packer, K. J.; Say, B. J. *Makromol. Chem.* **1981**, Suppl. 4, 117.
97. Möller, M.; Cantow, H.-J. *Polym. Bull.* **1981**, *5*, 119.
98. Möller, M.; Cantow, H.-J. *Macromolecules* **1984**, *17*, 733.
99. Sacchi, M. C.; Locatelli, P.; Zetta, L.; Zambelli, A. *Macromolecules* **1984**, *17*, 483.
100. Ferro, D. R.; Ragazzi, M. *Macromolecules* **1984**, *17*, 485.
101. Aujla, R. S.; Harris, R. K.; Packer, K. J.; Parameswaran, M.; Bunn, A.; Cudby, M. E. A. *Polym. Bull.* **1982**, *8*, 253.
102. Havens, J. R.; Koenig, J. L. *Polym. Commun.* **1983**, *24*, 194.
102a. Grenier-Loustalot, M.-F.; Bocelli, G. *Eur. Polym. J.* **1984**, *20*, 957.
103. Tashiro, K.; Nakai, Y.; Kobayashi, M.; Tadokoro, H. *Macromolecules* **1980**, *13*, 137.
104. Yokouchi, M.; Sakakibara, Y.; Chatani, Y.; Tadokoro, H.; Tanaka, T.; Yoda, K. *Macromolecules* **1976**, *9*, 266.
105. Earl, W. L.; VanderHart, D. L. *J. Magn. Reson.* **1982**, *48*, 35.
106. VanderHart, D. L. *Macromolecules* **1979**, *12*, 1232.
107. VanderHart, D. L. *J. Chem. Phys.* **1976**, *64*, 830.
108. Opella, S. J.; Waugh, J. S. *J. Chem. Phys.* **1977**, *66*, 4919.
109. Komoroski, R. A. *J. Polym. Sci., Polym. Phys. Ed.* **1983**, *21*, 2551.
110. Zumbulyadis, N. *J. Magn. Reson.* **1983**, *53*, 486.
111. Ostroff, E. D.; Waugh, J. S. *Phys. Rev. Lett.* **1966**, *16*, 1097.
112. Cheung, T. T. P.; Gerstein, B. C. *J. Appl. Phys.* **1981**, *52*, 5517.
113. Cheung, T. T. P.; Gerstein, B. C.; Ryan, L. M.; Taylor, R. E.; Dybowski, C. R. *J. Chem. Phys.* **1980**, *73*, 6059.
114. Torchia, D. A. *J. Magn. Reson.* **1978**, *30*, 613.
115. Havens, J. R.; Koenig, J. L. *Appl. Spectrosc.* **1983**, *37*, 226.
116. Caravatti, P.; Bodenhausen, G.; Ernst, R. R. *J. Magn. Reson.* **1983**, *55*, 88.
117. Wilson, M. A.; Pugmire, R. J.; Alemany, L. B.; Woolfenden, W. R.; Grant, D. M.; Given, P. H. *Anal. Chem.* **1984**, *56*, 933.
118. Alemany, L. B.; Grant, D. M.; Pugmire, R. J.; Stock, L. M. *Fuel* **1984**, *63*, 513.
119. Opella, S. J.; Frey, M. H. *J. Am. Chem. Soc.* **1979**, *101*, 5854.
120. Menger, E. M.; Veeman, W. S.; deBoer, E. *Macromolecules* **1982**, *15*, 1406.
121. Veeman, W. S.; Menger, E. M.; Ritchey, W.; deBoer, E. *Macromolecules* **1979**, *12*, 924.
122. Axelson, D. E.; Mandelkern, L.; Popli, R.; Mathieu, P. *J. Polym. Sci., Polym. Phys. Ed.* **1983**, *21*, 2319.
123. McBrierty, V. J.; Douglass, D. C.; Barham, P. J. *J. Polym. Sci., Polym. Phys. Ed.* **1980**, *18*, 1561.
124. Douglass, D. C.; McBrierty, V. J. *Polym. Eng. Sci.* **1979**, *19*, 1054.
125. Axelson, D. E. *J. Polym. Sci., Polym. Phys. Ed.* **1982**, *20*, 1427.
126. Semiquantitative calculations of the effect of $^{13}C-^{13}C$ natural abundance spin diffusion based on the polyethylene crystal structure yield the relationship $L = 11.9T_1^{1/2}$ Å, where L is the maximum diffusive path length.
127. Edzes, H. T.; Bernards, J. P. C. *J. Am. Chem. Soc.* **1984**, *106*, 1515.
128. Shirayama, K.; Kita, S.-I.; Watabe, H. *Makromol. Chem.* **1972**, *151*, 97.
129. Martuscelli, E. *J. Macromol. Sci. Phys.* **1975**, *B11*, 1.
130. Balta-Calleja, F. J.; Hosemann, R. *J. Polym. Sci., Polym. Phys. Ed.* **1980**, *18*, 1159.
131. Balta-Calleja, F. J.; Gonzalez Ortega, J. C.; Martinez de Salazar, J. *Polymer* **1978**, *19*, 1094.
132. Hosemann, R. *Endeavour* **1973**, *32*, 99.
133. Preedy, J. E. *Br. Polym. J.* **1973**, *5*, 13.

134. Swan, P. R. *J. Polym. Sci.* **1962**, *56*, 409.
135. Reding, F. P.; Lovell, C. M. *J. Polym. Sci.* **1960**, *46*, 147.
136. Richardson, M. J.; Flory, P. J.; Jackson, J. B. *Polymer* **1963**, *4*, 221.
137. Baker, C. H.; Mandelkern, L. *Polymer* **1966**, *7*, 71.
138. Elston, C. T. Canadian Patent No. 920,742, 1973; U.S. Patent No. 3,645,992, 1972.
139. McCrum, N. G.; Read, B. E.; Williams, G. "Anelastic and Dielectric Effects in Solids"; John Wiley: New York, 1967.
140. Sauer, J. A.; Richardson, G. C.; Morrow, D. R. *J. Macromol. Sci. Rev. Macromol. Chem.* **1973**, *C9*, 149.
141. Ashcraft, C. R.; Boyd, R. H. *J. Polym. Sci., Polym. Phys. Ed.* **1976**, *14*, 2153.
142. Meier, D. J., Ed. "Molecular Basis of Transitions and Relaxations"; Gordon and Breach: New York, 1978.
143. Popli, R.; Glotin, M.; Mandelkern, L.; Benson, R. S. *J. Polym. Sci., Polym. Phys. Ed.* **1984**, *22*, 407.
144. Illers, V. K. H. *Kolloid Z. Z. Polym.* **1973**, *251*, 394.
145. Stehling, F. C.; Mandelkern, L. *Macromolecules* **1970**, *3*, 242.
146. Takayanagi, M.; Matsuo, T. *J. Macromol. Sci. Phys.* **1967**, *B1*, 407.
147. Nielsen, L. E. *J. Polym. Sci.* **1960**, *42*, 357.
148. Hoffman, J. D.; Williams, G.; Passaglia, E. *J. Polym. Sci.* **1966**, *C14*, 173.
149. Snyder, R. G.; Scherer, J. R. *J. Polym. Sci., Polym. Phys. Ed.* **1980**, *18*, 421.
150. Schaufele, R. F.; Shimanouchi, T. *J. Am. Chem. Soc.* **1967**, *47*, 3605.
151. Strobl, G. R.; Hagedorn, W. *J. Polym. Sci., Polym. Phys. Ed.* **1978**, *16*, 1181.
152. Glotin, M.; Mandelkern, L. *Colloid Polym. Sci.* **1982**, *260*, 182.
153. Popli, R.; Mandelkern, L. *Polym. Bull.* **1983**, *9*, 260.
154. Dechter, J. J.; Axelson, D. E.; Dekmezian, A.; Glotin, M.; Mandelkern, L. *J. Polym. Sci., Polym. Phys. Ed.* **1982**, *20*, 641.
155. Axelson, D. E.; Mandelkern, L. *J. Polym. Sci., Polym. Phys. Ed.* **1978**, *16*, 1135.
156. Glotin, M.; Mandelkern, L. *J. Polym. Sci., Polym. Phys. Ed.* **1983**, *21*, 29.
157. Reneker, D. H.; Mazur, J. *Polymer* **1982**, *23*, 401.
158. VanderHart, D. L. *J. Magn. Reson.* **1976**, *24*, 4671.
159. Boyer, R. F. *Macromolecules* **1973**, *6*, 228.
160. Davis, G. T.; Eby, R. K. *J. Appl. Phys.* **1973**, *44*, 4274.
161. Hendra, P. J.; Jobic, H. P.; Moritz, K. H. *J. Polym. Sci., Polym. Lett. Ed.* **1975**, *13*, 363.
162. Ahmad, S. R.; Charlesby, A. *Eur. Polym. J.* **1975**, *11*, 91.
163. Eyring, H. *J. Chem. Phys.* **1936**, *4*, 283.
164. Kauzmann, W. *J. Chem. Rev.* **1942**, *14*, 12.
165. McCall, D. W. *U.S. Natl. Bur. Std.* Spec. Publ. 310, p. 475, 1969.
166. Mandelkern, L. *Pure Appl. Chem.* **1982**, *54*, 611.
167. Dekmezian, A.; Axelson, D. E.; Dechter, J. J.; Borah, B.; Mandelkern, L. *J. Polym. Sci., Polym. Phys. Ed.* **1985**, *23*, 367.
168. Williams, M.; Landel, R. J.; Ferry, J. D. *J. Am. Chem. Soc.* **1955**, *77*, 370.
169. Sefcik, M. D.; Schaefer, J.; Stejskal, E. O.; McKay, R. A. *Macromolecules* **1980**, *13*, 132.
170. Schaefer, J.: Stejskal, E. O.; McKay, R. A. *Macromolecules* **1977**, *10*, 384.
171. Steger, T. R.; Schaefer, J.; Stejskal, E. O.; McKay, R. A. *Macromolecules* **1980**, *13*, 1127.
172. Sefcik, M. D.; Schaefer, J.; Stejskal, E. O.; McKay, R. A. *Macromolecules* **1980**, *13*, 1132.

Appendix

Since the completion of the original manuscript there have been additional contributions to the literature concerning subjects discussed in this chapter. These studies will be briefly noted here for the sake of completeness.

Lauprêtre and colleagues[1] have reported a variable temperature study of poly(cyclohexyl methacrylate). Line broadening resulting from motional modulation of the carbon-proton dipolar coupling was analyzed in terms of a chair-chair inversion of the side chain between inequivalent conformations. The results were shown to be in accord with mechanical measurements and conformational energy calculations.

Tékély, Lauprêtre and Monnerie[2] studied blends of a glassy polymer, poly(methyl methacrylate) (PMMA) and a semicrystalline polymer, polyvinylidene fluoride (PVF$_2$) with respect to the precise composition of the amorphous phase, as well as heterogeneities that occur during the PVF$_2$ crystallization process. Proton T$_{1\rho}$ relaxation times (as measured through the resolved carbon resonances) were obtained. Pure PMMA, 40/60 and 50/50 PVF$_2$/PMMA blends exhibited single exponential decays with values of 13.4, 6.0 and 3.7 ms, respectively. It was concluded that within the spatial scale of the experiment the blends appeared homogeneous, the decrease in T$_{1\rho}$ as a function of PVF$_2$ content suggesting that mixing was intimate enough that PMMA relaxation was dominated by interaction with the nearby PVF$_2$, or that molecular motion was altered by blending. For 60/40 and 70/30 PVF$_2$/PMMA blends the magnetization decay exhibits two components with the long component T$_{1\rho}$ being very similar to that found for pure PMMA. The authors suggested that due to PVF$_2$ crystallization during sample quenching in these high PVF$_2$ content samples, the chains of PMMA that were intermixed before crystallization were subsequently rejected from the crystalline regions. In the amorphous phase next to the growing crystals this would lead to domains with a much higher content of PMMA chains as compared with the intimate mixing of the bulk amorphous phase. This premise was tested and verified by annealing and variable-cooling-rate experiments.

Kitamaru, Horii, and Murayama have reported ^{13}C spin-lattice relaxation time analyses of bulk and solution-crystallized polyethylene.[3, 4] They identified four components in the magnetization decay of the bulk crystallized samples, with three attributed to the orthorhombic crystalline phase and one attributed to the noncrystalline phase. (Some samples exhibited a resonance at 34.4 ppm corresponding to the monoclinic crystalline structure in polyethylene.) For instance, for the high molecular weight bulk crystallized sample (M$_v$ = 3×10^6) the relaxation times reported were 2560, 263, 1.7 and 0.37 s, the latter value corresponding to the interfacial and rubbery components of the noncrystalline phase. The authors also noted that the noncrystalline phase was characterized by two carbon spin-spin relaxation times with values of 0.044 ms and 2.4 ms (although only one T$_1$ value is found for this region at 31 ppm). The former T$_2$ value was assigned to the interfacial region and the latter to the rubbery region.

References

A1. Lauprêtre, F., Virlet, J., Bayle, J-P., *Macromolecules* **1985**, *18*, 1846.
A2. Tékély, P., Lauprêtre, F., Monnerie, L., *Polymer* **1985**, *26*, 1081.
A3. Kitamaru, R., Horii, F., Murayama, K., *Macromolecules*. In press.
A4. Kitamaru, R., 1984 International Chemical Congress of Pacific Basin Societies, Honolulu, Hawaii, Dec. 16–21, paper 10B05.

6

CONFORMATIONAL ANALYSIS OF POLYMERS BY SOLID-STATE NMR

David E. Axelson

ENERGY, MINES AND RESOURCES CANADA
CANADA CENTER FOR MINERAL AND ENERGY TECHNOLOGY
EDMONTON COAL RESEARCH CENTER
P.O. BAG 1280, DEVON, ALBERTA, CANADA, T0C 1E0

Introduction

Although relatively few papers have been published relating specifically to the ^{13}C cross-polarization, magic angle spinning (CPMAS) solid-state NMR characterization of conformation in synthetic polymers, there has been a growing literature concerning this topic, including polypeptides;[1-15] cellulose;[16, 17] silk fibroin;[18, 19] oxygenated hexalin, octalin, and decalin;[20] p-alkoxybenzoic acids;[21] hydroxybenzaldehyde;[22] and tetrahydronaphthoquinones.[23] Although these studies are not discussed in this chapter, they do contain important information concerning the sensitivity of solid-state NMR to various important structural and morphological features that will at the very least form the data base for future work in this area.

A number of phenomena may induce chemical-shift changes in the solid-state including molecular packing, hydrogen bonding, changes in bond angles and bond lengths, and anisotropic bulk susceptibility. These factors in some way also may relate to conformational changes. Conformationally related shift variations generally are reflected through two effects, the γ-gauche effect and the vicinal gauche effect. In the first instance, assuming a rotational isomeric

Komoroski (ed): High-Resolution NMR Spectroscopy of Synthetic Polymers in Bulk

state (RIS) model with three conformations for each bond, there are two magnetically distinguishable γ positions, the trans (or anti) and the gauche (Figure 6-1). Replacement of a trans by a gauche position leads to an upfield shift, the magnitude of which depends on the type of carbon involved, the number, and the relative orientations. With respect to the second phenomenon, the conformation of the α bonds affects the chemical shift if the positions of the α and β carbons are changed with respect to one another. It remains to be determined whether these parameters are sufficient, or to what extent they may be invoked, to rationalize structure–shift correlations. Examples of various calculations reported to date are used to illustrate this aspect of the research.

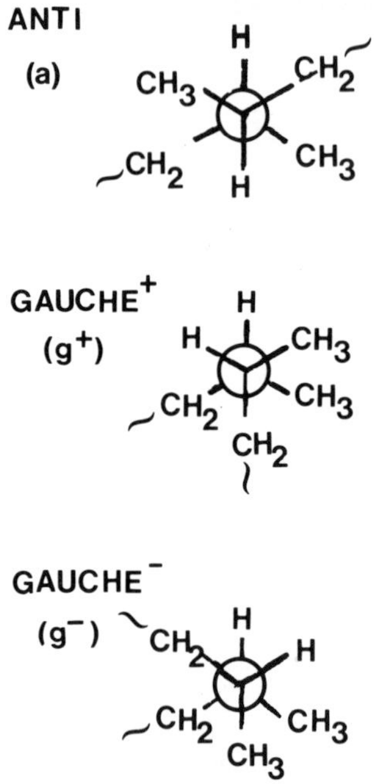

ANTI

(a)

GAUCHE$^+$

(g$^+$)

GAUCHE$^-$

(g$^-$)

Figure 6-1. Newman projections showing anti (trans), gauche-$(+)$, and gauche-$(-)$ conformations.

There are several areas of interest that can be probed by careful consideration of the chemical-shift changes induced by changes in conformation, including comparison of NMR and X-ray diffraction data to confirm a structure; characterization of conformation in complex multiphase systems (eg, semicrystalline polymers); and the study of the effects of stress, orientation, drawing, thermal history, and crystallization conditions on phase structure.

As a consequence of restricted internal mobility in molecules in the solid state, nuclei at different conformational sites but identical in other respects can give different signals. The analysis is not limited to crystalline regions. Signals of different conformers are resolved if the exchange is slow in comparison with the time scale of the NMR experiment. Temperature-dependent measurements offer a means of controlling or influencing this factor. Resolution enhancement experiments based on differences in relaxation offer additional benefits yet to be explored.

Polypeptides provide excellent evidence for the diagnostic ability of ^{13}C solid-state NMR in correlating structure and chemical-shift variations. For example, the chemical shifts of some carbons shift by as much as 7 ppm between a right-handed α-helix and the β-sheet forms as a consequence of variations in the dihedral angles describing bond orientation.[5] Kricheldorf and Muller[10] also have shown that quantitative analysis of α helices can be obtained in the presence of antiparallel pleated-sheet structures, the poly(glycine II) structure (3_1 helix) in the presence of poly(glycine I) (pleated sheet), and the poly(proline) I structure (10_3 helix) in the presence of poly(proline) II (3_1 helix).

Semicrystalline polymers provide a fertile area for study because of the differences in ordering and nature of the phases within a given sample. Chemical-shift differences between phases have been observed for a number of polymers. (See Chapter 5 for a more detailed discussion of this topic.) For example, the crystalline component of poly(methylene oxide) has an all-gauche conformation, whereas the amorphous phase has a distribution of gauche and trans bonds. Consistent with this, the amorphous resonance is found downfield from the crystalline resonance.[24]

Nuclear magnetic resonance may provide the key to the analysis of conformations in samples that have been the subject of much controversy over the years, ie, solution crystals. Contrary to some models of chain folding, in all cases to date the fold surface carbons of the amorphous phase exhibit the same chemical shifts as those observed for the amorphous carbons in bulk-crystallized samples. This indicates little or no difference in average conformation. However, the differences in chemical shift between the amorphous and crystalline components also have been the subject of study.

Examples of Solid-State NMR and Conformational Analysis

Poly(3-methyl-1-pentene)

Inferences may be made regarding the similarity of chain conformations of polymers in the solid state by careful analysis of chemical shifts and crystal structures known by X-ray analysis. This approach has been applied to the characterization of racemic and optically active isotactic poly(3-methyl-1-pentene) (poly3M1P).[25, 26]

Ferro and Ragazzi[26] described a semiquantitative scheme for the analysis of the chemical shifts in poly(3-methyl-1-pentene) as reported by Sacchi et al.[25] It involved a generalization of the reported substituent chemical-shift analysis formula of Lindeman and Adams[27] combined with terms that account for conformational effects, namely the conformer populations of the rotational states for the bonds of interest (determined on the basis of differences in the various potential energy minima obtained by means of a minimization procedure in Cartesian coordinates without any geometrical constraint). Differences between observed and calculated chemical shifts (averaging 1.4 ppm, Table 6-1) were ascribed to deviations from staggered conformations (an idealized helical, all-staggered structure was assumed) and specific solid-state effects.

TABLE 6-1. COMPARISON BETWEEN THE OBSERVED AND CALCULATED SHIFTS OF POLY[(RS)-3M1P][a]

| Carbon[b] | Pred.[c] | Calculated | | Observed[d] | |
		S	R	S	R
C-5	11.36	—	12.53	—	—	13.7(14.1)[e]	—
C-3'	16.64	12.44	—	17.47	13.7(14.1)[e]	—	17.9
C-4	27.16	29.15	—	21.81	28.9(29.0)[e]	—	22.8
C-1	34.10	—	31.18	—	—	32.4	—
C-3	37.06	—	34.60	—	—	36.7	—
C-2	37.45	34.74	—	37.72	36.7(38.0)[e]	—	40.6

[a] Reprinted with permission from Ferro, D. R.; Ragazzi, M. *Macromolecules* **1984**, *17*, 485. Copyright 1984, American Chemical Society. The designations R and S refer to the optical activity of the monomer. The designation RS indicates a racemic mixture.
[b] —^{1}CH$_2$^{2}CH(^{3}CH($^{3'}$CH$_3$)^{4}CH$_2$^{5}CH$_3$)—.
[c] Values predicted from substituent chemical shift effects reported by Lindeman, L. P.; Adams, J. Q. *Anal. Chem.* **1971**, *43*, 1245.
[d] From reference 25.
[e] Values in parentheses refer to poly[(S)-3M1P].

trans-Polybutadiene

The amorphous and crystalline chemical shifts observed in a solution crystal of trans-polybutadiene (TPBD) were interpreted by Schilling et al in terms of γ-gauche shielding effects and the accepted rotational isomeric state model of unperturbed 1,4-trans-polybutadiene.[28] For the purposes of this summary it is instructive to illustrate the calculations for one of the carbons (as given by Schilling et al). Specifically, the relationship between the amorphous olefinic carbon in fold surfaces and the corresponding crystalline component carbon should be considered.

The gauche arrangement of two olefinic carbons has been estimated to produce a shielding (upfield shift) of 4.15 ppm relative to the trans arrangement by the following arguments. The C-4 and C-7 olefinic carbons in 3,7-decadiene resonate 2.2-ppm upfield from the olefinic carbons in 3-hexene. From Mark's rotational isomeric state model of TBPD, bond 5 in 3,7-

decadiene is expected to be 47% trans and 53% gauche ($\pm$).[29] One therefore expects the full shielding effect of a gauche arrangement of two olefinic carbons to be 2.2/0.53 = 4.15 ppm. Olefinic carbons were found to be shielded by 5.6 ppm as a result of being cis to a methylene carbon three bonds removed. The C-2 olefinic carbon in 2-pentene resonates 1.8-ppm upfield from the olefinic carbons in 2-butene. From the RIS model the =C—C— bond in 2-pentene is 68% skew and 32% cis, so the full shielding effect on an olefinic carbon produced by a methylene carbon in the cis arrangement is expected to be 1.8/0.32 = 5.6 ppm.

An RIS model of TPBD (unperturbed randomly coiling chains) has the —CH$_2$—CH$_2$— bond 53% gauche-($\pm$) and 47% trans and the =CH—CH$_2$— bond 79% exact skew and 21% cis. The net effect of these gauche arrangements with olefinic carbons and cis arrangements with methylene carbons is to produce an expected upfield shift of (0.53 × 4.15) + (0.21 × 5.6) = 3.4 ppm for the amorphous olefinic carbons in TPBD relative to those olefinic carbons in the crystalline region. This compares with the observed value of 1.2 ppm. A similar calculation for the amorphous —CH$_2$— carbon leads to a calculated value of 2.8 ppm as opposed to the observed value of 2.3–2.4 ppm.[28] Additional discussion concerning TPBD appears in Chapter 5.

Polypropylene and Polybutene

The solid-state NMR results of Bunn et al have provided novel confirmation of the helical nature of syndiotactic polypropylene as deduced previously by X-ray diffraction.[30] X-Ray diffraction studies have suggested that the conformation of syndiotactic polypropylene is an involuted helix having the appearance of a figure eight when viewed down the helix axis (Figure 6-2A). Thus, there are two equally probable distinct sites for the methylene carbons, one lying on the axis of the helix and the other on the periphery of the helix, whereas only single sites for the methine and methyl carbons exist. A chemical shift difference of 8.7 ppm (Figure 6-2B) was observed for the methylene carbon, consistent with the fact that the internal methylene carbons have two gauche γ carbons (average γ-gauche shielding = 4.4 ppm), whereas the outer methylenes have two trans γ carbons. Therefore the 39.6-ppm peak was assigned to the internal methylene carbons (Table 5-10).

Bunn et al also investigated the effect of thermal history on the structure of isotactic polypropylene in some detail.[31] Three samples representing the α-crystalline (annealed and quenched) and the β-crystalline forms were found to exhibit differences in chemical shifts and linewidths that were related to differences in crystal structure (see Table 5-11 and Figure 5-7). Isotactic polypropylene adopts a helical conformation in the solid state with three monomer units per turn (3$_1$ helix). The crystalline region of the α form is depicted in Figure 6-3A. The view is a projection along the helix axis with the corners of the triangles representing the positions of the methyl groups. The circular

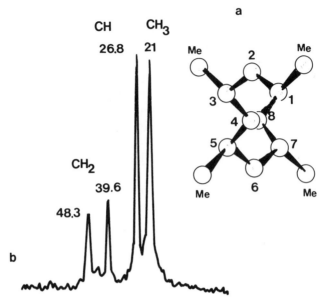

Figure 6-2. (A) View of the conformation of syndiotactic polypropylene looking down the helix axis. The backbone methine and methylene carbons are numbered sequentially, with the methyl carbons labeled Me. The two nonequivalent sites for the methylene carbons are typified by atoms 2 and 6, and 4 and 8, respectively. (B) Carbon-13 spectrum of syndiotactic polypropylene. Number of scans 50,000, contact time 1 ms, recycle delay 1 s, magic angle spinning of 2.3 kHz. (Reprinted with permission from Bunn, A.; Cudby, M. E. A.; Harris, R. K.; Packer, K. J.; Say, B. J. *Chem. Commun.* **1981**, 15, Royal Society of Chemistry.)

arrows indicate the handedness of the individual helices, the structure showing both left- and right-handed pairs in close proximity. The two distinct monomer environments that exist as a result of this interaction are identified by the labels A and B. Each corner of the triangles representing the chains can be equated in symmetry terms to the positions of all three carbon atoms of a monomer unit and so all three carbon resonances can be expected to be split into two lines with relative intensities 2 : 1.

The annealed α form shows resolved splittings for the methyl (22.6; 22.1 ppm) and methylene (45.2, 44.2 ppm) carbons, whereas the asymmetrical lineshape of the methine carbon resonance (26.8 ppm) indicates an unresolved splitting (Figure 5-7). Unlike the large splittings observed in syndiotactic polypropylene arising from inequivalent sites intrinsic to the conformation and the associated γ effect, the small splittings observed in α-isotactic polypropylene are consistent with their being intermolecular in origin. The β form of isotactic polypropylene is thought to differ in the way the helices are packed in the crystalline regions, although it is a 3_1 helix also. The proposed structure (Figure 6-3B) includes the presence of left- and right-handed helices arranged in groups of the same handedness, the packing not allowing the close approach of pairs of helices as found in the α form. As a consequence, inter-chain interactions may well be much smaller in the β form, consistent with the absence of resolvable splittings.

Similar effects have been observed in the study of the solid-state NMR spectra of polybutene (Figure 1-2) obtained under a variety of conditions by Belfiore et al.[32] and which are described in part in Chapter 5 of this book. Of particular interest here is the observation that the chemical shift variations could be rationalized on the basis of changes in conformation, as illustrated by the Newman projections in Figure 6-4. The key to this successful treatment was the utilization of an angle-dependent γ-gauche effect.

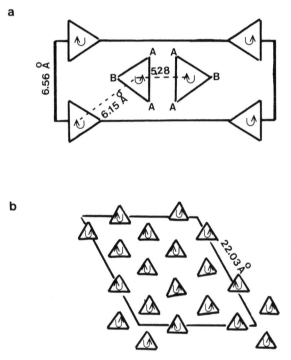

Figure 6-3. (A) Structure of the α form of solid isotactic polypropylene. The figure shows a projection perpendicular to the molecular axis. The triangles represent the molecules that have a 3_1 helical conformation. The arrows represent the handedness of each helix. The labels A and B identify the inequivalent sites discussed in the text, which, because the CH_2—CH bond is almost parallel with the c axis, are applicable to all three types of carbon site, ie, methyl, methine, and methylene. The structure consists of paired helices of opposite handedness with their center-to-center separation of 5.28 Å substantially less than the separation from other helices. (B) Structure of the β form of isotactic polypropylene. The conventions are as for (A). Note the absence of any pairing of helical chains of opposite handedness and the rotation of the three molecules in the bottom right of the unit cell with respect to the others. (Reprinted from Bunn, A.; Cudby, M. E. A.; Harris, R. K.; Packer, K. J.; Say, B. J. *Polymer* **1982**, *23*, 694, with permission of the publishers, Butterworth and Co. Ltd.)

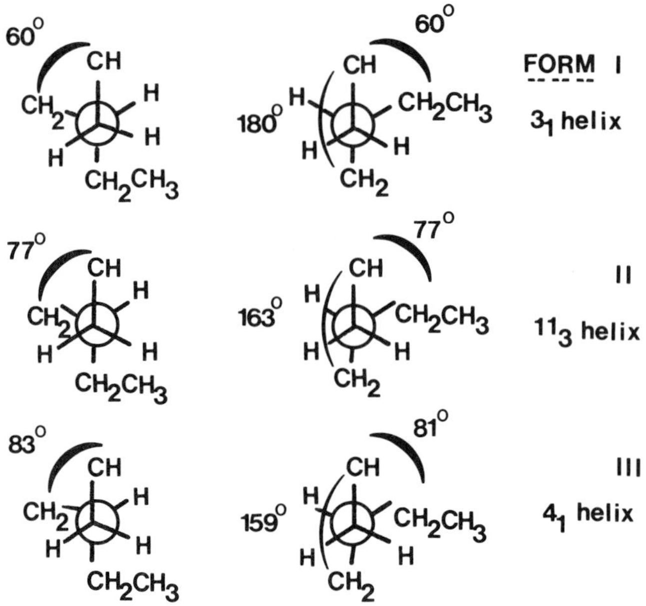

Figure 6-4. Newman projections for the various forms of poly(butene-1). The left side shows projections looking along the C_α—C_β bond in the direction of the gauche bond. The right side shows the projection in the opposite direction, ie, toward the trans bond. (Reprinted with permission from Belfiore, L. A.; Schilling, F. C.; Tonelli, A. E.; Lovinger, A.; Bovey, F. A. *Macromolecules* **1984**, *17*, 2561. Copyright 1984, American Chemical Society.)

Polyethylene and n-Alkane Models

Another interesting example of the effect of crystal structure on chemical shift values has been reported by VanderHart and Khoury.[33] It has been shown that the deformation of polyethylene under certain conditions leads to the formation of a crystallographically distinct, metastable crystalline phase (monoclinic crystalline phase, MCP) in addition to the usual orthorhombic crystalline phase (OCP). Indications of a monoclinic phase have been observed in drawn samples with residual strains in the range 0.2–1.7, samples that had been stretched and kept under tension at constant strain, and (in reduced amounts) in stretched samples that subsequently have been allowed to contract. The formation of this phase is partially reversible and stabilized by the presence of external stresses.[33]

In the monoclinic structure the planes of the zigzag, all-trans chains are parallel to one another, as in n-alkanes in n-C_yH_{2y+2} with $6 \leq y$ (even) < 26, whose unit cells are triclinic and contain one chain per unit cell. In contrast, the angle between the planes of the two all-trans chains that traverse the orthorhombic cell of polyethylene is close to 90°. This is also the situation in the orthorhombic subcell of both the monoclinic and/or the orthorhombic n-alkanes with $26 \leq y$ (even) ≤ 100 and in the orthorhombic n-alkanes with $11 \leq y$ (odd) ≤ 39.[33]

In a previous study VanderHart reported the chemical shift analysis for a series of n-alkanes (n-$C_{19}H_{40}$, n-$C_{20}H_{42}$, n-$C_{23}H_{48}$, n-$C_{32}H_{66}$) with different crystal structures including the pseudo-hexagonal (high-temperature), triclinic, orthorhombic, and monoclinic forms.[34] For all samples except the triclinic, the chemical shifts of the interior methylene carbons were the same. For the triclinic case the methylene carbon shifts were 1.3 ± 0.4 ppm downfield. This was taken to be diagnostic of n-alkanes in which the planes of the zigzag chains are parallel to one another. Therefore, the MCP methylene resonance for polyethylene was expected to be shifted downfield of the order of 1.3 ppm with respect to the orthorhombic phase. This is indeed the case, with the MCP resonance being observed about 1.4 ppm downfield from the OCP resonance.

Cycloalkane Models for Polyethylene

Next, consider in more detail the conformational effects that have been observed in the solid-state NMR characterization of several cyclic and long-chain hydrocarbons. The strategy that has been developed by Cantow et al[35-43] to correlate the NMR chemical shifts with specific conformations in the solid state has consisted of (i) verifying that low molecular weight analogs of the respective polymers were representative models, (ii) freezing out the lowest energy conformer in solution (slow-exchange conditions), and (iii) applying the derived chemical shift of the conformers for the interpretation of the spectra.

Cycloalkanes provide good models for the investigation of chemical-shift changes that arise from conformational effects, because a certain number of carbon–carbon bonds are constrained to gauche conformations by the ring structure. Anet et al originally reported the low-temperature, slow-exchange, spectra for cyclododecane, cyclotetradecane, and cyclohexadecane.[44-46] Ando et al reported the analysis of the CPMAS spectra of cyclic paraffins varying from $C_{24}H_{48}$ to $C_{80}H_{160}$, n-paraffins and crystalline polyethylene.[47] Table 6-2 summarizes their results (obtained at ambient temperature).

While $C_{24}H_{48}$ exhibits a single sharp peak, $C_{28}H_{56}$ and $C_{32}H_{64}$ have broader single peaks, all larger species showing high-field shoulders or small peaks which decrease in intensity as the number of carbons increases. The main signal and shoulders or small peaks are denoted by I, II, and III, respectively. The difference between I and II increases from 4 to 7 ppm as the number of carbons increases.

A methylene carbon appears at high field by 4–6 ppm if any carbon atom three bonds away is in a gauche, rather than a trans, conformation. On this premise it was suggested that peaks I, II, and III could be assigned to the methylene carbons arising from the trans zigzag (I) structure region and the folded structure region (II and III), respectively. Cyclic paraffins crystallize in a conformation characterized by two parallel, all-trans (a) planar zigzag strands

TABLE 6-2. Carbon-13 Chemical Shifts (ppm) of Cyclic Paraffins, n-Paraffins, and Solution-Grown Polyethylene in the Solid State[a]

Sample	Peak I, δ_I	Peak II, δ_{II}	Peak III, δ_{III}	$\delta_I - \delta_{II}$	$\delta_I - \delta_{III}$
Cyclic paraffin					
$C_{24}H_{48}$	30.0				
$C_{28}H_{56}$	29.7				
$C_{32}H_{64}$	30.7				
$C_{36}H_{72}$	34.0	30.0	26.0	4.0	8.0
$C_{40}H_{80}$	33.5	27.0	24.5	6.5	9.0
$C_{48}H_{96}$	33.7	27.0	22.0	6.7	11.7
$C_{64}H_{128}$	33.8	27.0	22.0	6.8	11.8
$C_{80}H_{160}$	34.0	~27	~22	~7	~12
n-Paraffin					
n-$C_{19}H_{40}$	33.4				
n-$C_{22}H_{46}$	33.0				
n-$C_{32}H_{66}$	33.0				
Polyethylene	33.0	30.7		2.3	

[a] Reprinted with permission from Ando, I.; Yamanobe, T.; Sorita, T.; Komoto, T.; Sato, H.; Deguchi, K.; Imanari, M. *Macromolecules* **1984**, *17*, 1955. Copyright 1984, American Chemical Society.

connected by two ggagg loops. Some loop carbons would be expected to resonate one γ effect upfield and other loop carbons two γ effects upfield from the all-trans carbons (and those loop carbons whose next nearest neighbor bonds are also both trans). For $C_{36}H_{72}$ one would expect 24 carbons with no γs, eight carbons with one γ, and four carbons with two γs, ie, a ratio of 6 : 2 : 1 in peak intensities. The measured ratio was 5.8 : 1.7 : 1. Similarly, for $C_{40}H_{80}$ the expected ratio from this type of analysis would be 7 : 2 : 1 as compared with the measured ratio of 7.2 : 1.8 : 1. The agreement is also good for the molecules $C_{48}H_{96}$ (calculated 9 : 2 : 1, measured 8.5 : 2.3 : 1) and $C_{64}H_{128}$ (calculated 13 : 2 : 1, measured 14 : 2.6 : 1).

Choosing a γ effect of 4–6 ppm results in calculated shifts for $C_{36}H_{72}$ of 34 (I), 28–30 (II), and 22–26 ppm (III). The chemical shifts for peak I in the C_{24}, C_{28}, and C_{32} alkanes appear at higher field by 3–4 ppm than those for C_{36}, C_{40}, C_{48}, and C_{80} and n-paraffins. This was attributed to the observation that the former series of molecules are conformationally more mobile at room temperature.

Information regarding trans–gauche populations can be derived from the equation: $\delta_{II} = \delta_0 - 2\gamma f_g$, where f_g is the equilibrium fraction of gauche bonds, δ_0 is the chemical shift for peak I in cyclic paraffins and n-paraffins or that for the all trans zigzag structure in polyethylene, and the value of γ is attributed entirely to the trans to gauche change. The probability that any internal bond in a long n-paraffin in the liquid state is gauche is 0.357. Each carbon would therefore experience $2(0.357) \times \gamma = 0.714$ γ-gauche interactions and would resonate 2.9–4.3 ppm upfield from its solid-state, all-trans conformation. Therefore the noncrystalline or fold carbons in polyethylene should appear upfield from the crystalline resonance. If the fold carbons are sharply folded, relatively immobile, structures in a lozenge-shaped polyethylene, they should appear 5–10 ppm upfield from the all-trans carbons, as shown in the cyclic paraffins. The observed shift is 2.3 ppm. Therefore, the fold carbons are not in a sharply folded structure but are in the mobile state with slightly less gauche character than truly amorphous or molten polyethylene.

Sorita et al have reported additional data relating to saturated hydrocarbon chemical shifts in the solid state.[48]

Möller and colleagues[35] have examined the CPMAS, variable-temperature behavior of the chemical shifts of cyclododecane, $(CH_2)_{12}$; cyclotetraeicosane, $(CH_2)_{24}$; and cyclohexatriacontane, $(CH_2)_{36}$ (Table 6-3). The 298 K CPMAS spectrum of cyclododecane exhibited a single resonance at 25.72 ppm which in the slow-exchange limit (≤ 155 K) split into two resonances of relative intensities 1 : 2 with shifts of 28.73 ppm and 24.40 ppm, respectively. An unambiguous assignment of the signals was made on the basis of the X-ray diffraction results, which indicated a conformation of the type (gag)$_4$. There are two different conformational sites for the carbons: four carbons in the corners of the molecule are within the center of an agga sequence; the eight carbons in between are within a gagg sequence. This is in agreement with the observed relative intensities and also allows the signal at 28.73 to be assigned to the carbons within the agga sequence and the signal at 24.40 to be assigned to the center of gagg segments.

Cyclotetraeicosane (Figure 6-5) exhibits a single resonance at 31.32 ppm at 299 K because all carbon atoms occupy all conformational sites in the time average with equal probability. The spectrum splits into five peaks at 248 K with chemical shifts of 35.4, 30.6, 28.9, 28.1, and 23.6 ppm with relative intensities of 2 : 1 : 1 : 1 : 1, respectively. The structure of this strained molecule is that of two all-anti strands bridged by two ggagg loops (Figure 6-6). Five conformational sequences can be identified: eight carbons are in the center of an aaaa segment, four carbons at a time are in the center of an agga, an aagg, a gagg, and an aaag segment. However, the resonances could not be explained solely on the basis of γ-gauche interactions.

The use of cyclic hydrocarbons as models for polymer conformation must be tempered by the fact that ring molecule chemical-shift observations do not necessarily agree exactly with linear molecule shifts for the same sequences. Consider, for example, that the aaaa signal of the cycloalkanes differs by 1.8 ppm from the all-anti segments of the crystalline component of polyethylene.

Head-to-Head Polypropylenes

Head-to-head–tail-to-tail polypropylenes [poly(1,2-dimethyltetramethylene)s (DMTM)] have been the subject of intensive study of the rotational isomeric states within the solid state by Cantow and coworkers.[37–43] The advantages offered by these polymers include the facts that they exhibit two centers of asymmetry and three types of nonequivalent bonds CH—CH, CH—CH$_2$, and CH$_2$—CH$_2$, leading to a range of configurations (Figure 6-7).

TABLE 6-3. CARBON-13 CHEMICAL SHIFTS (ppm) OF CYCLOALKANES[a]

Conformational sites	(CH$_2$)$_{36}$[b]	(CH$_2$)$_{24}$[b]	(CH$_2$)$_{16}$[c,d]	(CH$_2$)$_{14}$[c,d]	(CH$_2$)$_{12}$[b]	(CH$_2$)$_{12}$[c]	(CH$_2$)$_{12}$[c,d]
gagg	23.8	23.6		21.3	24.4	22.1	21.8
	23.6						
gaag	27.6	28.1	22.8	23.3			
ggaa	28.5	28.9	26.8	26.2	28.7	26.8	26.6
agga	30.2	30.6	27.0	26.8			
gaaa							
aaaa	35.5	35.4					

[a] Reprinted with permission from Möller, M.; Gronski, W.; Cantow, H.-J.; Höcker, H. *J. Am. Chem. Soc.* **1984**, *106*, 5093.
[b] Solid state.
[c] In solution.
[d] Data from references 44–46.

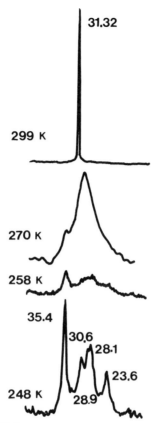

Figure 6-5. Carbon-13 CPMAS spectra of cyclotetraeicosane at various temperatures. (Reprinted with permission from Möller, M.; Gronski, W.; Cantow, H.-J.; Höcker, H. *J. Am. Chem. Soc.* **1984**, *106*, 5093. Copyright 1984, American Chemical Society.)

Several aspects of this important work are noted here in order to illustrate the tremendous potential of solid-state NMR in this area.

Investigations of ditactic poly(1,2-dimethyltetramethylene)s and their low molecular weight analogs have been reported; it has been concluded that meso-4,5-dimethyloctane is the appropriate model for erythrodiisotactic poly(1,2-dimethyltetramethylene). The temperature-dependent populations of the conformers in the macromolecule were related to the fast-exchange chemical shifts, in agreement with slow-exchange data measured on the low molecular weight models.[37-39]

The CPMAS spectrum of semicrystalline erythrodiisotactic poly(1,2-dimethyltetramethylene) is shown in Figure 6-8.[40-43] Peaks from the amorphous regions are found slightly upfield from the resonances in the solution spectrum, 38.0 for CH, 32.74 for CH_2, and 16.88 ppm for CH_3 (Table 5-13). The peak at 30.18 ppm was assigned to the CH_2 sequences. Crystalline-region signals are found at 41.15, 40.8, 36.3, 27.85, 20.78, and 12.74 ppm. The signals

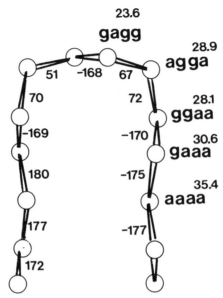

Figure 6-6. Conformational model of cyclotetraeicosane from X-ray diffraction data. Only the top part of the cyclic compound is shown here along with bond angles and the specific conformations about certain bonds. (Reprinted with permission from Möller, M.; Gronski, W.; Cantow, H.-J.; Höcker, H. *J. Am. Chem. Soc.* **1984**, *106*, 5093. Copyright 1984, American Chemical Society.)

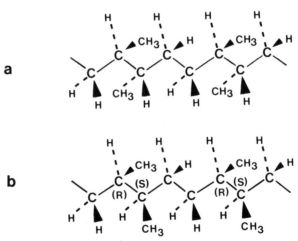

Figure 6-7. Poly(1,2-dimethyltetramethylene) in all-anti conformation: (A) erythro-diisotactic chain, (B) threodiisotactic chain. (Reprinted from Gronski, W.; Möller, M.; Cantow, H.-J. *Polym. Bull.* **1982**, *8*, 503, with permission of the publisher, Springer-Verlag.)

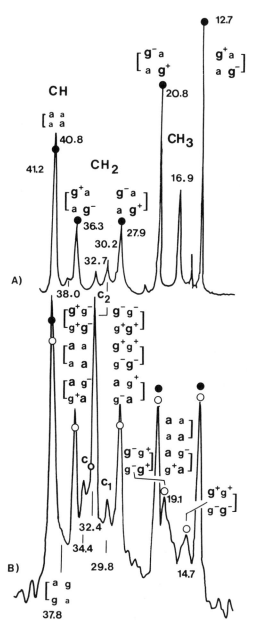

Figure 6-8. Solid-state ^{13}C CPMAS spectra of erythrodiisotactic poly(1,2-DMTM) (A) spectrum at 303 K, 75.42 MHz; (B) spectrum at 233 K, 25.14 MHz. ● crystalline component, ○ glassy component. (Reprinted from Gronski, W.; Möller, M.; Cantow, H.-J. *Polym. Bull.* **1982**, *8*, 503, with permission of the publisher, Springer-Verlag.)

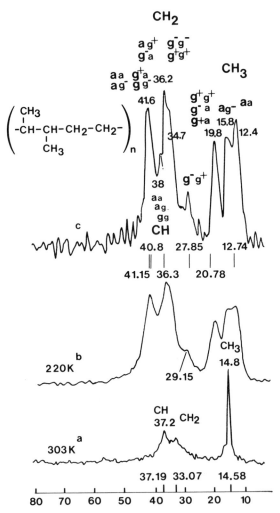

Figure 6-9. Solid-state ^{13}C CPMAS spectra of amorphous threodiisotactic poly(1,2-DMTM). (A) Spectrum at 303 K, 75.42 MHz; chemical shifts of the polymer in CDCl$_3$ solution are marked by vertical lines. (B) Spectrum at 220 K, 25.14 MHz. (C) Resolution-enhanced spectrum of (B) with assignment of conformations (big letters for the CH—CH bond, medium for the CH—CH$_2$, and small letters for the CH$_2$—CH$_2$ bonds); the vertical lines mark the chemical shifts of the equivalent carbons in the g^+aaag$^-$aaag$^+$ conformation of the crystalline phase of erythrodiisotactic poly(1,2-DMTM). (Reprinted from Gronski, W.; Hasenhindl, A.; Limbach, H. H.; Möller, M.; Cantow, H.-J. *Polym. Bull.* **1981**, *6*, 93, with permission of the publisher, Springer-Verlag.)

at 36.3 and 27.85 are shifted symmetrically upfield and downfield versus the CH_2 resonance of the methylene carbon within the amorphous phase, 32.74 ppm. Similarly, the crystalline component CH_3 signals, 20.78 and 12.74, ppm center the amorphous phase signal at 16.88 ppm. They show the same splitting pattern as the CH_2 and CH_3 signals of meso-4,5-dimethyloctane for the gauche conformation of the CH—CH bond in the slow-exchange solution spectrum.

The signals at 41.15 and 40.8 ppm were assigned to the crystalline component CH carbons. Their shift relative to the CH amorphous carbon resonance position of 38 ppm results only in part from rotational isomerism of the CH—CH bond, which does not influence the position of the C atoms situated γ with respect to these nuclei. The γ-gauche effects for the methine atoms are caused by the rotational isomerism around the CH—CH_2 and CH_2—CH_2 bonds.

Simultaneous downfield shifts of both the CH signals was interpreted as arising from the anti position of both of these bonds. The NMR spectrum demonstrates that $(g^{\pm}aaa)_n$ is the conformation of the chain within the crystalline region. It is plausible to argue that the g^+ and g^- positions must alternate within the sequence in order to achieve an ordered lattice.[40–43]

A similar analysis of the chemical shifts observed in amorphous threodiisotactic poly(1,2-dimethyltetramethylene) has been reported[40–43] (Figure 6-9). Consider the assignment of the methyl resonances. The methyl carbon exhibits three resonances. In principle there are nine conformational pairs of the CH—CH and the CH—CH_2 bonds, although some are of low statistical probability. The g^- conformation of the CH—CH bond involves energetically unfavorable syn-axial positions of the two methyl groups. Near T_g the fraction of g^- conformation is estimated to be 0.08 as compared with the more likely a and g^+ conformations characterized by probabilities of 0.63 and 0.29, respectively. (These estimates were based on ^{13}C analysis of d,l-4,5-dimethyloctane under slow-exchange conditions.) The energy difference between the g^+ and a conformations was estimated to be 1.6 kJ mol^{-1} (4 kJ mol^{-1} between g^- and a). The high-field resonance was assigned to the aa conformation as it involves a double γ-gauche interaction from one side and a simple γ-gauche interaction from the other. The conformational change from aa to ag$^-$ involves a transition of a γ-CH_2 group from gauche to anti, leading to a shift to lower field by about 3.4 ppm. The shift of 7.4 ppm from aa to $g^{\pm}a$ corresponds to a double γ-gauche interaction being replaced by a simple γ-gauche interaction.

Summary

High-resolution ^{13}C solid-state spectra obtained with magic angle spinning and dipolar decoupling for resolution enhancement and cross polarization for sensitivity enhancement have proved sensitive to a variety of inter- and intramolecular interactions, conformational changes, crystal packing effects, bond angle distortions, and thermal history effects. The additional information provided by careful variable-temperature studies should enable a detailed understanding of chemical shifts and the nature of the associated interactions for particular rotational isomeric states.

References

1. Mueller, D.; Stulz, J.; Kricheldorf, H. R. *Makromol. Chem.* **1984**, *185*, 1739.
2. Shoji, A.; Ozaki, T.; Saito, H.; Tabeta, R.; Ando, I. *Macromolecules* **1984**, *17*, 1472.
3. Ando, I.; Saito, H.; Tabeta, R.; Shoji, A.; Ozaki, T. *Macromolecules* **1984**, *17*, 457.
4. Kricheldorf, H. R.; Muller, D. *Polym. Bull.* **1983**, *10*, 513.
5. Saito, H.; Tabeta, R.; Ando, I.; Ozaki, T.; Shoji, A. *Chem. Lett.* **1983**, 1437.
6. Flippen-Anderson, J. L.; Gilardi, R.; Karle, I. L.; Frey, M. H.; Opella, S. J.; Gierasch, L. M.; Goodman, M.; Madison, V.; Delaney, N. G. *J. Am. Chem. Soc.* **1983**, *105*, 6609.
7. Kessler, H.; Bermel, W.; Forster, H. *Angew. Chem.* **1982**, *94*, 703.
8. Muller, D.; Kricheldorf, H. R. *Polym. Bull.* **1981**, *6*, 101.
9. Taki, T.; Yamashita, S.; Satoh, M.; Shibata, A.; Yamashita, T.; Tabeta, R.; Saito, H. *Chem. Lett.* **1981**, 1803.
10. Kricheldorf, H. R.; Muller, D. *Macromolecules* **1983**, *16*, 615.
11. Kricheldorf, H. R.; Muller, D. *Int. J. Biol. Macromol.* **1983**, *5*, 171.
12. Saito, H.; Tabeta, R.; Shoji, A.; Ozaki, T.; Ando, I. *Macromolecules* **1983**, *16*, 1050.
13. Kricheldorf, H. R.; Muller, D. *Polym. Bull.* **1982**, *8*, 495.
14. Kricheldorf, H. R.; Muller, D.; Forster, H. *Polym. Bull.* **1982**, *8*, 487.
15. Tonelli, A. E. *Biopolymers* **1984**, *23*, 819.
16. VanderHart, D. L.; Atalla, R. H. *Macromolecules* **1984**, *17*, 1465.
17. Horii, F.; Hirai, A.; Kitamaru, R. *Polym. Bull.* **1983**, *10*, 357.
18. Saito, H.; Tabeta, R.; Asakura, T.; Iwanaga, Y.; Shoji, A.; Ozaki, T.; Ando, I. *Macromolecules* **1984**, *17*, 1405.
19. Saito, H.; Iwanaga, Y.; Tabeta, R.; Narita, M.; Asakura, T. *Chem. Lett.* **1983**, 427.
20. Ariel, S.; Scheffer, J. R.; Trotter, J.; Wong, Y. F. *Tetrahedron Lett.* **1983**, *24*, 4555.
21. Hays, G. R. *J. Chem. Soc., Perkin Trans. 2* **1983**, 1049.
22. Imashiro, F.; Maeda, S.; Takegoshi, K.; Terao, T.; Saika, A. *Chem. Phys. Lett.* **1983**, *99*, 189.
23. McDowell, C. A.; Naito, A.; Scheffer, J. R.; Wong, Y. F. *Tetrahedron Lett.* **1981**, *22*, 4779.
24. Cholli, A. L.; Ritchey, W. M.; Koenig, J. L. *Spectrosc. Lett.* **1983**, *16*, 21.
25. Sacchi, M. C.; Locatelli, P.; Zetta, L.; Zambelli, A. *Macromolecules* **1984**, *17*, 483.
26. Ferro, D. R.; Ragazzi, M. *Macromolecules* **1984**, *17*, 485.
27. Lindeman, L. P.; Adams, J. Q. *Anal. Chem.* **1971**, *43*, 1245; see also Dalling, D. K.; Grant, D. M. *J. Am. Chem. Soc.* **1974**, *96*, 1827, for a discussion of the origin of the γ effect.
28. Schilling, F. C.; Bovey, F. A.; Tonelli, A. E.; Tseng, S.; Woodward, A. E. *Macromolecules* **1984**, *17*, 728.
29. Mark, J. E. *J. Am. Chem. Soc.* **1967**, *89*, 6829.
30. Bunn, A.; Cudby, M. E. A.; Harris, R. K.; Packer, K. J.; Say, B. J. *Chem. Commun.* **1981**, 15.
31. Bunn, A.; Cudby, M. E. A.; Harris, R. K.; Packer, K. J.; Say, B. J. *Polymer* **1982**, *23*, 694.
32. Belfiore, L. A.; Schilling, F. C.; Tonelli, A. E.; Lovinger, A.; Bovey, F. A. *ACS Polym. Prepr.* **1984**, *25*, 351; *Macromolecules* **1984**, *17*, 2561.
33. VanderHart, D. L.; Khoury, F. *Polymer* **1984**, *25*, 1589.
34. VanderHart, D. L. *J. Magn. Reson.* **1981**, *44*, 117.
35. Möller, M.; Gronski, W.; Cantow, H.-J.; Höcker, H. *J. Am. Chem. Soc.* **1984**, *106*, 5093.
36. Emeis, D.; Cantow, H.-J.; Möller, M. *Polym. Bull.* **1984**, *12*, 557.
37. Möller, M.; Ritter, W.; Cantow, H.-J. *Polym. Bull.* **1981**, *4*, 609.
38. Möller, M.; Cantow, H.-J. *Polym. Bull.* **1980**, *3*, 579.
39. Möller, M.; Ritter, W.; Cantow, H.-J. *Polym. Bull.* **1980**, *2*, 543.
40. Möller, M.; Cantow, H.-J. *Polym. Bull.* **1981**, *5*, 119.

41. Gronski, W.; Hasenhindl, A.; Limbach, H. H.; Möller, M.; Cantow, H.-J. *Polym. Bull.* **1981**, *6*, 93.
42. Gronski, W.; Möller, M.; Cantow, H.-J. *Polym. Bull.* **1982**, *8*, 503.
43. Möller, M. *Adv. Polym. Sci.* **1985**, *66*, 60.
44. Anet, F. A.; Cheng, A. K.; Wagner, J. J. *J. Am. Chem. Soc.* **1972**, *94*, 9250.
45. Anet, F. A. L.; Cheng, A. K. *J. Am. Chem. Soc.* **1975**, *97*, 2420.
46. Anet, F. A. L.; Rowdah, T. N. *J. Am. Chem. Soc.* **1978**, *100*, 7166.
47. Ando, I.; Yamanobe, T.; Sorita, T.; Komoto, T.; Sato, H.; Deguchi, K.; Imanari, M. *Macromolecules* **1984**, *17*, 1955.
48. Sorita, T.; Yamanobe, T.; Komoto, T.; Ando, I.; Sato, H.; Deguchi, K.; Imanari, M. *Makromol. Chem., Rapid Commun.* **1984**, *5*, 657.

7

POLYMER MOTION IN THE SOLID STATE

Alan A. Jones

DEPARTMENT OF CHEMISTRY
CLARK UNIVERSITY
WORCESTER, MA 01610

Introduction

Nuclear magnetic resonance provides a new tool for characterizing chain dynamics in bulk polymers. It complements other experimental approaches, such as dielectric and dynamic mechanical response, with the advantage of greater structural specificity versus these more traditional approaches. Nuclear magnetic resonance can probe motion at several places in a repeat unit by utilizing either one or several experiments. Carbon-13 experiments often can probe several sites[1] simultaneously, whereas systems labeled with either carbon-13[2] or deuterium[3,4] (see Chapter 10) probe only one site per labeled polymer. Even proton spectroscopy can be used to probe specific sites under favorable circumstances.[2,5,6]

If the goal is to develop a repeat-unit-level characterization of local motion, certain aspects of the motion must be defined. Arbitrarily, one can start with the geometry of the motion. Is it isotropic? If not, is there an axis of anisotropic rotation? Does the motion take place by large-angle jumps between well-defined minima or does it take place by stochastic diffusion over a relatively flat potential surface? If the motion is anisotropic, is it restricted in angular amplitude at low temperatures, becoming a complete anisotropic rotation about a given axis at higher temperatures?

A second aspect of the motion to be characterized is time scale. In the time

Komoroski (ed): High-Resolution NMR Spectroscopy of Synthetic Polymers in Bulk

domain this is best summarized by a correlation function and in the frequency domain by a spectral density. In this regard, polymer motions are often complex, with two general factors contributing to this complexity. First, motions taking place in an isolated polymer backbone are complex relative to small molecules because of the nature of a chain system. Conformational changes, such as conformational exchange in a chain molecule, lead to Bessel function correlation functions[7-8] as opposed to an exponential correlation function for rotational diffusion in a small molecule. The second factor leading to complexity in solid polymers is intermolecular interactions. In bulk polymers, especially glasses,[9-11] these interactions can lead to correlation functions characterized in terms of very broad distributions of exponential correlation times.

Another aspect of the motion is the temperature dependence or energetics. As just mentioned, the amplitude of various motions can change with temperature, although more commonly one thinks of changes in the time scale.[12] The time scale of a given motion is often found to have an Arrhenius dependence on temperature; and therefore this dependence can be summarized conveniently by an apparent activation energy.

A last issue is often extremely important to the materials scientist studying chain dynamics. After motions are detected and characterized by NMR, a relationship to the results of other dynamics experiments is sought as well as a relationship to material properties. For instance, the glass transition is an important concept in the study of polymeric solids. This transition can be observed by NMR[13] and the influence of this transition on local motions can be observed also (see Chapter 4). In the end, an improved understanding of local motion above and below the glass transition can serve to better define the nature of the glass transition itself.

Experimental Techniques for Characterizing Local Motions

Spin–Lattice Relaxation

Spin–lattice relaxation measurements have been used widely to characterize chain dynamics in solution[14] and the same general approach can be extended to bulk polymers,[1] although some more difficulties are encountered. Several points should be considered in designing a spin–lattice relaxation study.

First, each spin–lattice relaxation time samples motion at one frequency or one combination of frequencies. In a pure spin–lattice measurement, the frequencies are related to the Larmor frequencies of the nuclei involved.[13] If spin–lattice relaxation in the rotating frame is measured, the relevant frequency is given by the magnitude of the spin-lock field[13] (see Chapter 2).

To vary the circumstances of the experiment, temperature commonly is swept, which changes the nature of the motion sampled by the spin–lattice relaxation experiment. Whereas temperature variation is a valuable experimental control, it is even more useful to vary the frequency of the measurement.[15,16] Here a drawback to the NMR approach is encountered. To vary frequency in a spin–lattice relaxation measurement, the magnetic field must be varied. With common commercial instrumentation, magnetic fields are fixed so an individual spectrometer is required for each frequency. Another possible approach to varying frequency is to study different nuclei such as protons and carbon in the same polymer[17] because they have different gyromagnetic ratios and therefore different Larmor frequencies at the same field strength. For spin–lattice relaxation in the rotating frame the spin-lock field can be varied, but frequently the range of useful variation only extends over a factor of two or three.[18] However, relaxation in the rotating frame nicely complements normal spin–lattice relaxation because the former is in the kilohertz region and the latter is in the megahertz region.

The particular experimental approach used to observe spin–lattice relaxation in bulk polymers depends on the state of the polymer. The easiest experimental situation to study is a rubber at a temperature somewhat above the glass transition[19] (see Chapter 4). Here ^{13}C spectra can be obtained with low-power or scalar proton decoupling, allowing for relaxation studies almost entirely analogous to solution studies. The presence of nearly isotropic segmental motion[12] removes dipolar couplings and chemical-shift anisotropy, leaving a spectrum comparable to a dissolved macromolecule. A standard measurement of spin–lattice relaxation time, T_1, by a 180-τ-90 pulse sequence as a function of temperature and Larmor frequency should be sufficient to characterize local motion in a rubber.

In a glassy polymer at least two general types of experimental approaches may be employed. The first is wide-line relaxation studies. Here the best system is one where only one nucleus contributes to the spectrum because the presence of several nuclei leads to signal overlap. One therefore must first select a polymer system where only one chemical position contributes to the spectrum, and this is usually achieved through labeling, although occasionally a judicious choice provides an appropriate case.[6] In typical wide-line relaxation studies, the lineshape is governed by the orientation of the molecule relative to the externally applied field. The sources of this orientation dependence may be dipolar coupling, quadrupole coupling, or chemical-shift anisotropy. In proton dipolar studies, relaxation of all points in a lineshape are equal because of spin diffusion (homogeneous broadening).[6] However for deuterium quadrupolar and ^{13}C chemical shift anisotropy cases, relaxation of the lineshape changes across the lineshape and can depend on the orientation of the molecule relative to the externally applied field (heterogeneous broadening). Exploitation of this latter situation has just begun.

High-resolution ^{13}C spectra can be obtained in glassy polymers through a combination of cross-polarization, high-power dipolar decoupling and magic angle spinning[1] (see Chapters 2 and 3). The resulting spectra are almost comparable to liquid spectra in resolution, so studies of unlabeled systems in natural abundance are straightforward. Both spin–lattice relaxation and spin–lattice relaxation in the rotating frame can be observed. However several complicating factors are often present. The return of the magnetization to equilibrium is often nonexponential. Several possible sources of the nonexponential behavior have been proposed, including a distribution of correlation times[1] and spin–spin relaxation contributions.[18,20] In any case, ^{13}C relaxation times in glassy solids have not been susceptible to quantitative interpretation with a correlation function comparable to the interpretations developed for relaxation of dissolved polymers. Qualitative interpretations and comparisons of mobility between polymers is possible and has proved most useful in a number of cases.[1]

In any application of the spin–lattice relaxation approach, the information generated primarily reflects the time scale of the motion. Some information about the geometry of the motion can be determined by performing spin–lattice relaxation experiments at several positions in the repeat unit.[17] Some information about the amplitude of anisotropic restricted rotation can be ascertained also.[21] However, in solid polymers the best information about the geometry of local motions comes from the study of lineshape.

Lineshape

Examination of lineshape and linewidth also provides insight into chain dynamics. Following the analogy to solution studies, one might consider employing the spin–spin relaxation time, T_2, as a source of information. In the rubbery state, T_2 can be determined from the linewidth but careful examination by several authors shows this relaxation effect to be dominated by long-range motions and not by local motion.[12,22,23] This situation is also true for concentrated solutions and interpretations based on motions, such as the Rouse-Zimm modes and reptation, have been proposed.[24] The validity of these interpretations has been challenged[12] but it is clear that linewidths or other measures of spin–spin relaxation are not sources of information about local chain dynamics. This view is presented here in spite of early work by this author[25] and others[19] attempting to use spin–spin relaxation as a source of information on local chain dynamics in rubbers.

In glassy polymers under conditions of magic angle spinning and dipolar decoupling, ^{13}C linewidths are also not a function of local motions. Here the linewidth results from isotropic chemical-shift heterogeneity, which reflects the heterogeneity of the glass itself.[1] In cases where chemical exchange occurs, local chain dynamics can be studied in a manner comparable to chemical exchange in small, dissolved molecules.[26] The major difference between the glassy polymer and the small molecule in solution is again in the complexity of the correlation function reflecting the strong intermolecular interactions in the glass. High-resolution lineshape studies on phenomena loosely categorized as chemical exchange yield primarily information about the time scale of the chain dynamics, although some inferences about geometry can be drawn by considering the observed lineshape changes.

The best information about the geometry of motions below the glass transition comes from wide-line NMR. Until recently the only wide-line spectra commonly obtained were those of proton and fluorine nuclei. Proton spectra of glassy solids are always dominated by dipolar interactions. The presence of this strong coupling in addition to a very limited chemical-shift range yields spectra with extensive overlap of resonances from chemically distinct points in the repeat unit. Under such circumstances, very little geometric information is available. In certain cases only one type of proton is present in a repeat unit and then geometric information can be extracted.[2,5] Fluorine spectra also are dominated by dipolar interactions in the solid state, but the chemical-shift anisotropy is large and can be observed more easily under coherent averaging conditions[27].

Major advances in solid-state ^{13}C[2,28] and deuterium[3,4] spectroscopy have made detailed geometric information available. Depending on the experimental approach, the lineshape reflects either chemical-shift anisotropy,[2] quadrupole,[3,4] or dipolar interactions.[2,5] Whatever the interaction, the observed lineshape depends on the orientation of the repeat unit with the externally applied field. If molecular motion is present, orientations are exchanged and the lineshape becomes partially averaged when the motion becomes rapid with respect to the frequency separation between the points in the lineshape under exchange.[29] Lineshape changes are very distinctive and provide a fairly decisive test of various motional possibilities.

The time scale of the motion also can be examined when the rate of the motion is comparable to the frequency of the lineshape.[30] The analysis of the lineshape is quite comparable to the analysis of chemical exchange in high-resolution spectra of dissolved molecules. The major differences are the summation over all orientations present in the glassy state and the possible complexity of the correlation function from strong intermolecular interactions. Only a restricted range of frequency is accessible in a lineshape experiment, which limits the ability of this approach to characterize a complex correlation function. The lineshape studies can be performed in conjunction with spin–lattice relaxation measurements to extend the sampling in the frequency domain.[31]

Another method for extending the range in time or frequency of lineshape studies is to employ the Jeener-Brockaert pulse sequence.[3] This method has yet to be applied fully to a polymer glass because the analysis of the raw data requires a knowledge of the correlation function. Currently, correlation functions for motions in glassy polymers are not readily available or even known to be of a particular form. This general approach will no doubt be more useful as the understanding of local chain dynamics in bulk polymers grows.

Polycarbonates as an Example

The chain dynamics of the polycarbonate of bisphenol A (BPA) has been studied widely by solid-state NMR.[1-3,5,6] The polymer itself is a commercially successful engineering plastic, highly touted for impact resistance. This polymer is also known to be very mobile below the glass transition and this mobility often is associated with the property of impact resistance.[32,33] Many traditional dynamic studies have been performed on BPA, but until the recent NMR work a repeat-unit-level understanding of the motions was not agreed upon. Even now some aspects of the motion are under review, although a reasonable consensus exists about other aspects.

Isolated Chain Studies

It is easier to consider local chain motion for a single polymer backbone, both experimentally and theoretically. In the isolated chain, intermolecular effects are absent and these effects can dominate some aspects of motional properties in the glass. On the other hand, a characterization of the motions in an isolated chain is still helpful in understanding motion in the bulk polymer. If a specific local conformational process is not possible in a chain in the absence of intermolecular interactions, it is not likely to occur in the presence of such additional contributions, which increase the barriers restricting motion. It seems more reasonable to find which motions are possible in the absence of large intermolecular interactions and then see how these motions are altered when the interactions are present.

Following this line of thought, dilute solution spin–lattice relaxation studies can be employed effectively to characterize local chain motion in a situation where intermolecular contributions are relatively small. High-resolution proton and ^{13}C relaxation studies have been performed on dissolved BPA polycarbonate and a number of related polymers.[17,21,34,35] Figure 7-1 shows the repeat units of the polycarbonates studied to date.

In the most recent solution studies,[21,35] proton-relaxation measurements are conducted at at least two field strengths and frequently partially deuterated forms are employed to reduce dipole–dipole interactions among protons in different chemical environments in the repeat unit. Two partially deuterated repeat units employed are shown in Figure 7-2. The proton spin–lattice relaxation time can be written in terms of the spectral density, J, the internuclear distance, r, and constants γ_H and $\hbar$ as follows:

$$1/T_1 = \sum_j (9/8)\gamma_H^4 \hbar^2 r^{-6}[(2/15)J_1(\omega_H) + (8/15)J_2(2\omega_H)] \qquad (7\text{-}1)$$

Here ω_H is the proton Larmor frequency. If partially deuterated forms are employed, the relevant distances in this equation usually are known to within a few hundredths of an Ångstrom.

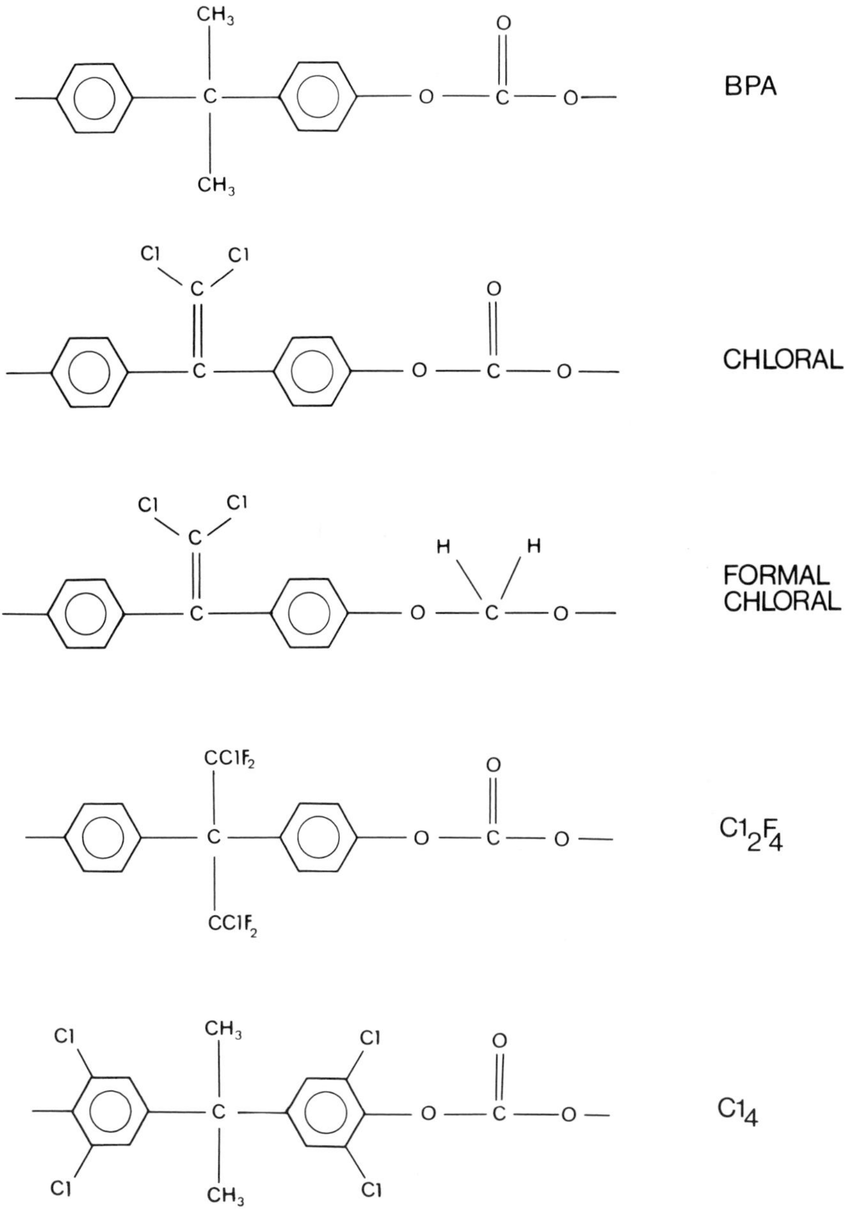

Figure 7-1. Structures of the repeat units and abbreviations for the polycarbonates and polycarbonate analogs.

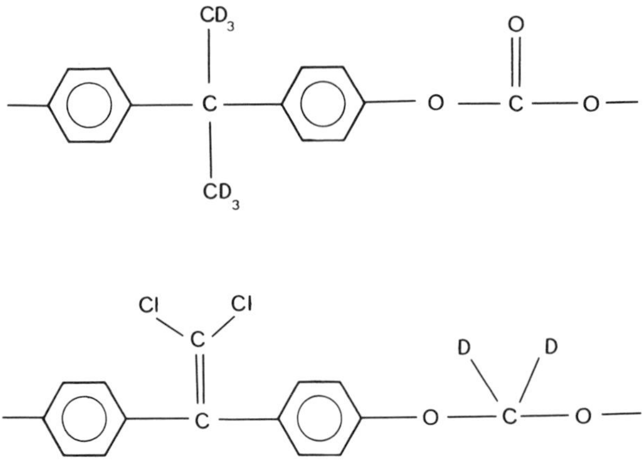

Figure 7-2. Structures of the repeat units of partially deuterated polymers used in proton relaxation studies.

To obtain data at other frequencies and with different geometric orientations in the repeat unit, ^{13}C spin–lattice relaxation times usually are measured as well. Only carbons with directly bonded protons are measured because the relaxation mechanism is usually dipole–dipole relaxation and known carbon–proton internuclear distances are required (see Chapter 4). Again, measurements are performed at several frequencies to obtain a data base capable of supporting an informative analysis. The equation relating the spin–lattice relaxation time to the spectral density for ^{13}C is:

$$1/T_1 = W_0 + 2W_{1C} + W_2 \qquad (7\text{-}2)$$

where

$$W_0 = \sum_j \gamma_C^2 \gamma_H^2 \hbar^2 J_0(\omega_0)/20 r_j^6$$

$$W_{1C} = \sum_j 3\gamma_C^2 \gamma_H^2 \hbar^2 J_1(\omega_C)/40 r_j^6$$

$$W_2 = \sum_j 3\gamma_C^2 \gamma_H^2 \hbar^2 J_2(\omega_2)/10 r_j^6$$

$$\omega_0 = \omega_H - \omega_C, \qquad \omega_2 = \omega_H + \omega_C$$

The dynamic information is contained in the spectral densities and differences among the spectral densities at different points in the repeat unit. In dilute solution, local chain motions have been divided generally between segmental motions and anisotropic rotation of functional groups.[25] Segmental motions correspond to changes of direction of backbone bonds. These motions are thought to involve one or several backbone units but not long portions of the chain. Hall and Helfand[7] have developed a correlation function for segmental motion based on observations made during computer simulations of segmental motion in polyethylene. Two subclasses of segmental motion are observed. First, there are conformational changes produced by rotation about a single backbone bond. This process is described by a single exponential correlation time, τ_0. The second process is correlated motions where a second rotation about a backbone bond is observed to follow a first rotation closely in time. The second rotation compensates for distortion of the backbone produced by the first conformational event. Relaxation caused by this correlated motion is characterized by a Bessel function of order zero (I_0) with a time constant τ_1. The Bessel function is a standard solution for one-dimensional diffusion problems.[36] The complete Hall-Helfand correlation function is written:

$$\phi(t) = \exp\left(-t/\tau_0\right) \exp\left(-t/\tau_1\right) I_0(t/\tau_1) \qquad (7\text{-}3)$$

which may be Fourier transformed to yield the spectral density:

$$J(\omega) = 2\{[(\tau_0^{-1})(\tau_0^{-1} + 2\tau_1^{-1}) - \omega^2]^2 + [2(\tau_0^{-1} + \tau_1^{-1})\omega]^2\}^{-1/4}$$

$$\times \cos\{1/2 \arctan 2(\tau_0^{-1} + \tau_1^{-1})\omega/[\tau_0^{-1}(\tau_0^{-1} + 2\tau_1^{-1}) - \omega^2]\} \quad (7\text{-}4)$$

For consideration of substituent groups attached to the backbone, the Hall-Helfand correlation function may be combined with anisotropic internal rotation, as originally described by Woessner[37] under the assumption of independence of the two motions. The correlation time for internal rotation of the substituent group is τ_{ir} and the geometry of the motion is characterized by the angle Δ between the axis of internal rotation and the internuclear interaction. The complete spectral density for segmental motion and internal rotation of functional groups is:

$$J(\omega) = AJ_a(\tau_0, \tau_1, \omega) + BJ_b(\tau_{b0}, \tau_1, \omega) + CJ_c(\tau_{c0}, \tau_1, \omega) \qquad (7\text{-}5)$$

where

$$A = (3 \cos^2 \Delta - 1)^2/4$$

$$B = 3 (\sin^2 2\Delta)/4$$

$$C = 3 (\sin^4 \Delta)/4$$

For stochastic diffusion:

$$\tau_{b0}^{-1} = \tau_0^{-1} + \tau_{ir}^{-1}$$

$$\tau_{c0}^{-1} = \tau_0^{-1} + (\tau_{ir}/4)^{-1}$$

For a threefold jump:

$$\tau_{b0}^{-1} = \tau_{c0}^{-1} = \tau_0^{-1} + \tau_{ir}^{-1}$$

The form of J_a, J_b, and J_c is the same as in Eq. (7-4) with τ_0 replaced by τ_0, τ_{b0}, τ_{c0}, respectively.

The results of proton and ^{13}C spin–lattice relaxation studies on polycarbonates can be summarized in terms of correlation times for segmental motion, methyl group rotation, and phenylene group rotation.[17,21,34,35] The time scale of each motion can be separated from the others based on repeat unit structure and the orientation of the dipole–dipole interactions in the repeat unit. A complete discussion of the analysis is presented in the original work.[17] Because measurements are made as a function of temperature, the temperature dependence of the correlation times is usually expressed in terms of the Arrhenius equation. Table 7-1 lists the activation energies and prefactors for the polycarbonates shown in Figure 7-1.

The uncertainty in the activation energies is usually about 5 kJ/mol, although the measurements made in the solvent $C_2D_2Cl_4$ are more accurate than the measurements made in $CDCl_3$. A larger frequency base was employed in the $C_2D_2Cl_4$ study, as well as partially deuterated forms to reduce cross relaxation. Also, the $CDCl_3$ studies were interpreted with an earlier segmental motion model and the outcome of that analysis has been approximately converted to the Hall-Helfand approach using a published conversion factor.[36]

In solution, BPA is found to be a rather mobile chain, especially considering the complexity of the repeat unit. Segmental motion is dominated by cooperative backbone motions as opposed to single backbone bond reorientations.[35] Phenylene group rotation is rapid and rather similar to the cooperative backbone motions in time scale. This similarity persists as concentration is raised from 5 to 30%. Methyl group rotation is also facile, although it is not coincident with the time scale of cooperative segmental motion or phenylene group motion as concentration is raised from 5 to 30%.

The chloral polycarbonate is very similar to BPA polycarbonate in all regards[35] except that it contains no methyl group. The polyformal, structurally analogous to the chloral polycarbonate, has similar phenylene group rotation, although segmental motions are different.[21] In polyformal, the cooperative segmental motions dominate at high temperatures and the single backbone reorientations dominate at low temperatures.

The last two polycarbonates listed, Cl_2F_4 and Cl_4, are rather different. Segmental motion is somewhat slower in these two polymers but phenylene group rotation is reduced greatly. Because of the limited data base, the phenylene group rotation was modeled as slow, complete anisotropic rotation, which yields low apparent activation energies and large Arrhenius prefactors. A more complete study and simulation might well have shown the motion to be restricted anisotropic diffusion or, in other words, an incomplete anistropic rotation. In any case phenylene group motion is either reduced in amplitude or slower than in the other three polymers because rather little additional motion beyond segmental motion is indicated by the ^{13}C T_1s.

This change in phenylene group motion was linked empirically to the low-temperature or γ dynamic mechanical loss peak. The first three polymers all have similar dynamic mechanical spectra below the glass transition, dominated by the presence of a substantial loss peak near 173 K. All three of these polymers also have comparable phenylene group rotation in solution. In the last two polymers, the γ relaxation peak is shifted to much higher tem-

TABLE 7-1. Apparent Activation Energies and Arrhenius Prefactors

Polymer[a]	Solvent[b]	Apparent activation energies (kJ/mol)				Arrhenius prefactor ($\tau_\infty \times 10^{14}$ s)			
		Correlated segmental τ_1	Single backbone rotation	Phenylene rotation	Methyl rotation	Correlated segmental τ_1	Single backbone rotation	Phenylene rotation	Methyl rotation
BPA	$C_2D_2Cl_4$	19	16	22	24	28	1603	6	1
	$CDCl_3$	~19	—	13	22	—	—	—	—
Chloral	$C_2D_2Cl_4$	17	18	18	—	94	409	40	—
	$CDCl_3$	~13	—	15	—	—	—	—	—
Formal chloral	$C_2D_2Cl_4$	30	9	21	—	1	9700	30	—
Cl_2F_4	$CDCl_3$[c]	~16	—	13	14	—	—	—	—
Cl_4	$CDCl_3$[c]	~18	—	15	19	—	—	—	—

[a] See Figure 7-1 for repeat-unit structures corresponding to abbreviations.
[b] 10 wt% solutions.
[c] The studies performed in $CDCl_3$ are less accurate because they are based on a poorer data set.

peratures and the phenylene group rotation in solution is reduced or restricted greatly. These empirical correlations were noted and phenylene group rotation was linked phenomenologically to the low-temperature loss peak.

These dilute solution studies on BPA show the intramolecular barriers to motion to be quite low: 15–25 kJ. Theoretical calculations on an isolated BPA chain also indicate very low barriers for local chain motions.[32,33,38] For instance, rotation about the backbone CO bonds is found to be 10 kJ by one investigator[38] and 12 kJ by another.[32,33] Somewhat higher values would be found in solution because the effects of solvent viscosity are added to the intramolecular barrier.

Bulk Relaxation Measurements

Solution studies and theoretical calculations show conformational changes to be easy in isolated polycarbonate chains. Also, some correlation between phenylene group motion and the low-temperature loss peak is indicated. However, direct measurement of relaxation in the bulk polymer is the best way to delineate further the nature of the motions below the glass transition.

Spin–lattice relaxation times can be used to set the time scale of the motion in the bulk polymer. McCall determined the proton T_1 and $T_{1\rho}$ on BPA polycarbonate, but the observed relaxation behavior is dominated by the methyl protons and methyl group rotation.[13] To determine motion at the phenylene group, proton relaxation times were measured on the chloral polycarbonate[6] shown in Figure 7-1. The dynamic mechanical spectrum[39] and the solution relaxation results show this polymer to be quite similar to BPA polycarbonate.[35] Because it contains only phenylene protons, the proton relaxation times will reflect phenylene group motion. Either phenylene group rotation or translation modulates intermolecular dipole–dipole interactions to produce relaxation. Figure 7-3 shows the proton T_1 at 20 and 90 MHz as well as the $T_{1\rho}$ at 43 kHz.

The relaxation measurements at all three frequencies show the presence of one broad, asymmetrical minimum. The $T_{1\rho}$ data as a function of temperature show the shape of the minimum best and all these data can be fitted with a correlation function in a fashion similar to the solution data. The proton T_1s have the same functional form as shown in Eq. (7-1); in solids, however, it is traditional to write the equation in terms of the second moment, S:

$$1/T_1 = (2/3)\gamma^2 S[J_1(\omega_H) + 4J_2(2\omega_H)] \qquad (7\text{-}6)$$

The proton $T_{1\rho}$ can be written in a similar fashion, but here the relaxation depends on the spectral density at the frequency corresponding to the strength of the radiofrequency field, ω_e, as well as the Larmor frequency, ω_H:

$$1/T_{1\rho} = (2/3)\gamma^2 S[1.5J_e(2\omega_e) + 2.5J_1(\omega_H) + J_2(2\omega_H)] \qquad (7\text{-}7)$$

As before, the spectral density is the Fourier transform of the correlation function, which should reflect the nature of the local motion in the bulk

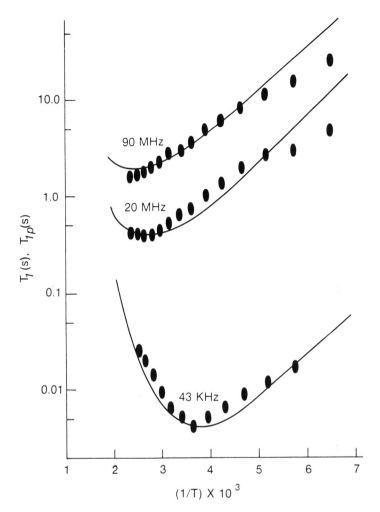

Figure 7-3. Proton spin–lattice relaxation time and spin–lattice relaxation time in the rotating frame versus inverse temperature for chloral polycarbonate. The solid line corresponds to a fit of the relaxation data with a Williams-Watts-Ngai fractional exponential correlation function. Abscissa is in K^{-1}. (Reprinted with permission from Jones, A. A.; O'Gara, J. F.; Inglefield, P. T.; Bendler, J. T.; Yee, A. F.; Ngai, K. L. *Macromolecules* **1983**, *16*, 658. Copyright 1983, American Chemical Society.)

polymer. Correlation functions of the type used in solution based on segmental motion and anisotropic rotation of functional groups do not account for the data in Figure 7-3.[6] A variety of correlation functions was tested, but the only successful one was the Williams-Watts fractional exponential,[11] commonly written:

$$\phi_{ww}(t) = \phi(0) \exp\left[-(t/\tau_{ww})^\beta\right] \qquad (7\text{-}8)$$

Ngai[9,10] has proposed an explanation of the occurrence of this function in glassy materials and writes the correlation function in the form:

$$\phi(t)_{CSM} = \phi(0) \exp\left[-(t/\tau_p)^{1-n}\right] \tag{7-9}$$

where $0 < n < 1$ and the loss peak frequency is located at $\omega_p \propto \tau_p^{-1}$, where

$$\tau_p = [(1-n)e^{nu}E_c^n\tau_0]^{1/(1-n)}$$

The model yielding this correlation function is based on a correlated states model; hence the subscript CSM. The symbol u designates Euler's constant (0.5722). The two key parameters used to fit the data are τ_p and n. As just shown, τ_p is a function of n, the cutoff energy, E_c, and the microscopic correlation time, τ_0. This correlation cannot be determined from a fit of the data but the activation energy for the microscopic process, E_A, can be determined for the apparent activation energy, E_A^*, according to the equation:

$$E_A^* = E_A/(1-n) \tag{7-10}$$

In practice one writes the correlation time τ_p in an Arrhenius form:

$$\tau_p = \tau_\infty^* \exp(E_A^*/RT) \tag{7-11}$$

where

$$\tau_\infty^* = [(1-n)\exp(nu)E_c^n\tau_\infty]^{1/(1-n)}$$

The parameters E_A^*, τ_∞^*, and n are varied to fit the data as a function of temperature. Figure 7-3 shows the fit of the CSM model to the T_1 and $T_{1\rho}$ data using the parameters $E_A = 10$ kJ, $n = 0.8$, and $\tau_\infty^* = 2.29 \times 10^{-16}$ s. The activation energy, E_A, corresponds to the barrier encountered in the isolated chain and is comparable to the barriers calculated for simple rotations.[32,33,38] The apparent activation energy E_A^* is much higher and reflects the contribution to the barrier height from intermolecular interactions encountered in the glass. The parameter n controls the effective width of the relaxation process and accounts for the breadth and asymmetry of the T_1 and $T_{1\rho}$ minima. The parameter n also relates E_A and E_A^* as written in Eq. (7-10) above.

The fit of the data argues for the utility of the fractional exponential correlation function but it is also worth noting the realistic value obtained for E_A. Also, the correlated states model rationalizes the physically unrealistic value obtained for τ_∞^* because it is not simply an Arrhenius prefactor but a combination of several parameters.

Recently a comparable proton-relaxation study has been completed on a partially deuterated form of BPA where the methyl protons have been replaced with deuterons.[40] This system will reflect phenylene group motion just as chloral did, and, indeed, rather similar T_1 and $T_{1\rho}$ data were obtained. With a summary of the time scale of the motion contained in the Williams-Watts-Ngai correlation function it becomes quite interesting to consider the geometry of the motion.

First it is important to consider the ^{13}C $T_{1\rho}$ reported for BPA poly-carbonate.[1] Although these relaxation times can be written in terms of the spectral density just as the proton times, the $T_{1\rho}$ decay curves cannot be char-acterized by a single time constant. The decay curves reflect the presence of at least two time constants and possibly more. In addition, there may be some spin–spin contributions to the decay curves as well.[18,20] The net effect of these complications is enough to discourage quantitative data fitting comparable to the proton interpretation. The ^{13}C $T_{1\rho}$s indicate the presence of considerable local chain motion in agreement with the proton interpretation and with approximately the same time scale. However, the ^{13}C data may reflect an inhomogeneous distribution of correlation times,[26] yielding the observed $T_{1\rho}$ dispersions, whereas spin diffusion equalizes the proton relaxation data, yield-ing a single relaxation time. Another source of the dispersion may be aniso-tropic motion leading to relaxation dependent on orientation in a glass. In any case, an analysis of the ^{13}C relaxation data in terms of a correlation function is not yet in hand.

Solid-State Lineshape Studies

Proton lineshape measurements provided the first direct information on the geometry of the motion of the phenylene group. To obtain significant informa-tion, the chloral polycarbonate was selected for the first proton lineshape study because it contains only one type of proton, phenylene protons.[5] There-fore this system contains geometrically significant information, whereas most proton spectra of polymers consist of overlapping resonances of several chemi-cal types of protons with little chance of extracting information.

Figure 7-4 shows several proton spectra observed as a function of tem-perature for the chloral polycarbonate as well as a frozen solution of the chloral polycarbonate. At low temperatures, the spectra are featureless bell-shaped curves with no obvious information content. At about 193 K, the line narrows and at a temperature of 273 K the lineshape becomes that of a Pake doublet with substantial broadening. The splitting of the Pake doublet is 2.5 G which corresponds to a proton–proton internuclear distance of 2.6 Å. The only intramolecular interaction comparable to this is between the 2 and 3 protons on the phenylene ring.

As temperature is raised from 273 K up to the glass transition, the line continues to narrow gradually but the Pake doublet persists with an undiminished splitting. These observations have significant implications for the motions of the phenylene group in polycarbonates. The substantial line narrowing shown in Figure 7-5 in terms of the decrease of the second moment of the proton lineshape as the temperature is raised is indicative of the pre-sence of considerable motion in the glassy polymer. This point has been gener-ally recognized. However, the observation of the Pake doublet and the persistence of the doublet up to the glass transition greatly restricts the motions that may be proposed for the phenylene group. The interaction between the 2,3 protons of the phenylene group is parallel to the C_1C_4 axis of the phenylene group. This axis must not be reoriented substantially if the Pake doublet persists as temperature is raised. Considerable motion of the phenylene group must be present because intermolecular dipole–dipole interactions between phenylene groups are averaged. The only intramolecular motion consistent with averaging of intermolecular interactions while preserving the Pake doublet is rotation or rotational oscillation about the C_1C_4 axis of the phenylene group. While proton NMR is most effective at locating the axis about which motion must be occurring, it cannot identify the exact nature of the

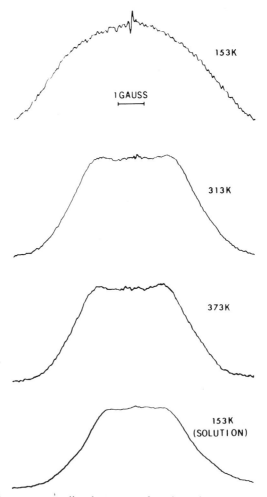

Figure 7-4. Solid-state proton lineshapes as a function of temperature for chloral polycarbonate. The bottom spectrum is for a 15 wt% frozen solution of the polymer in $C_2D_2Cl_4$. (Reprinted with permission from Inglefield, P. T.; Jones, A. A.; Lubianez, R. P.; O'Gara, J. F. *Macromolecules* **1981**, *14*, 288. Copyright 1981, American Chemical Society.)

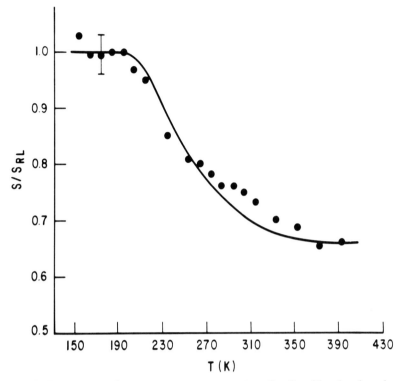

Figure 7-5. Proton second moment versus temperature for the chloral polycarbonate. The solid line is a fit of the second moment data using the same parameters used to account for spin–lattice relaxation shown in Figure 7-3. (Reprinted with permission from Jones, A. A.; O'Gara, J. F.; Inglefield, P. T.; Bendler, J. T.; Yee, A. F.; Ngai, K. L. *Macromolecules* **1983**, *16*, 658. Copyright 1983, American Chemical Society.)

motion about the axis because the motion does not affect the predominant intramolecular interaction contributing to the proton lineshape.

The change in proton second moment, S, from the rigid lattice value, S_{RL}, can be written in terms of the spectral density just as the spin–lattice relaxation time was earlier [Eq. (7-6)].[6]

$$S/S_{RL} = a + m \int_{-\delta v}^{+\delta v} J_0(v)\, dv \qquad (7\text{-}12)$$

Here, a and m are constants and the integration is carried out over the frequency range of the absorption line. Because a correlation function has already been developed to characterize the time scale of motions contributing to T_1 and $T_{1\rho}$, this same correlation function can be employed to predict the change in proton second moment. Figure 7-5 shows this prediction, which leads to an adequate interpretation of the second moment without further adjustment of the parameters used in the Williams-Watts-Ngai correlation function.

Although the axis of motion is clear from the proton lineshape, the nature of the motion about this axis is not apparent. Possibilities include free rotation, restricted rotation, or jumps between distinct minima. To distinguish between these possibilities additional lineshape studies were performed. The first report came from Spiess[3] and employed deuterium spectroscopy. Phenylene protons were replaced with deuterons in BPA and deuterium wide-line spectra were reported at room temperature and at 380 K. The spectra were simulated by the combination of two motions: jumps between two minima separated by 180° and restricted rotation over an rms angular amplitude of ±15° at room temperature. The jumps between two minima separated by 180° are referred to as π flips and the presence of this motion initially was disputed. Because the deuterium quadrupolar interaction is axially symmetrical, the high-temperature spectra given by Spiess could be simulated fairly well by large-amplitude restricted rotation.

A second lineshape approach confirmed the motion as π flips and removed the possibility of large-amplitude oscillation as the only motion.[2,31] A ^{13}C labeled BPA polycarbonate was prepared as shown in Figure 7-6 with 90% of

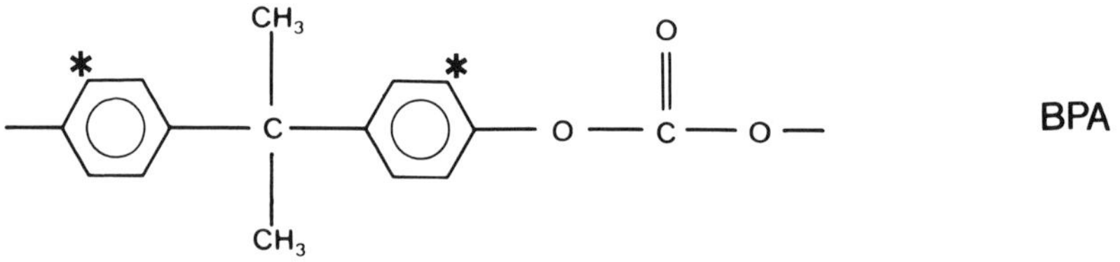

Figure 7-6. Structure of the ^{13}C-labeled repeat unit for BPA polycarbonate. The asterisks indicate the position labeled to a level of more than 90%.

the carbons in the aromatic ring at the position ortho to the carbonate replaced with [13]C. Carbon-13 spectra were then taken under conditions of cross polarization and dipolar decoupling. The resulting spectra are shown in the low- and high-temperature limit in Figure 7-7. There is a pronounced change in the lineshape as temperature is raised from 113 to 393 K. Also the low-temperature lineshape shows the classical asymmetric chemical shift anisotropy pattern (see Figure 2-2C). This pattern can be fit with the Bloembergen and Rowland[41] equation:

$$I(\sigma; \sigma_{11}, \sigma_{22}, \sigma_{33}, \Delta\sigma) = \int_{-\infty}^{+\infty} I^0(\sigma - \xi; \sigma_{11}, \sigma_{22}, \sigma_{33})F\,d\xi \qquad (7\text{-}13)$$

For $\sigma_{33} \leq \sigma < \sigma_{22}$:

$$I^0(\sigma; \sigma_{11}, \sigma_{22}, \sigma_{33}) = \pi^{-1}K(x)(\sigma_{11} - \sigma)^{-0.5}(\sigma_{22} - \sigma_{33})^{-0.5}$$

and

$$x = (\sigma_{11} - \sigma_{22})(\sigma - \sigma_{33})/[(\sigma_{22} - \sigma_{33})(\sigma_{11} - \sigma)]$$

For $\sigma_{22} < \sigma \leq \sigma_{11}$

$$I^0(\sigma; \sigma_{11}, \sigma_{22}, \sigma_{33}) = \pi^{-1}K(x)(\sigma - \sigma_{33})^{-0.5}(\sigma_{11} - \sigma_{22})^{-0.5}$$

and

$$x = (\sigma_{11} - \sigma)(\sigma_{22} - \sigma_{33})/[(\sigma - \sigma_{33})(\sigma_{11} - \sigma_{22})]$$

For $\sigma < \sigma_{11}$, and $\sigma \geq \sigma_{33}$

$$I^0(\sigma; \sigma_{11}, \sigma_{22}, \sigma_{33}) = 0$$

to yield the three values of the shielding tensor in the principal axis system. In all cases

$$K(x) = \int_0^{\pi/2} (1 - x^2 \sin^2 \Psi)^{-1/2} d\Psi$$

$$F = f(\sigma; \Delta\sigma) = 1/[1 + (2\sigma/\Delta\sigma)^2]$$

where F describes the Lorentzian line broadening of the chemical-shift dispersion, I^0.

The principal shielding values are $\sigma_{11} = 17 \pm 1$, $\sigma_{22} = 52 \pm 1$, and $\sigma_{33} = 175 \pm 1$ ppm relative to liquid CS_2 scale. Positive values are upfield in the direction of increasing shielding. In the principal axis system, the shielding interaction usually is written as a diagonal tensor.

$$\boldsymbol{\sigma}_{123} = \begin{bmatrix} \sigma_{11} & 0 & 0 \\ & \sigma_{22} & 0 \\ 0 & 0 & \sigma_{33} \end{bmatrix} \tag{7-14}$$

To consider the effects of motion, it is necessary to transform the shielding tensor in the principal axis system to the axis system of the molecular motion. For the case at hand, the shielding tensor can be transformed to the new coordinate system by a rotation matrix $\tilde{\mathbf{R}}$ defined by a set of Euler angles α, β, and γ using the convention of Rose.[42] The angle α is a rotation about the original z axis, β is about the new y axis, and γ is about the final z axis.

$$\mathbf{R}(\alpha\beta\gamma) = \begin{bmatrix} \cos\alpha\cos\beta\cos\gamma & \sin\alpha\cos\beta\cos\gamma & -\sin\beta\cos\gamma \\ -\sin\alpha\sin\gamma & +\cos\alpha\sin\gamma & \\ -\cos\alpha\cos\beta\sin\gamma & -\sin\alpha\cos\beta\sin\gamma & \sin\beta\sin\gamma \\ -\sin\alpha\cos\gamma & +\cos\alpha\cos\gamma & \\ \cos\alpha\sin\beta & \sin\alpha\sin\beta & \cos\beta \end{bmatrix} \tag{7-15}$$

The effects of motion on the shielding tensor in the motional axis system are analogous to chemical exchange. Because the chemical shift experienced by a

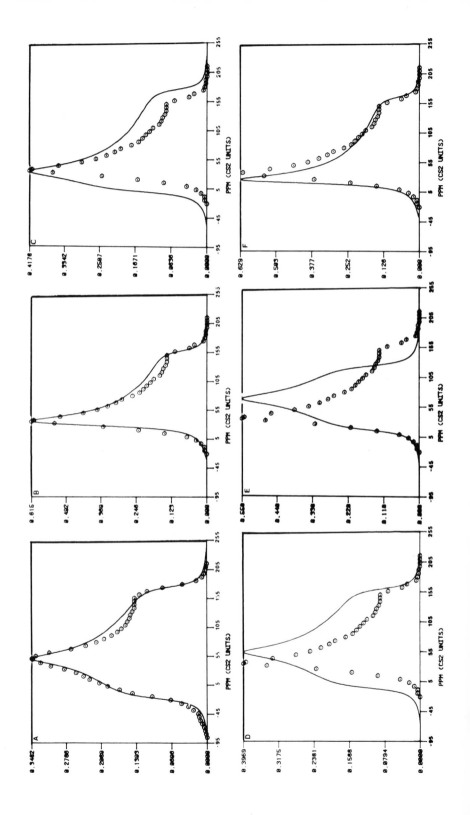

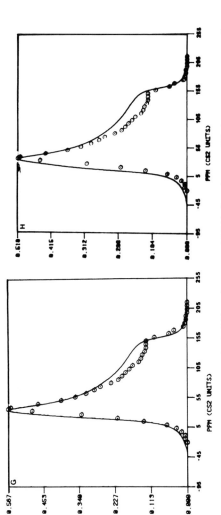

Figure 7-7. Carbon-13 chemical-shift anisotropy (CSA) lineshapes: (A) low-temperature data and simulation based on powder pattern function; (B) high-temperature data and simulation based on powder function; (C) high-temperature data and simulation based on π flips about the C_1C_4 axis; (D) high-temperature data and simulation based on restricted rotation about the C_1C_4 axis over a 60° range; (E) high-temperature data and simulation based on restricted rotation about the C_1C_4 axis over a 120° range; (F) high-temperature data and simulation based on restricted rotation over a 180° range about an axis inclined at 70° to C_1C_4 (parallel to the O—CO bond axis); (G) high-temperature data and simulation based on flips between 0 and 140° about the C_1C_4 axis; (H) high-temperature data and simulation based on C_1C_4 axis π flips plus restricted rotation over a 60° range. The points represent experimental data and the line the theoretical simulation. (Reprinted with permission from Inglefield, P. T.; Amici, R. M.; O'Gara, J. F.; Hung, C.-C.; Jones, A. A. *Macromolecules* **1983**, *16*, 1552. Copyright 1983, American Chemical Society.)

^{13}C nucleus depends on the orientation of the phenylene group relative to the field, as the orientation is exchanged by molecular motion so is the chemical shift. The main complicating factor is the superposition of all chemical shifts for all orientations corresponding to a powder lineshape.

The simplest case to deal with is rapid chemical exchange, which is achieved at high temperatures. For the data taken on labeled BPA at 22.6 MHz, the high-temperature limit for π flips is achieved at about room temperature. The chemical-shift anisotropy (CSA) pattern observed at these temperatures can be simulated by an approach outlined by Slotfeldt-Ellingsen and Resing.[29] Here the x axis of the molecular motion is chosen to be coincident with the C_1C_4 axis of the phenylene group and the shielding tensor is transformed to this axis system by using a rotation matrix $\mathbf{R}(\delta)$ as follows:

$$\sigma_{xyz} = \mathbf{R}(\delta) \cdot \sigma_{123} \cdot \mathbf{R}(\delta)^{-1} \qquad (7\text{-}16)$$

where

$$\mathbf{R}(\delta) = \begin{bmatrix} \cos\delta & -\sin\delta & 0 \\ \sin\delta & \cos\delta & 0 \\ 0 & 0 & 1 \end{bmatrix}$$

and δ is the angle between x and y axes of the motion system and the x and y axes of the principal shielding system. With the shielding tensor in the molecular motional axis system, various possible motions about the C_1C_4 axis can be modeled. Rotating the molecule about the x axis by an angle α can be defined as follows:

$$\sigma_{xyz}(\alpha) = \mathbf{R}(\alpha) \cdot \sigma_{xyz} \cdot \mathbf{R}(\alpha)^{-1} \qquad (7\text{-}17)$$

where

$$\mathbf{R}(\alpha) = \begin{bmatrix} 1 & 0 & 0 \\ 0 & \cos\alpha & -\sin\alpha \\ 0 & \sin\alpha & \cos\alpha \end{bmatrix}$$

Using this approach, jumps between various minima can be modeled by averaging the shift experienced in each of the minima. For π flips, the shielding is averaged over two values separated by a 180° rotation about the x axis ($\alpha = 180°$).

$$\sigma_{flip} = (1/2)[\sigma_{xyz} + \sigma_{xyz}(\alpha)] \qquad (7\text{-}18)$$

Another possible type of motion about the x axis is rapid rotation through all angular positions in a specified angular range. This can be modeled by averaging over α in the equation:

$$\langle\sigma_{xyz}\rangle_\alpha = \langle\mathbf{R}(\alpha) \cdot \sigma_{xyz} \cdot \mathbf{R}(\alpha)^{-1}\rangle_\alpha \qquad (7\text{-}19)$$

The preceding average depends on the potential function and if one assumes a square well with walls at $+\alpha/2$ and $-\alpha/2$, then the average shift is:

$$\langle \sigma_{xyz} \rangle_\alpha = \int_{-\alpha/2}^{+\alpha/2} \mathbf{R}(\alpha) \cdot \sigma_{xyz} \cdot \mathbf{R}(\alpha)^{-1} \, d\alpha \Bigg/ \int_{-\alpha/2}^{+\alpha/2} d\alpha \qquad (7\text{-}20)$$

These examples are only a few of the possibilities. Calculated chemical-shift anisotropy lineshapes are compared with the observed high-temperature limit in Figure 7-7. Neither the π-flip lineshape nor restricted rotation can simulate the shape individually. The π-flip lineshape has the correct general shape but is too broad. The restricted rotation lineshape does not even approximately match the observed shape when the angular amplitude is made large, so this possibility may be ruled out. Another arbitrary axis inclined at $70°$ to the C_1C_4 yields a lineshape close to the observed but this choice of axis is inconsistent with the solid-state proton spectrum of BPA-d_6 shown in Figure 7-8. A reasonably good simulation of the observed lineshape at high temperatures can be obtained by combining π flips with restricted rotation over a modest

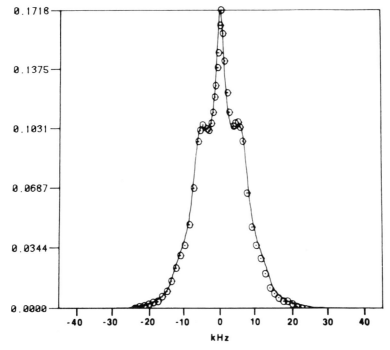

Figure 7-8. Proton dipolar lineshapes of BPA-d_6 in perdeutero-BPA. The points represent experimental data and the line the theoretical simulations in the high-temperature limit simulation. The lineshape consists of a Pake doublet plus a sharp line from a small percentage of protons remaining in deuterated sites. (Reprinted with permission from Inglefield, P. T.; Amici, R. M.; O'Gara, J. F.; Hung, C.-C.; Jones, A. A. *Macromolecules* **1983**, *16*, 1552. Copyright 1983, American Chemical Society.)

angular amplitude. The simulated spectrum for a restricted rotation over a range of $\pm 36°$ in addition to π flips yields a lineshape comparable to the observed lineshape at 393 K.

The proton and ^{13}C lineshape data in conjunction with the deuterium lineshape results specify the nature of the geometry of the motion of the phenylene group. Carbon–proton dipolar rotational spin-echo lineshape experiments (a technique discussed in Chapters 2 and 3) are also consistent with the π flips and limited restricted rotation.[43] With the geometry of the motion determined by comparing the spectra in low-temperature and high-temperature limits, the rates and amplitudes of the motions can then be considered.

If the rate of motion is comparable to the frequency width of the lineshape, a more elaborate simulation procedure must be undertaken. Equations are available for multiple site exchange, leading to the collapse of CSA lineshape.[30] The general lineshape equation for N sites is:

$$g(\omega) = (1/N)(L/(1 - KL)) \tag{7-21}$$

where

$$L = \sum_{j=1}^{N} [i(\omega - \omega_j) - (1/T_{2j}) + NK]^{-1}$$

T_{2j} = the spin–spin relaxation time, $K = \tau^{-1}$ = exchange or flipping rate, N = number of sites, and ω_j = frequency of site j:

$$\omega_j = \sigma_{iso} + (\sigma_{33} - \sigma_{iso})[P_2(\cos \beta)P_2(\cos \theta)$$

$$+ (3/4) \sin^2 \theta \sin^2 \beta \cos^2 (\phi + \gamma) - 3 \sin \theta \cos \theta \sin \beta \cos \beta \cos (\phi + \gamma)]$$

$$+ [(\sigma_{11} - \sigma_{22})/2][\sin^2 \theta \cos^4 \beta \cos [2(\phi + \gamma + \alpha)]$$

$$+ \sin^2 \theta \sin^4 \beta \cos [2(\phi + \gamma - \alpha)] + \sin \theta \cos \theta \sin \beta$$

$$\times \{(\cos \beta^{+1}) \cos (\phi + \gamma + 2\alpha) + (\cos \beta^{-1}) \cos (\phi + \gamma + 2\alpha)\}$$

$$+ P_2(\cos \theta) \sin^2 \beta \cos (2\alpha)]$$

Other terms in the equation are the Legendre polynomial, P_2; the Euler angles of the flip axis with respect to the principal axis system of the shift tensor, (α, β, γ); and the Euler angles of the magnetic field with respect to the flip axis, (θ, ϕ). For an NMR lineshape, $I(\omega)$, the real part of $g(\omega)$ is required. The preceding formulation is for a single orientation with respect to the magnetic field and an average over all orientations must be taken:

$$I(\Omega) = \int d\Omega P(\Omega)I(\omega, \Omega) \tag{7-22}$$

where $P(\Omega) = (1/4\pi)$ and $d\Omega = \sin \theta \, d\theta \, d\phi$. Very suitable computer programs for these lineshape simulations as a function of rate are given in the Ph.D. thesis of Wemmer.[44]

While the π-flip process is simulated as a specific example of the general N-site exchange, the restricted rotation is simulated not with a temperature-dependent rate but with a temperature-dependent amplitude. At temperatures above 293 K, the π-flip process is in the rapid limit and all further narrowing is attributed to an increase in an amplitude of the restricted rotation. The root mean square of the amplitude of the restricted rotation is found to be linear versus the square root of temperature above 293 K. This linear dependence is assumed for lower temperatures as well, which allows for a complete simulation of the spectra at all temperatures as a result of both motions. The amplitudes of the restricted rotation as a function of temperature are summarized in Figure 7-9 and the comparable π flip rate information in Figure 7-10. The experimental spectra are shown in Figure 7-11 and comparisons between the simulations and the experimental spectra are shown in Figure 7-12.

An apparent activation energy can be calculated from the temperature

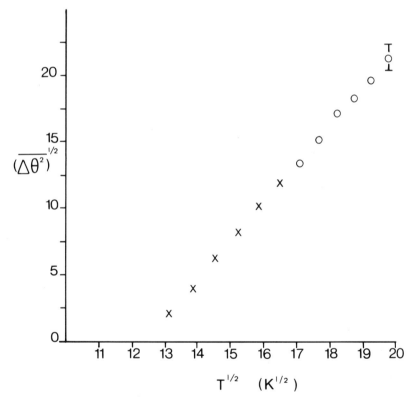

Figure 7-9. Root-mean-square amplitude of restricted phenylene group rotation about the C_1C_4 axis versus temperature to the one half power. The rms amplitude is determined from simulating the ^{13}C CSA lineshape. (Reprinted with permission from O'Gara, J. F.; Jones, A. A.; Hung, C.-C.; Inglefield, P. T. *Macromolecules* **1985**, *18*, 1117. Copyright 1985, American Chemical Society.)

dependence of the π-flip rate. The rather low value of 11 ± 5 kJ is found that is not consistent with the apparent activation energy of 50 kJ determined from proton spin–lattice relaxation data. The discrepancy reflects the use of an exponential correlation function for the simulation of the CSA spectra, whereas a highly nonexponential correlation function was used in the simulation of the T_1 and $T_{1\rho}$ data. The lineshape data can be simulated fairly well with a single exponential correlation function because only a limited range of frequencies near 1 kHz is sampled and the temperature dependence of π flips can be observed only from 183 to 273 K. On the other hand, the T_1 and $T_{1\rho}$ data extended from 90 MHz to 43 kHz and the temperature dependence could be determined from 153 to 393 K.

One other set of relevant lineshape data is available on the BPA unit in an epoxy resin.[26] The spectroscopic technique employed was variable-temperature ^{13}C magic angle sample spinning. At low temperatures, the protonated phenylene carbon lines become doublets reflecting two slightly different chemical environments. As temperature is raised the high-resolution lines obtained under these experimental conditions collapse to the single sharp line usually observed near room temperature. The two lines for a single type of carbon present at low temperature are attributed to two rotational conformations of the phenylene ring although at the time of the report it was not clear that the process involved was π flips. A correlation function was used to interpret the spectral collapse. In this case, as for the proton-relaxation data, a Williams-Watts-Ngai fractional exponential correlation function was employed with a correlation time, apparent activation energy, and fractional exponent comparable to the values used in the chloral polycarbonate interpretation. (See Chapter 3.)

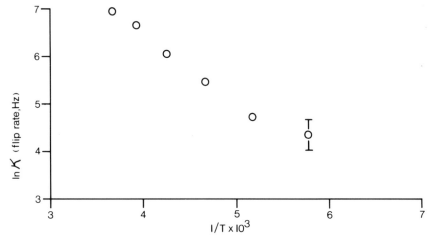

Figure 7-10. The natural logarithm of the π-flip rate versus inverse temperature. The π-flip rate is determined from simulation of the ^{13}C CSA lineshape. Abscissa is in K^{-1}. (Reprinted with permission from O'Gara, J. F.; Jones, A. A.; Hung, C.-C.; Inglefield, P. T. *Macromolecules* **1985**, *18*, 1117. Copyright 1985, American Chemical Society.)

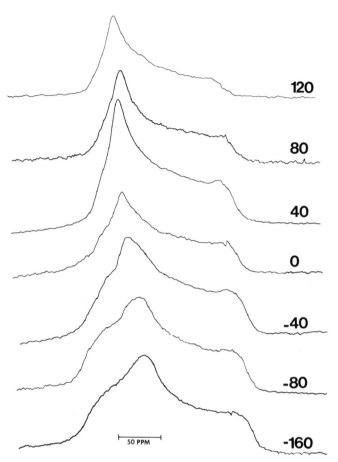

120

80

40

0

-40

-80

50 PPM

-160

Figure 7-11. Carbon-13 CSA lineshapes at several temperatures. (Reprinted with permission from O'Gara, J. F.; Jones, A. A.; Hung, C.-C.; Inglefield, P. T. *Macromolecules* **1985**, *18*, 1117. Copyright 1985, American Chemical Society.)

Carbon-13 chemical-shift anisotropy lineshape spectra are also available on the carbonate unit in BPA polycarbonate.[28] These spectra were obtained by Henrichs on a labeled sample under conditions comparable to the phenylene labeled system. The carbonate carbon CSA lineshape is quite broad, about 150 ppm, but shows rather little change with temperature. This result is surprising because a large dielectric loss is present in BPA polycarbonate, which can only reflect motion of the carbonate unit. Henrichs interprets the small change in the CSA tensor to be oscillation of the carbonate unit with an angular amplitude of about 40° at room temperature. The interpretation is not decisive because the principal axis system of the shielding tensor is not entirely clear in the carbonate unit and the lineshape change is rather small. However this information cannot be ignored as a complete dynamic picture of polycarbonate is developed.

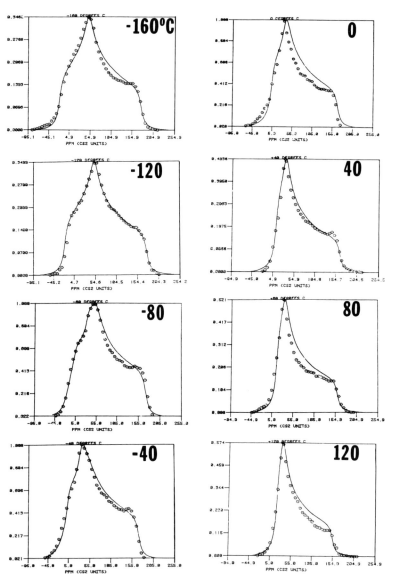

Figure 7-12. The lines are simulations of the ^{13}C CSA lineshapes at several temperatures, and the points are taken from the spectra in Figure 7-11. (Reprinted with permission from O'Gara, J. F.; Jones, A. A.; Hung, C.-C.; Inglefield, P. T. *Macromolecules* **1985**, *18*, 1117. Copyright 1985, American Chemical Society.)

Summary of Polycarbonate Dynamics from NMR

Polycarbonate is intrinsically a very mobile polymer chain in spite of a rather complex repeat unit. In dilute solution, segmental motion, phenylene group rotation and methyl group rotation are all rapid and have rather low apparent barriers to surmount. However, even in dilute solution, segmental motion cannot be characterized by a single correlation time, although phenylene group rotation and methyl group rotation can still be treated in the simplest manner. Presently, the best model for segmental motion appears to be the approach by Hall and Helfand[7] based on computer simulations of chain dynamics in polyethylene.

In the bulk, segmental motion of the type that creates new backbone directions is present above the glass transition. Below the glass transition, the backbone bonds represented by the BPA unit are not reorientated significantly, as evidenced by the proton spectra. Carbon-13 and deuterium spectroscopy show phenylene group rotation persists in the glass in the form of π flips and restricted rotation. Carbon-13 spectroscopy is less definitive with respect to the motion of the carbonate unit.

Proton-relaxation experiments show the correlation function describing the motion to be complex. In the frequency domain, there is significant spectral density over five decades. The lineshape, time-scale analysis does not require as complex a correlation function, although it is not inconsistent with the presence of it either. The only correlation function found that is capable of summarizing the relaxation data is the fractional exponential form commonly referred to as the Williams-Watts function. Justifications of this form have been offered by Ngai[9, 10] and by Schlesinger and Montroll.[45]

Relationship Between NMR and Other Dynamical Methods

Nuclear magnetic resonance has clearly established the geometry of the phenylene motion in BPA polycarbonate and proton relaxation yields the correlation function. It is fair to ask whether the motions seen by proton and ^{13}C spectroscopy are the same and then whether these motions are related to dynamic mechanical response and dielectric relaxation.

First a relaxation map can be constructed by plotting the logarithm of the frequencies versus temperature for the T_1 minima, the $T_{1\rho}$ minima, the average coalescence point of ^{13}C lineshape collapse, and maximum point of dynamic mechanical loss. In Figures 7-13 and 7-14, this plot is seen to be linear for chloral polycarbonate and BPA polycarbonate. The line on these plots is derived from the Williams-Watts-Ngai fit of the proton-relaxation data shown in Figures 7-3 and 7-5. The conclusion to be drawn is that the NMR data are linked in time and they are linked to the time scale of the mechanical loss.

Furthermore, the breadth as well as the position of the mechanical loss peak is fairly well represented by the correlation function developed from proton-relaxation data as shown in Figure 7-15.

The frequency–temperature superposition of the NMR data and the dynamic mechanical loss data raises questions. It is difficult to see how jumps of a symmetrical group between two equivalent minima can produce a large mechanical loss. Also motion of the phenylene group certainly cannot produce a large dielectric loss nor does it seem likely that the large dielectric loss can be attributed entirely to carbonate group oscillation. The dynamic mechanical and dielectric loss peaks are rather large for such sub-glass transition peaks and a mechanism reflecting this fact is required.

A motion can be proposed that is consistent with the NMR data and the presence of large low-temperature loss peaks.[46] This proposal is offered in the

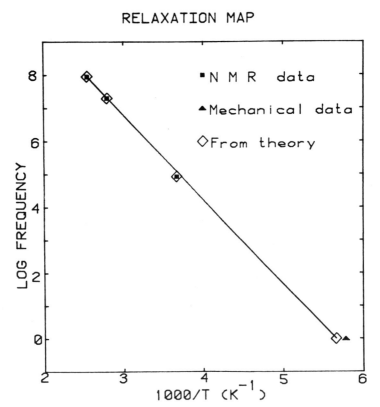

Figure 7-13. Relaxation map for chloral polycarbonate. The three NMR points are spin–lattice relaxation minima at 90 MHz, 50 MHz, and 43 kHz corresponding to the data in Figure 7-3. The mechanical datum is the maximum of a dynamic mechanical loss peak at 1 Hz. The line is a Williams-Watts-Ngai fit based on parameters determined from the proton-relaxation data in Figure 7-3. (Reprinted with permission from Jones, A. A.; O'Gara, J. F.; Inglefield, P. T.; Bendler, J. T.; Yee, A. F.; Ngai, K. L. *Macromolecules* **1983**, *16*, 658. Copyright 1983, American Chemical Society.)

RELAXATION MAP

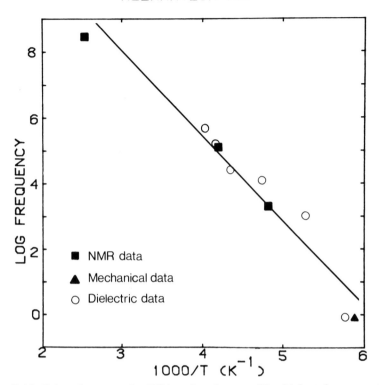

Figure 7-14. Relaxation map for BPA polycarbonate. The highest frequency NMR point is the 90 MHz proton T_1 minimum. The next highest frequency NMR point is the 43 kHz $T_{1\rho}$ minimum. The lowest frequency NMR point is the ^{13}C chemical-shift anisotropy lineshape. The open circles are maxima of various dielectric loss curves taken at different frequencies. The positions of all points have an uncertainty of the order of 10° because of the breadth of the loss peaks and relaxation minima. (Reprinted with permission from O'Gara, J. F.; Jones, A. A.; Hung, C.-C.; Inglefield, P. T. *Macromolecules* **1985**, *18*, 1117. Copyright 1985, American Chemical Society.)

spirit of stimulating further investigation because, although it is consistent with the data in hand, it is not proved by it. No doubt further experimental and theoretical tests can be developed to evaluate the proposal.

The basic motion is shown in Figure 7-16 and consists of an interchange of a cis–trans carbonate conformation with a neighboring trans–trans carbonate conformation. This is not a simultaneous interchange but a correlated conformational change of the type observed by Helfand[7] in computer simulations of polyethylene chains. The carbonate groups are reoriented by this process but the BPA groups are not reoriented significantly, although they are translated. The carbonate groups are thought to be mostly in the trans–trans conformation, with a smaller number of cis–trans units present. The conformational interchange diffuses the cis–trans conformation down the chain of largely

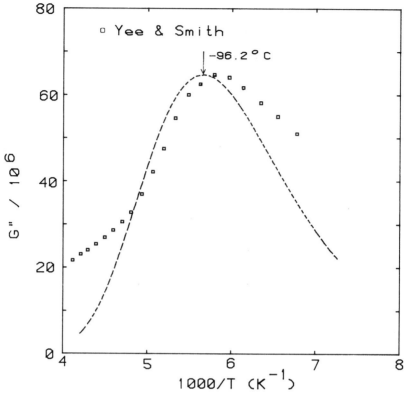

Figure 7-15. Dynamic mechanical spectrum for chloral polycarbonate. The dashed line is the simulation employing the Williams-Watts-Ngai fractional exponential with the parameters set from the proton relaxation data and fit shown in Figure 7-3. (Reprinted with permission from Jones, A. A.; O'Gara, J. F.; Inglefield, P. T.; Bendler, J. T.; Yee, A. F.; Ngai, K. L. *Macromolecules* **1983**, *16*, 658. Copyright 1983, American Chemical Society.)

trans–trans units. As this diffusional process occurs, BPA units are translated. The occurrence of π flips is coupled to this motion through intramolecular or intermolecular interactions. The presence of mobile phenylene groups may make the barrier to conformational interchange smaller by providing some fluctuations in the surrounding glassy matrix and vice versa.

The proposed motion is in agreement with NMR data because the BPA unit is not reoriented significantly although the phenylene groups undergo π flips and oscillation about the C_1C_4 axis, which corresponds to a virtual backbone bond direction. The proton lineshape data show these virtual backbone bonds do not change direction whereas the ^{13}C and deuterium results show π flips and oscillation about the C_1C_4 axis. The ^{13}C carbonate lineshape may not show significant changes because of coincidental geometric relationships or because most of the carbonate units are trans–trans and a minority are cis–trans, so interchange produces an average shape not greatly different from the trans–trans case. In the proposed model, the shear dynamic mechanical loss and the dielectric loss result from the reorientation of the carbonate group. A larger bulk mechanical loss results from the translation of the larger BPA unit during the conformational interchange. The translation produces volume fluctuations between chains. The bulk loss, shear loss, and dielectric loss are all coincident in time because the same cooperative motion is involved. The π flips are linked in time because they are controlled by the fluctuations in the glassy environment. The correlation function describing this motion is complex because the motion is a segmental rearrangement and also because of the strong intermolecular interactions in the glass. A complex correlation function would lead to broad loss peaks and relaxation minima. This segmental motion is different from the segmental motions of the glass transition because it creates no new backbone bond directions but merely interchanges carbonate group conformations. Because the proposed motion is a segmental motion, albeit not as general as those of the glass transition, large dielectric and mechanical loss peaks may result. The phenylene group and the rest of the chain undergo oscillations reflecting the general flexibility of this polymer but these oscillations do not produce the large loss peaks. The

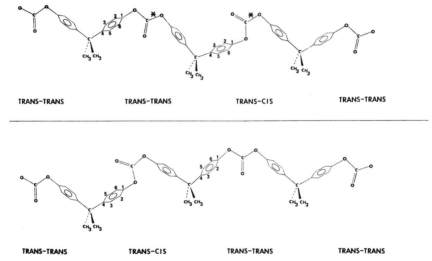

Figure 7-16. BPA polycarbonate chains and the local motion corresponding to the loss maxima and relaxation minima in the glass. The carbonate CO bonds with asterisks indicate points of bond rotation. The phenylene rings undergoing flips in association with the CO bond rotations are numbered. The correlated conformational change from the top chain to the lower chain involves two neighboring carbonate groups and is produced by the CO bond rotations which interchange the trans–trans and cis–trans conformations. Conventional bond angles of 109° are used for all backbone bonds except the carbonate bonds, which are set at 120°. These choices lead to an 11° change in the $C_1 C_4$ axis of the phenylene groups in the BPA unit between the carbonate unit undergoing conformational change. (Reprinted with permission from Jones, A. A. *Macromolecules* **1985**, *18*, 902. Copyright 1985, American Chemical Society.)

general flexibility of polycarbonate may lower the transition state during the conformational interchange, thus playing an important secondary role.

This dynamic model came out of the new information provided by NMR and the picture of segmental motion developed by Hall and Helfand.[7] The model agrees quantitatively with the phenylene group NMR data and is at least consistent with other dynamic information on polycarbonate. Quantitative estimates of the magnitude of the dielectric and mechanical loss peaks based on this proposal would be helpful in evaluating it but represent substantial projects in themselves. In any case, NMR spectroscopy has interjected new ideas into the discussion of motion in polycarbonate.

Acknowledgment

This research was carried out with the financial support of National Science Foundation Grant DMR-790677, of National Science Foundation Equipment Grant No. CHE 77-09059, of National Science Foundation Grant No. DMR-8108679, and of U.S. Army Research Office Grant DAAG 29-82-G-0001.

References

1. Schaefer, J.; Stejskal, E. O.; Buchdahl, R. *Macromolecules* **1977**, *10*, 384.
2. Inglefield, P. T.; Amici, R. M.; O'Gara, J. F.; Hung, C.-C.; Jones, A. A. *Macromolecules* **1983**, *16*, 1552.
3. Spiess, H. W. *Colloid Polym. Sci.* **1983**, *261*, 193.
4. Jelinski, L. W.; Dumais, J. J.; Engel, A. K. *Macromolecules* **1983**, *16*, 492.
5. Inglefield, P. T.; Jones, A. A.; Lubianez, R. P.; O'Gara, J. F. *Macromolecules* **1981**, *14*, 288.
6. Jones, A. A.; O'Gara, J. F.; Inglefield, P. T.; Bendler, J. T.; Yee, A. F.; Ngai, K. L. *Macromolecules* **1983**, *16*, 658.
7. Hall, C. K.; Helfand, E. *J. Chem. Phys.* **1982**, *77*, 3275.
8. Weber, T. A.; Helfand, E. *J. Phys. Chem.* **1983**, *87*, 2881.
9. Ngai, K. L.; White, C. T. *Phys. Rev. B* **1979**, *20*, 2475.
10. Ngai, K. L. *Phys. Rev. B* **1980**, *22*, 2066.
11. Williams, G.; Watts, D.C. *Trans. Faraday Soc.* **1970**, *66*, 80.
12. English, A. D.; Dybowski, C. R. *Macromolecules* **1984**, *17*, 446.
13. McCall, D. W. *Acc. Chem. Res.* **1971**, *4*, 223.
14. Heatley, F. "Progress in NMR Spectroscopy", Vol. 13; Pergamon Press, Ltd.: Oxford, 1979, p. 47.
15. Jones, A. A.; Robinson, G. L.; Gerr, F. E.; Bisceglia, M.; Shostak, S. L.; Lubianez, R. P. *Macromolecules* **1980**, *13*, 95.
16. Levy, G. C.; Axelson, D. E.; Schwartz, R.; Hochmann, J. *J. Am. Chem. Soc.* **1978**, *100*, 410.
17. Jones, A. A.; Bisceglia, M. *Macromolecules* **1979**, *12*, 136.
18. Garroway, A. N.; VanderHart, D. L. *J. Chem. Phys.* **1979**, *71*, 2772.
19. Schaefer, J. *Macromolecules* **1973**, *6*, 882.
20. Fleming, W. W.; Lyerla, J. R.; Yannoni, C. S. *ACS Symp. Ser.* **1984**, *247*, 83.
21. Tarpey, M. F.; Lin, Y.-Y.; Jones, A. A.; Inglefield, P. T. *ACS Symp. Ser.* **1984**, *247*, 67.
22. Cohen-Addad, J. P. *J. Chem. Phys.* **1974**, *60*, 2440.
23. Cohen-Addad, J. P.; Faure, J. P. *J. Chem. Phys.* **1974**, *61*, 1571.
24. Cohen-Addad, J. P.; Guillermo, A. *J. Polym. Sci., Polym. Phys. Ed.* **1984**, *22*, 931.

25. Jones, A. A.; Robinson, G. L.; Gerr, F. E. *ACS Symp. Ser.* **1974**, *103*, 271.
26. Garroway, A. N.; Ritchey, W. M.; Moniz, W. B. *Macromolecules* **1982**, *15*, 1051.
27. Vega, A. J.; English, A. D. *Macromolecules* **1980**, *13*, 1635.
28. Henrichs, P. M.; Linder, M.; Hewitt, J. M.; Massa, D.; Isaacson, H. V. *Macromolecules* **1984**, *17*, 2412.
29. Slotfeldt-Ellingsen, D.; Resing, H. A. *J. Phys. Chem.* **1980**, *84*, 2204.
30. Mehring, M. "Principles of High Resolution NMR in Solids"; Springer Verlag: Berlin, 1983.
31. O'Gara, J. F. Ph.D. Thesis, Clark University, Worcester, MA, 1984.
32. Tonelli, A. E. *Macromolecules* **1972**, *5*, 558.
33. Tonelli, A. E. *Macromolecules* **1973**, *6*, 503.
34. O'Gara, J. F.; Desjardins, S. G.; Jones, A. A. *Macromolecules* **1981**, *14*, 64.
35. Connolly, J. J.; Gordon, E.; Jones, A. A. *Macromolecules* **1984**, *17*, 722.
36. Lin, Y.-Y.; Jones, A. A.; Stockmayer, W. H. *J. Polym. Sci., Polym. Phys. Ed.* **1984**, *22*, 2195.
37. Woessner, D. E. *J. Chem. Phys.* **1962**, *36*, 1.
38. Bendler, J. T. *Ann. N.Y. Acad. Sci.* **1981**, *371*, 299.
39. Yee, A. F.; Smith, S. A. *Macromolecules* **1981**, *14*, 54.
40. O'Gara, J. F.; Jones, A. A.; Hung, C.-C.; Inglefield, P. T. *Macromolecules* **1985**, *18*, 1117.
41. Bloembergen, N.; Rowland, T. J. *Acta Metall.* **1955**, *1*, 731.
42. Rose, M. "Elementary Theory of Angular Momentum"; John Wiley and Sons: New York, 1957.
43. Schaefer, J.; Stejskal, E. O.; McKay, R. A. *Macromolecules* **1984**, *17*, 1479.
44. Wemmer, D. E. Ph.D. Thesis, University of California, Berkeley, CA, 1979.
45. Schlesinger, M. F.; Montroll, E. W. *Proc. Nat. Acad. Sci. (USA)* **1984**, *81*, 1280.
46. Jones, A. A. *Macromolecules* **1985**, *18*, 902.

8

HIGH-RESOLUTION NMR STUDIES OF ORIENTED POLYMERS

Anita J. Brandolini

MOBIL CHEMICAL COMPANY
RESEARCH AND DEVELOPMENT
EDISON, NJ 08818

Cecil Dybowski

DEPARTMENT OF CHEMISTRY
UNIVERSITY OF DELAWARE
NEWARK, DELAWARE 19716

Orientation in Polymers

The rather unusual, but extremely useful, properties of polymeric materials are a direct consequence of their complex molecular structure. The exceptional length of polymer chains allows a wide variety of conformations, both crystalline and amorphous. When polymer materials are processed into useful objects, the properties of the final product are often not the same as those of the original material, not as a result of chemical change, but as a result of a change in the supramolecular organization of these polymer chains.[1,2]

© 1986 VCH Publishers, Inc.
Komoroski (ed): High-Resolution NMR Spectroscopy of Synthetic Polymers in Bulk

The Orientation Phenomenon

One particularly important type of reorganization occurs when a polymer is subjected to a mechanical stress. A simplistic representation of the orientation process is shown in Figure 8-1.[3] The imposition of a stress (in this case, a uniaxial tensile stress) causes a molecule-fixed set of axes to tilt toward the stress axis (or draw axis) as the sample is stretched. After the stress is removed, the polymer does not return to its original state. It retains some degree of permanent elongation, characterized by the draw ratio, $\lambda = L/L_0$. More importantly, the molecules and molecular segments also retain some orientation.[1,2]

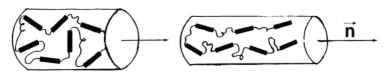

Figure 8-1. Schematic representation of the orientation process. A uniaxial tensile stress along ñ causes polymer molecule segments (designated by solid blocks) to align in the direction of the imposed stress.

The structure of a deformed polymer is obviously anisotropic, and so it is not surprising that its mechanical properties are also anisotropic. If the stress is uniaxial, then such properties as tensile strength and stiffness will be greatest along the draw direction. This effect is exploited in the manufacture of fibers. When high strength is required in two directions, as in a film, biaxial orientation may be induced by drawing in two different directions.[1,2]

Many different experimental approaches have been used to characterize orientation in polymers: birefringence, ultraviolet–visible light spectroscopy, infrared dichroism, sonic modulus measurements, various X-ray techniques, nuclear magnetic resonance, and others.[1,2,4] Nuclear magnetic resonance spectroscopy can provide a detailed description of the oriented state in both the crystalline and amorphous phases of polymers. Orientation-dependent NMR interactions[5–7] (homonuclear dipole–dipole, quadrupole, and chemical shift) have all been used to study order in deformed polymers.[1,8,9] After an introduction to the mathematical description of orientation, applications of NMR to these polymeric systems are surveyed, focusing on the anisotropy of chemical-shift interactions. The primary example for these studies is uniaxially deformed polytetrafluoroethylene (PTFE), a semicrystalline polymer (ie, one with both crystalline and amorphous domains).

Quantitative Description

The position of any polymer chain (or a segment of that chain) in an oriented sample can be described by the Euler angles θ, ϕ, and ψ[1,2,8–10] relative to some coordinate axes. In the most general case, there is a distribution of orientations, $N(\theta, \phi, \psi)$, describing the sample anisotropy. When the deformation is uniaxial, the distribution is cylindrically symmetrical about the draw axis, and $N(\theta, \phi, \psi)$ reduces to a function of one variable, $N(\theta)$, as shown in Figure 8-2.

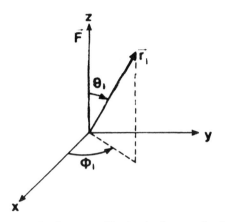

Figure 8-2. Position vector $\vec{r}_i$ of a crystallite in the frame of reference of an oriented polymer (x, y, z). The draw direction is designated by $\vec{F}$.

The exact form of this orientation distribution function is not necessarily known. However, $N(\theta)$ can be expanded as a series in the even-ordered associated Legendre polynomials P_l of cos θ, where[10,11]:

$$P_2(\cos \theta) = \tfrac{1}{2}(3 \cos^2 \theta - 1) \tag{8-1a}$$

$$P_4(\cos \theta) = \tfrac{1}{8}(35 \cos^4 \theta - 30 \cos^2 \theta + 3) \tag{8-1b}$$

and so forth. The expansion of $N(\theta)$ is written[11]:

$$N(\theta) = \sum_{\substack{l=0 \\ l \text{ even}}}^{\infty} a_l P_l(\cos \theta) \tag{8-2}$$

where

$$a_l = \frac{2l + 1}{4\pi} \langle P_l(\cos \theta) \rangle \tag{8-3}$$

and $\langle P_l(\cos \theta) \rangle$ represents the average value of the lth Legendre polynomial (denoted hereafter simply by $\langle P_l \rangle$). The sum in Eq. (8-2) includes only even-ordered terms because of symmetry considerations.[11] Because the $\langle P_l \rangle$ are given by:

$$\langle P_l \rangle = \sum_{\substack{i=0 \\ i \text{ even}}}^{l} b_i \langle \cos^i \theta \rangle \tag{8-4}$$

the expansion of N(θ) entails calculation of the moments of the distribution ($\langle\cos^2\theta\rangle$, $\langle\cos^4\theta\rangle$, ...) where:

$$\langle\cos^n\theta\rangle = \frac{\int_0^\pi N(\theta)\cos^n\theta\sin\theta\,d\theta}{\int_0^\pi N(\theta)\sin\theta\,d\theta} \tag{8-5}$$

These quantities can be used to make certain conclusions about the shape of N(θ). One cannot, of course, evaluate the sum in Eq. (8-2) over an infinite number of terms. Fortunately, the truncation error can be made small for low to moderate values of $\langle\cos^i\theta\rangle$ by including only a few terms.[12] For relatively high degrees of orientation, it may be simpler to make certain assumptions about the shape of N(θ), testing this assumed shape against the experimental results.[13]

Simple Theoretical Models

Several models have been proposed that attempt to describe the development of orientation in drawn polymers. One such scheme is the affine model, in which the polymer is assumed to be composed of a network of flexible chains. As the material deforms, the macroscopic change in length is reflected in the rotation of network segments.[1,14,15] The affine model leads to a fairly slow initial increase in $\langle\cos^2\theta\rangle$, as shown by the dashed line in Figure 8-3.[1] This model holds fairly well for rubber-like and some amorphous, glassy polymers, but it fails for semicrystalline polymers, which typically exhibit quite different behavior, as shown by the solid line in Figure 8-3.

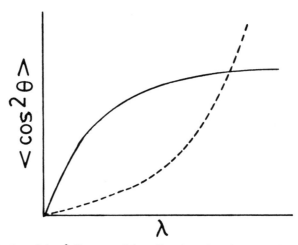

Figure 8-3. A plot of $\langle\cos^2\theta\rangle$ versus λ for affine (– – –) and pseudoaffine (—) deformation models.

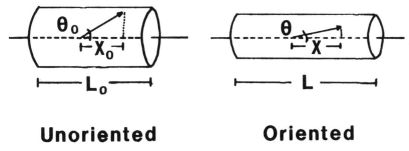

Unoriented   Oriented

Figure 8-4. Pseudoaffine deformation model. The length of the sample increases from L_0 to L, causing the crystallite to rotate such that $X/X_0 = L/L_0$.

If one assumes that the crystallites in the polymer act as rigid rods embedded in a viscous matrix, then $\langle \cos^2 \theta \rangle$ does vary with the draw ratio λ in the manner expected for the crystalline material. This deformation scheme, the pseudoaffine, assumes that rigid crystallites rotate freely to account for the macroscopic increase in length, as illustrated in Figure 8-4.[1,16,17] The moments of the orientation distribution function for this type of behavior are given in terms of λ by[18]:

$$\langle \cos^2 \theta \rangle = \frac{\lambda^3}{(\lambda^3 - 1)^{1/2}} \left\{ \frac{1 - \tan^{-1} (\lambda^3 - 1)^{1/2}}{(\lambda^3 - 1)^{1/2}} \right\} \tag{8-6}$$

and

$$\langle \cos^k \theta \rangle = \frac{\lambda^3}{(k - 2)(\lambda^3 - 1)} \left\{ (k - 1)\langle \cos^{k-2} \theta \rangle - 1 \right\} \tag{8-7}$$

Anisotropic NMR Interactions

The NMR spectrum of an oriented system, such as a deformed polymer, is subject to interactions that arise from the anisotropic distribution of molecular orientations.[5,7] Strong homonuclear dipolar coupling dominates the spectrum of an abundant spin-1/2 nucleus (such as ^1H or ^{19}F) in the solid state, just as quadrupolar coupling is the principal interaction observed when the nuclear spin is greater than 1/2 (for example, for ^2H). The chemical shift dominates the spectrum of rare spin-1/2 nuclei, such as ^{13}C, under conditions of high-power heteronuclear decoupling (see Chapter 2), and special homonuclear dipolar decoupling experiments permit observation of ^1H or ^{19}F chemical-shift spectra (see Chapter 9). Each of these nuclear interactions exhibits a dependence on the state of orientation of the molecular framework relative to the magnetic field direction, a dependence that can be analyzed to characterize the sample anisotropy.

Homonuclear Dipolar Coupling

The NMR spectrum of abundant spin-1/2 nuclei in a solid is broadened considerably by homonuclear dipole–dipole interactions, resulting in spectra whose linewidths typically exceed 10 kHz.[5-7] The strength of the homonuclear couplings between two nuclei i and j depends on, among other variables, the length of the internuclear vector $\vec{r}_{ij}$ and on the angle ξ_{ij} between $\vec{r}_{ij}$ and the static magnetic field, $\vec{B}_0$. For a many-spin system, the magnitude of this coupling is reflected in the Van Vleck second moment of the spectrum[1,19]:

$$\overline{\Delta\omega_2} \propto \sum_{i>j} r_{ij}^{-6}(3\cos^2\xi_{ij} - 1)^2 \qquad (8\text{-}8)$$

The quantity $\overline{\Delta\omega_2}$ is calculated from the observed spectrum $I(\omega)$:[1]

$$\overline{\Delta\omega_2} = \frac{\displaystyle\int_{-\infty}^{\infty} I(\omega)(\omega - \omega')^2 \, d(\omega - \omega')}{\displaystyle\int_{-\infty}^{\infty} I(\omega) \, d(\omega - \omega')} \qquad (8\text{-}9)$$

where ω' is the center of the resonance. Because these broad-line spectra are symmetrical about ω', all odd moments are zero. For a randomly oriented sample, $\overline{\Delta\omega_2}$ does not depend on the position of the sample relative to $\vec{B}_0$. In an anisotropic system, on the other hand, $\overline{\Delta\omega_2}$ varies with β, the angle between the sample's unique axis and $\vec{B}_0$:[1]

$$\overline{\Delta\omega_2} \propto \sum_{l=0,2,4} a_l P_l(\cos\beta)\langle P_l(\cos\theta)\rangle \qquad (8\text{-}10)$$

The factors a_l include both normalization constants and the "lattice sums," which relate the position of $\vec{r}_{ij}$ to the orienting unit (crystallite or chain segment). Both $\langle\cos^2\theta\rangle$ and $\langle\cos^4\theta\rangle$ can be evaluated from Eq. (8-10), and similar calculations based on the fourth moment, $\overline{\Delta\omega_4}$, yield $\langle\cos^6\theta\rangle$ and $\langle\cos^8\theta\rangle$.

An important application of the analysis of dipolar-coupled spectra has been the testing of the theoretical deformation schemes discussed in the Simple Theoretical Models section. McBrierty and Ward[20] illustrated that the pseudoaffine model best described the orientational behavior of polyethylene, and Kashiwagi and Ward showed that this scheme also applied to poly(vinyl chloride).[21] Poly(methyl methacrylate), on the other hand, exhibited affine behavior.[22]

The quantitative analysis of orientation outlined above assumes that both $\vec{r}_{ij}$ and ξ_{ij} are static on the NMR time scale. This condition pertains only at low temperature, in the so-called rigid-lattice limit.[1] At higher temperatures, molecular motions partially average both $\vec{r}_{ij}$ and ξ_{ij}. The second and fourth moments of the spectrum reflect this dipolar averaging under those conditions[1]:

$$\overline{\Delta\omega_2} \propto \left\langle \frac{3\cos^2\xi_{ij} - 1}{r_{ij}^3} \right\rangle^2 \qquad (8\text{-}11)$$

In this temperature regime, the exact dependence of $\overline{\Delta\omega_n}$ on β will vary according to the specific type of molecular motion taking place. In other words, the angular dependence contains dynamic as well as orientational information. Olf and Peterlin[23] and Folkes and Ward[24] developed models for motion in polyethylene based on such studies of spectral moments.

Broad-line NMR studies have been reviewed in greater detail elsewhere.[1] Many other polymers have been investigated by this approach: polyethylene,[25-36] polypropylene,[26,32,36,37-41] polyisoprene,[42] poly(vinyl alcohol),[28,36] polytetrafluoroethylene,[28,36,43-47] poly(vinylidene fluoride),[48] polyoxymethylene,[36,49-51] poly(ethylene terephthalate),[31,36,52-57] polyformaldehyde,[58,59] polytetrahydrofuran,[60] Nylon 6,[36,57] Nylon 66,[61,62] and polycaproamide.[63]

Other ^1H and ^{19}F NMR parameters, such as T_1, T_2, and $T_{1\rho}$, are orientation dependent and can be utilized to examine nonrandomness in samples. Studies of tetrafluoroethylene–hexafluoropropylene copolymers, for example, have shown that T_1 has only minimal dependence on β, whereas T_2 and $T_{1\rho}$ are strongly anisotropic.[64] One difficulty of using relaxation analysis is in being able to extract the appropriate contributions to the relaxation function. This area is discussed in great detail by McBrierty and Douglass.[64]

Quadrupolar Coupling

Nuclei with spin $> 1/2$ experience quadrupolar coupling arising from the interaction of the nuclear quadrupole moment with the electric-field gradient of the surroundings. In polymer studies, the isotope of primary interest is ^2H in natural abundance or, more commonly, in selectively enriched samples. A deuteron subject to a quadrupolar interaction has a doublet structure, the spacing of which depends on orientation[65]:

$$\delta\omega = \omega_Q[1/2(3\cos^2\xi - 1) + 1/2\ \eta\sin^2\xi\cos 2\chi] \qquad (8\text{-}12)$$

where ω_Q is the static quadrupolar coupling constant and ξ and χ are the polar angles in the quadrupolar principal axis system of $\vec{B}_0$.[65] The parameter η is the asymmetry of the quadrupolar coupling. Frequently, the materials observed are organic, for which the quadrupole coupling tensor is axially symmetrical ($\eta \approx 0$). The unique axis is then assumed to be along the C—^2H bond, and ξ represents the angle between this bond and the magnetic field direction.

Deuterium NMR spectroscopy has not been applied widely to the study of orientation in polymers until recently. Elastomers have been investigated through the use of deuterated swelling agents[66] and of site-specifically labeled rubbers, very small degrees of order being measured in these experiments.[65] The complex orientational behavior of highly drawn semicrystalline polyethylene was studied by Spiess et al, who illustrated the dependence of the ^2H lineshape on β, reproduced in Figure 8-5.[13] Rather than determine the orientational distribution function directly, they chose to model the function as a

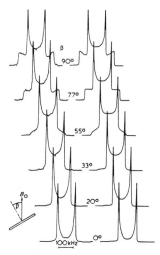

Figure 8-5. The β-dependence of ^2H lineshape for oriented polyethylene. (From reference 13, by permission. Copyright 1981 by Butterworth and Co., Ltd.)

Gaussian in $\sin \theta$:[13]

$$N(\theta) = \frac{\int_0^\pi \exp\left(-\sin^2 \theta / \sin^2 \bar{\theta}\right) \cos \theta \sin \theta \, d\theta}{\int_0^\pi \exp\left(-\sin^2 \theta / \sin^2 \bar{\theta}\right) \sin \theta \, d\theta} \qquad (8\text{-}13)$$

and to calculate the best fit to the observed spectrum by varying the width parameter $\bar{\theta}$. In addition to the orientational information, careful analysis of the change in spectral features with temperature can infer mechanisms and rates of motion in these materials. Because the deuteron spectra are uniquely affected by a single interaction (the quadrupolar coupling) that is directly related to chemical structure (the C—^2H bond), deuterium NMR is a probe of geometry and structure in polymers that can be expected to be exploited more in the future (see Chapter 10).[9,13,66]

Chemical-Shift Interaction

The chemical shift is the interaction most familiar to those who have experience with high-resolution NMR of solutions. The precise resonant frequency of a nucleus is determined by the electron density around that nucleus.[5,7] This interaction is also anisotropic. In liquids and solutions, rapid molecular motions average the orientation-dependent components of the chemical-shift tensor, so the resonance is observed only at the isotropic value. In a solid, however, chemical bonds lie at fixed orientations relative to $\vec{B}_0$, and a wide variety of chemical shifts are observed from nuclei in identical chemical environments.[5-7] For a randomly ordered sample (eg, a powder), the lineshape depends on the symmetry of the chemical-shift tensor, as shown in Figure 8-6. For the case of an axially symmetric chemical-shift tensor, there is a one to one correspondence between ξ and the observed shift[5]:

$$\sigma = \sigma_{\parallel} \cos^2 \xi + \sigma_{\perp} \sin^2 \xi \qquad (8\text{-}14)$$

Orientation of the sample results in a lineshape that is a form of the powder spectrum convoluted with a distribution function.[67]

Just as for the case of the dipolar-coupled ^{1}H broad-line spectra and the quadrupolar-split ^{2}H spectra discussed above, orientational information can be extracted from chemical-shift spectra by calculating moments of the

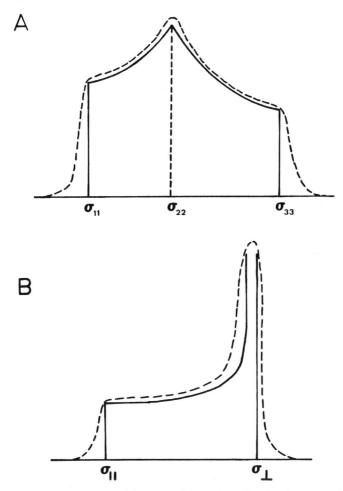

Figure 8-6. Theoretical (——) and broadened (– – –) powder lineshapes resulting from: (A) nonaxially symmetrical chemical-shift interaction with principal tensor components σ_{11}, σ_{22}, and σ_{33}; (B) axially symmetrical chemical-shift interaction with principal tensor components $\sigma_{11} = \sigma_{22} = \sigma_{\perp}$ and $\sigma_{33} = \sigma_{\parallel}$.

spectra.[67,68] In this chapter the chemical-shift spectra are analyzed in terms of moments relative to the isotropic value.

$$\overline{\Delta\sigma_n} = \frac{\displaystyle\int_{\text{all }\sigma} (\sigma - \sigma_{\text{iso}})^n \, I(\sigma) \, d\sigma}{\displaystyle\int_{\text{all }\sigma} I(\sigma) \, d\sigma} \tag{8-15}$$

Because the chemical-shift spectrum is not symmetric about σ_{iso}, both odd and even moments are nonzero. When the unique axis of the chemical-shift tensor lies along the chain axis, the nth moment, $\overline{\Delta\sigma_n}$, of the spectrum is given by an expansion in the powers of the sine of the angle between the draw axis and the magnetic field direction[64]:

$$\overline{\Delta\sigma_n} = (\sigma_\perp - \sigma_\parallel)^n \sum_{i=0}^{n} B_{in} \sin^{2i} \beta \tag{8-16}$$

where

$$B_{in} = \sum_{j=0}^{n} C_{ji}^n \langle \cos^{2j} \theta \rangle \tag{8-17}$$

The set of coefficients C_{ji}^n has been published elsewhere.[67] The coefficients are related directly to moments of the distribution of crystallite axes relative to the draw axis. The axes of the chemical-shift tensor need not be colinear with crystallite axes. In the general case, the NMR spectrum gives the distribution function of the chemical-shift axis system. To obtain the distribution function of the crystallite axes (a more meaningful parameter for the polymer chemist), a transformation must be made that accounts for the orientation of the chemical-shift axes relative to the system describing the crystallite.

The anisotropy of the chemical shift can be observed for rare-spin nuclei, such as ^{13}C, in spectra obtained employing high-power ^{1}H-decoupling without magic angle spinning.[5-7] In the usual cross-polarization, magic angle spinning (CPMAS) experiment, decoupling suppresses the effects of ^{13}C–^{1}H heteronuclear coupling, whereas MAS suppresses the effects of the chemical-shift dispersion resulting from anisotropy (CSA). To observe this orientational effect, the sample must be stationary in the magnetic field. The dependence of such nonspinning spectra on orientation has been illustrated for polyimide[69] and polycarbonate.[69,70] Detection of the effects of chemical-shift anisotropy for spectra of abundant-spin nuclei, such as ^{1}H or ^{19}F, requires the use of homonuclear dipolar-decoupling pulse sequences, such as WAHUHA-4, MREV-8, or BR-24[5-7] (see Chapter 9). A report has been made of the ^{1}H spectrum of oriented Nylon 6,[71] but most investigations have concentrated on the ^{19}F NMR of polytetrafluoroethylene (PTFE).[67,68,72-76] The type of analysis used and the information gained from these studies of PTFE are also applicable to broad-line, deuterium, or ^{13}C studies of oriented polymers. Our ^{19}F results discussed in the next section illustrate the wealth of information

NMR spectroscopy can provide about the structure of oriented polymers, but the general technique is applicable to a variety of problems where a chemical-shift spectrum of a nucleus imbedded in an ordered sample can be observed.

NMR Spectroscopy of Oriented PTFE

Polytetrafluoroethylene (PTFE) is a semicrystalline polymer that exhibits NMR characteristics that make the resulting solid-state ^{19}F chemical shift spectra relatively simple to analyze. There is, of course, only one type of ^{19}F nucleus present, neglecting end groups, and the chemical-shift dispersion of PTFE is broad (113.6 ppm).[68,73] The chemical-shift tensor of the crystalline portions is axially symmetric above 30°C and the unique axis of the chemical-shift tensor comes closely coincident with the crystallite axis.[72]

Orientation Distribution Functions for Drawn and Compressed PTFE

Chemical-shift ^{19}F spectra obtained with suppression of dipole–dipole coupling (see Chapter 9) of a compressed ($\lambda = 0.80$) and a drawn ($\lambda = 1.40$) sample are compared in Figure 8-7. Each is shown for two orientations indicated by β, the angle between the draw axis and the magnetic field. One spectrum is shown with the direction of the applied stress along the magnetic field

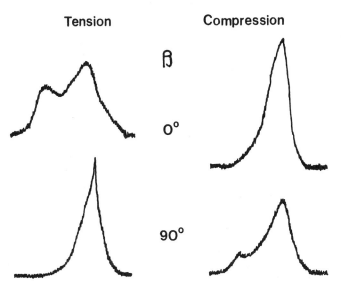

Figure 8-7. Comparison of ^{19}F spectra of polytetrafluoroethylene oriented under tensile ($\lambda = 1.40$) and compressive ($\lambda = 0.80$) stress at two angles β. (From reference 67 by permission. Copyright 1983 by John Wiley and Sons.)

and one with it perpendicular to $\vec{B}_0$. Crystallites lying parallel to the field contribute to the downfield region of the spectrum, whereas those oriented perpendicular to $\vec{B}_0$ contribute to the upfield region. All of these spectra show a broad symmetric resonance in the center arising from fluorine nuclei in CF_2 groups on amorphous chains. The observed lineshape is dependent on β, and comparison of these spectra qualitatively illustrate that the direction of the stress has a very significant effect on the way in which the crystallites orient. Drawing produces alignment of the crystallite axes parallel to the draw axis, and compression causes them to orient perpendicular to the stress.

A more detailed study of the β dependence of the ^{19}F NMR spectrum for a drawn sample ($\lambda = 2.40$) is shown in Figure 8-8. The moments of these spectra, $\Delta\sigma_n$, calculated according to Eq. (8-15), do exhibit the expected functional dependence on $\sin^2 \beta$ (linear for $\Delta\sigma_1$, quadratic for $\Delta\sigma_2$, as seen in Figures 8-9 and 8-10.[67,68] The moments $\langle \cos^{2j} \theta \rangle$ of the orientational distribution func-

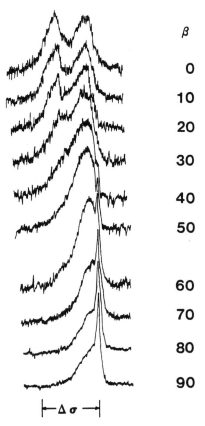

β

0

10

20

30

40

50

60

70

80

90

$\vdash \Delta\sigma \dashv$

Figure 8-8. Fluorine-19 chemical-shift spectra of drawn ($\lambda = 2.40$) PTFE at various angles β. Spectra are not all drawn to the same vertical scale. The anisotropy of the chemical shift, $\Delta\sigma$, is 113.6 ppm. (From reference 67 by permission. Copyright 1983 by John Wiley and Sons, Inc.)

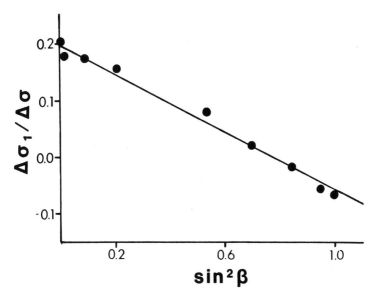

Figure 8-9. A plot of $\overline{\Delta\sigma}_1$ versus $\sin^2 \beta$ for spectra shown in Figure 8-8.

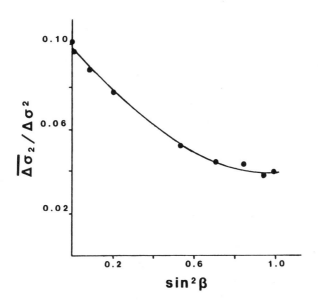

Figure 8-10. A plot of $\overline{\Delta\sigma}_2$ versus $\sin^2 \beta$ for spectra shown in Figure 8-8. (From reference 67 by permission. Copyright 1983 by John Wiley and Sons.)

TABLE 8-1. RESULTS OF MOMENT ANALYSIS FOR A DRAWN
PTFE SAMPLE ($\lambda = 2.40$)

n	$\langle \cos^{2n} \theta \rangle$	$\langle P_{2n}(\cos \theta) \rangle$
1	0.67	0.50
2	0.51	0.09
3	0.41	−0.03
4	0.34	0.03

tion may be calculated from the nonintercept terms of Eq. (8-16), and the results for this sample are summarized in Table 8-1. The distribution function is approximated from the $\langle P_1 \rangle$ values through Eqs. (8-4) and (8-2). Because this approximation to $N(\theta)$ contains only a finite number of terms, the truncated series must be renormalized so that:

$$\int_0^\pi N(\theta) \sin \theta \, d\theta = 1 \qquad (8\text{-}18)$$

The resulting $N(\theta)$ are shown in Figure 8-11 for several samples that have different extension ratios.

To assess the validity of the calculated $N(\theta)$, the reverse analysis may be performed to simulate the NMR spectrum that would result from a particular distribution of orientations:

$$I(\sigma) = \int \int N(\theta) \, \delta[\sigma' - \sigma(\theta)] B(\sigma - \sigma') \sin \theta \, d\theta \, d\sigma' \qquad (8\text{-}19)$$

where $B(\sigma - \sigma')$ is an appropriate broadening function.[67] Figure 8-12 shows a comparison of the simulation according to Eq. (8-19) to the experimentally observed spectra of one sample ($\lambda = 2.40$) for four orientations of the draw axis relative to the magnetic field. The simulated spectra do not, by intention, account for the amorphous resonance.

This analysis has assumed that there is little or no permanent orientation in the amorphous domains of drawn PTFE. The nth moment of each spectrum certainly contains contributions from both domains:

$$\overline{\Delta \sigma_n} = \phi_{cr} \, \overline{\Delta \sigma_n^{cr}} + (1 - \phi_{cr}) \, \overline{\Delta \sigma_n^{am}} \qquad (8\text{-}20)$$

where ϕ_{cr} is the degree of crystallinity. However, if the amorphous chains are truly unordered, then $\overline{\Delta \sigma_n^{am}}$ does not vary with β, and it does not contribute to any values of $\langle P_1 \rangle$ calculated from nonintercept terms. The resonance from nuclei on chains in the amorphous region in Figure 8-8 does not appear to change with β. Although the crystalline contribution to the spectrum can be simulated quite closely by neglecting it, direct spectroscopic evidence is still desirable. The difference in molecular mobility between these two domains (as manifested in T_2) may be exploited by inserting a short delay into the eight-pulse experiment (see Chapter 9). This modified sequence, shown schematically in Figure 8-13B, allows the signal from nuclei on the crystalline chains (and,

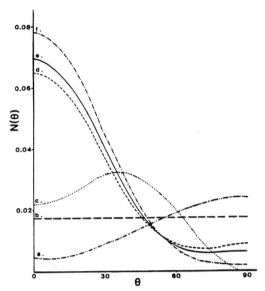

Figure 8-11. Orientation distribution functions $N(\theta)$ for several PTFE samples: (A) $\lambda = 0.82$; (B) $\lambda = 1.00$; (C) $\lambda = 1.40$; (D) $\lambda = 1.75$; (E) $\lambda = 2.10$; (F) $\lambda = 2.40$. (From reference 67, by permission. Copyright 1983 by John Wiley and Sons, Inc.)

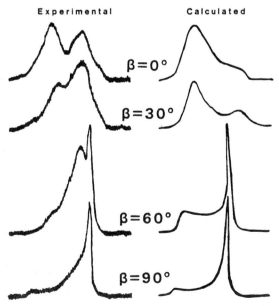

Figure 8-12. Simulation of spectra for drawn PTFE ($\lambda = 2.40$) based on $N(\theta)$ shown in Figure 8-11. The simulation program does not contain a contribution from the amorphous regions of the polymer. (From reference 67, by permission. Copyright 1983 by John Wiley and Sons, Inc.)

hence, with shorter T_2) to decay because of dipolar dephasing before the start of dipolar decoupling. The resulting spectra represent only the spins on chains in the amorphous region. As anticipated, the lineshape from spins in the amorphous regions is independent of β as seen in Figure 8-14, indicating that the amorphous material is relatively unoriented.[76] This result had also been predicted on the basis of broad-line ^{19}F spectra.[45] It is important to note that this lack of amorphous orientation at low to moderate draw ratios may not apply to all polymers, and it may not apply to PTFE at higher elongations.[1,2]

The orientation distributions obtained for the deformed samples displayed in Figure 8-11 deviate from the distribution for a random sample (curve B). The compressed sample (curve A) exhibits a maximum in N(θ) around 90°, whereas the highly drawn samples have maxima at 0°. There are relatively minor changes in the shape of N(θ) for the three most highly drawn samples (curves D, E, and F), but the function of curve C is markedly different. These data imply that some change in the orientation process has occurred between $\lambda = 1.40$ and $\lambda = 1.75$. This behavior may be caused by structural changes that take place around the "yield point."[1,2] At strains below this point the crystalline deformation is primarily intracrystalline, so that orientation is somewhat restricted. Above the yield point, lamellar slip occurs as the layers of crystalline material begin to separate. Orientation proceeds readily under these conditions, and large changes in N(θ) occur for small changes in λ. For PTFE drawn at room temperature, the onset of yield has been reported to occur at strains in the range of 40–65%.[77] The draw ratios, λ, for these PTFE samples represent the permanent elongations of the samples, not necessarily the maximum strains to which the materials were subjected. Therefore, the exact point at which yield occurred cannot be determined based solely on

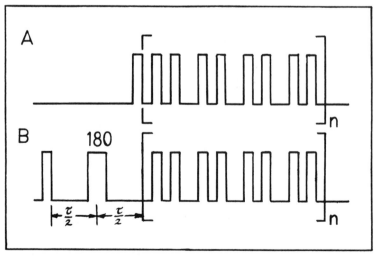

Figure 8-13. (A) Homonuclear dipole-decoupling sequence; (B) pulse sequence used to suppress resonance from relatively rigid crystalline domains.

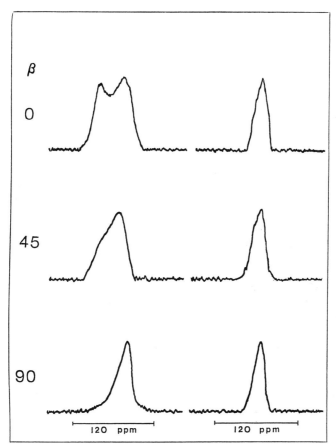

Figure 8-14. Comparison of ^{19}F spectra obtained with normal (on left) and modified (on right) line-narrowing pulse sequences. The isotropic character of the spectra on the right indicates that there is little or no permanent orientation in the amorphous regions of this drawn ($\lambda = 2.10$) PTFE sample.

these data. The distribution functions of Figure 8-11 do, however, reflect some of the microscopic processes that occur during the drawing process.

The three most highly drawn samples (curves D, E, and F) are similar in shape. They may, at first, appear to be simply Gaussian, but best agreement with the experimental data is obtained when the model function is a super-position of a partially ordered and a random distribution function[67]:

$$N(\theta) = \frac{N_r + \exp\left(-\theta^2/2\bar{\theta}^2\right)}{\displaystyle\int_0^\pi \left[N_r + \exp\left(-\theta^2/2\bar{\theta}^2\right)\right] \sin\theta \; d\theta} \tag{8-21}$$

where $\bar{\theta}$ is the width parameter of the Gaussian component, and N_r is the random contribution. The orientational distribution that results from Eq. (8-21) is compared with the experimentally determined $N(\theta)$ in Figure 8-15.

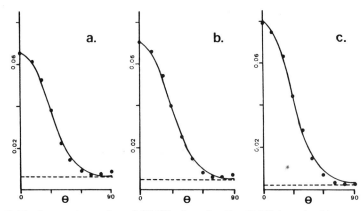

Figure 8-15. Comparison of model N(θ) given by Eq. (8-21) (——) with experimental data (●): (A) $\lambda = 1.75$, $\bar{\theta} = 26°$, $N_r/N = 0.35$; (B) $\lambda = 2.10$, $\bar{\theta} = 26°$, $N_r/N = 0.20$; (C) $\lambda = 2.40$, $\bar{\theta} = 26°$, $N_r/N = 0.15$. The dashed line indicates the contribution from N_r. (From reference 67, by permission. Copyright 1983 by John Wiley and Sons, Inc.)

The width of the Gaussian component ($\bar{\theta} = 26°$) remains the same as λ increases, but the fraction N_r/N of randomly oriented material decreases as the sample is drawn. It is important to note that this random component is crystalline, not amorphous. The physical description that emerges from these model calculations is that, as PTFE is drawn, more and more of this randomly oriented material becomes ordered in a way characterized by a particular Gaussian distribution.

The distribution function's moments $\langle \cos^{2n} \theta \rangle$ derived from the spectra of oriented PTFE can be compared to the values predicted by the affine and pseudoaffine models discussed earlier. Figure 8-16 shows the dependence of

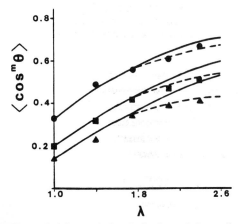

Figure 8-16. Comparison of values of $\langle \cos^m \theta \rangle$ (m = 2n) predicted by pseudoaffine deformation model with experimental data: (A) ●, n = 1; (B) ■, n = 2; (C) ▲, n = 3. The dashed line indicates the trend of the experimental data. (From reference 67, by permission. Copyright 1983 by John Wiley and Sons, Inc.)

the experimental $\langle \cos^{2n} \theta \rangle$ values on λ. These data are clearly closer to the pseudoaffine than to the affine model. This agreement is to be expected because PTFE is a semicrystalline material. Orientation does, however, appear to be developing somewhat more slowly than predicted for a polymer exhibiting pseudoaffine behavior. There is a further discrepancy; the distribution function resulting from purely pseudoaffine deformation is given by[78]:

$$N(\theta) = \frac{\lambda^3 \sin \theta \, d\theta}{[1 + (\lambda^3 - 1) \sin^2 \theta]^{3/2}} \tag{8-22}$$

which gives rise to a continuously narrowing distribution, $N(\theta)$, with increasing λ. As discussed above, this type of distribution is not observed from the NMR studies of drawn PTFE. It appears that, although the pseudoaffine model predicts approximate $\langle \cos^{2n} \theta \rangle$ values, it does not completely describe the complex molecular reorientations taking place when PTFE is subject to a uniaxial elongation.

Orientation Distribution Functions for Annealed PTFE

During the processing of polymeric materials, annealing at temperatures near the melting point is often done to relieve fabrication-induced internal stresses. This kind of processing improves several properties of the unannealed material: impact strength, dimensional stability, and heat resistance.[79] When a drawn semicrystalline polymer, such as PTFE, is annealed, the sample shrinks and the strained tie molecules that connect the crystalline lamellae relax rapidly.[79, 80] This relaxation process causes rapid disorientation of crystallites,[79] and the accompanying crystallite randomization can be analyzed by observing changes in the NMR spectra.[75] Figure 8-17 compares spectra of drawn PTFE ($\lambda_0 = 2.10$) before and after annealing at 300°C. It is clear that considerable crystallite disorientation has taken place during the annealing process. The annealed sample's lineshapes at various angles, β, are similar (as compared to those of the unannealed sample), indicating that the distribution of crystallite orientations is more nearly random after heating. There is still some residual orientation, however, as evidenced by the fact that the lineshape for the annealed polymer does change slightly as a function of β. The values of $\langle P_2 \rangle$ calculated from these spectra depend strongly on the degree of shrinkage, whereas the values of $\langle P_4 \rangle$ and $\langle P_6 \rangle$ are relatively independent of shrinkage.[75] A plot of $\langle P_2 \rangle$ versus shrinkage is shown in Figure 8-18 and, in agreement with other results,[12] the relationship between $\langle P_2 \rangle$ and λ', the draw ratio after annealing, is linear.

As indicated by the spectra of Figure 8-17 or the derived distribution functions $N(\theta)$ of the annealed samples, the randomization process that occurs during annealing is incomplete. Distribution functions for several annealed samples whose original draw ratio was 2.10 are given in Figure 8-19.[75] This figure also shows the unoriented case for comparison. It is obvious that the crystallites have suddenly relaxed to a distribution that is very nearly random

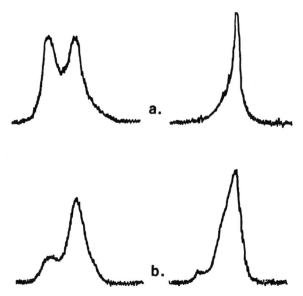

Figure 8-17. Comparison of spectra of unannealed (A) and annealed (B) PTFE ($\lambda_0 = 2.10$) shown for $\beta = 0°$ (on left) and $\beta = 90°$ (on right). (Reprinted with permission from Brandolini, A. J.; Rocco, K. J.; Dybowski, C. *Macromolecules* **1984**, *17*, 1455. Copyright 1984, American Chemical Society.)

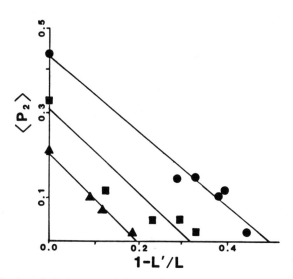

Figure 8-18. A plot of $\langle P_2 \rangle$ versus shrinkage for deformed PTFE samples annealed at 300°C. (●) $\lambda_0 = 2.10$; (■) $\lambda_0 = 1.75$; (▲) $\lambda_0 = 1.40$. (Reprinted with permission from Brandolini, A. J.; Rocco, K. J.; Dybowski, C. *Macromolecules* **1984**, *17*, 1455. Copyright 1984, American Chemical Society.)

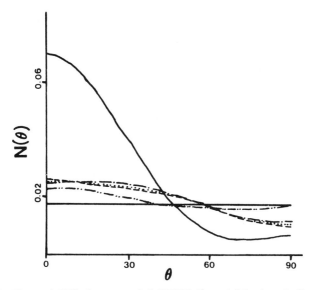

Figure 8-19. Plots of N(θ) for annealed PTFE ($\lambda_0 = 1.75$): $(- \cdot -)$ $\lambda' = 1.17$; $(- \cdot \cdot)$ $\lambda' = 1.24$; $(- - -)$ $\lambda' = 1.34$; $(\cdot \cdot \cdot)$ $\lambda' = 1.53$; (———) $\lambda' = 1.75$ (unannealed). The solid horizontal line indicates the value for a random distribution. (Reprinted with permission from Brandolini, A. J.; Rocco, K. J.; Dybowski, C. *Macromolecules* **1984**, *17*, 1455. Copyright 1984, American Chemical Society.)

but that still retains some slight anisotropy. The small deviations from the isotropic distribution cannot be attributed simply to errors in the calculated moments of N(θ) because all spectra clearly exhibit some residual orientation, just as Figure 8-17 does. The subtle deviations from isotropy are indicative of the kinds of variation observable with NMR spectroscopy. Indeed, it should be possible to measure kinetics of disorientation in thermal processes by careful NMR examination.

Summary

Deformation—drawing, compression, and annealing—of a polymer, can be investigated using any anisotropic NMR parameter. A number of examples have been reported in the literature, and it is clear that a great deal of geometric and motional information is present in the NMR spectrum of a deformed polymeric specimen. We have delved into examples from our own work using the ^{19}F chemical-shift interaction to obtain information on structure in the prototype polymer, PTFE, under one specific set of conditions. There are, of course, many other aspects of the mechanical behavior of polymer materials that can be investigated by solid-state NMR spectroscopy. The way in which orientation develops as a function of strain as the material is drawn at constant strain rate could be studied more closely by varying the test conditions or the polymer's properties. Other types of deformation—creep, stress relaxation, and cyclic loading, for example—may produce quite different orientational behavior, the results of which could be examined with NMR spectroscopy.

Solid-state NMR has been a valuable tool for the study of oriented polymers, particularly because of the linewidth anisotropy observed in the broadline spectra. The advent of magic angle spinning has made possible analyses comparable to those obtained with NMR spectroscopy of solutions. The chemical-shift and quadrupolar interactions, however, are sensitive to orientation, because in these cases the lineshape, not just the width, is characteristic of the orientational distribution function. The analysis described in this discussion can be used to study, quantitatively, fairly subtle and not so subtle deformation-induced changes in the polymer structure. More sophisticated studies on polymer materials using NMR techniques will give a wealth of information on geometrically deformed polymers, on the effects of changes in processing, on the response of material to an applied stress, on the effects of chemical constitution of the material on the structures assumed under deformation and on a variety of other ways in which the orientational response of a polymeric material to an applied stress may be altered.

Acknowledgment

The work presented here has been supported by a Crown-Zellerbach Grant of the Research Corporation, and by the Division of Materials Research of the National Science Foundation.

References

1. Ward, I. M. "Structure and Properties of Oriented Polymers"; John Wiley: New York, 1975.
2. Samuels, R. J. "Structured Polymer Properties"; John Wiley: New York, 1974.
3. Wilchinsky, Z. W. *Polymer* **1964**, *5*, 271.
4. Rabek, J. F. "Experimental Methods in Polymer Chemistry"; John Wiley: Chichester, 1980.
5. Mehring, M. "High Resolution NMR in Solids"; Springer Verlag: Berlin, 1983.
6. Haeberlen, U. In "High Resolution NMR in Solids: Selective Averaging"; Adv. Magn. Reson. Ser., Waugh, J. S., Ed., Suppl. 1; Academic Press: New York, 1976.
7. Gerstein, B. C.; Dybowski, C. "Transient Techniques in NMR of Solids: An Introduction to the Theory and Practice"; Academic Press: New York, 1985.
8. McBrierty, V. J.; Douglass, D. C. *Phys. Rep.* **1980**, *63*, 6.
9. Spiess, H. W. In "NMR: Basic Principles and Progress", Diehl, P.; Fluck, E.; Kosfeld, R., Eds., Vol. 15; Springer-Verlag: Berlin, 1978.
10. McBrierty, V. J. *J. Chem. Phys.* **1974**, *61*, 872.
11. Bower, D. I. *J. Polym. Sci., Polym. Phys. Ed.* **1983**, *19*, 93.
12. Nomura, S.; Kawai, H.; Kimura, I.; Kagiyama, N. *J. Polym. Sci.* **1980**, *A2, 8*, 383.
13. Hentschel, R.; Sillescu, H.; Spiess, H. W. *Polymer* **1981**, *22*, 1516.

14. Treolar, L. R. G. "Physics of Rubber Elasticity"; Clarendon Press: Oxford, 1958.
15. Kuhn, W.; Gruen, F. *Kolloid-Z.* **1942**, *101*, 248.
16. Ward, I. M. "Mechanical Properties of Polymers"; Wiley-Interscience: New York, 1971.
17. Kratky, O. *Kolloid-Z.* **1933**, *64*, 213.
18. Hennecke, M.; Fuhrmann, J. *Polymer* **1982**, *23*, 797.
19. Van Vleck, J. H. *Phys. Rev.* **1948**, *74*, 1168.
20. McBrierty, V. J.; Ward, I. M. *Brit. J. Appl. Phys. (J. Phys. D)* **1968**, *1*, 1529.
21. Kashiwagi, M.; Ward, I. M. *Polymer* **1972**, *13*, 145.
22. Kashiwagi, M.; Folkes, M. J.; Ward, I. M. *Polymer* **1971**, *12*, 1529.
23. Olf, H. G.; Peterlin, A. *J. Polym. Sci.* **1970**, *A2, 8*, 753.
24. Folkes, M. J.; Ward, I. M. *J. Mater. Sci.* **1971**, *6*, 582.
25. McCall, D. W.; Slichter, W. P. *J. Polym. Sci.* **1957**, *26*, 171.
26. Hyndman, D.; Origlio, G. F. *J. Polym. Sci.* **1959**, *36*, 556.
27. Hyndman, D.; Origlio, G. F. *J. Polym. Sci.* **1960**, *46*, 259.
28. Yamagata, K.; Hirota, S. *J. Appl. Phys. Jpn* **1961**, *30*, 261.
29. Chujo, R.; Sudzuki, T. *Buss. Kenkyu* **1961**, *10*, 159.
30. Yamagata, K.; Hirota, S. *Rep. Progr. Polym. Phys. Jpn* **1962**, *5*, 261.
31. Boye, C. A.; Goodlet, V. W. *J. Appl. Phys.* **1963**, *34*, 59.
32. Fisher, E. W.; Peterlin, A. *Makromol. Chem.* **1964**, *74*, 1.
33. Slonim, I. Y.; Lyubimov, A. N. "The NMR of Polymers"; Plenum Press: New York, 1970.
34. McBrierty, V. J.; McDonald, I. R.; Ward, I. M. *J. Phys. D., Appl. Phys.* **1971**, *4*, 88.
35. Arai, N.; Hayakawa, N.; Tamura, N.; Kuriyama, I. *J. Polym. Sci., Polym. Phys. Ed.* **1977**, *15*, 1697.
36. Egorov, E. A.; Zhikhenkov, V. V. *J. Polym. Sci., Polym. Phys. Ed.* **1982**, *20*, 1089.
37. Cackovic, H.; Hoseman, R.; Laboda-Cackovic, J. *J. Polym. Sci., Polym. Lett. Ed.* **1978**, *16*, 129.
38. Wool, R. P.; Lohse, M. I.; Rowland, T. J. *J. Polym. Sci., Polym. Lett. Ed.* **1979**, *17*, 385.
39. Murin, J. *Czech. J. Phys.* **1981**, *331*, 62.
40. Murin, J.; Olcak, D.; Rakos, M.; Simo, K. R. *Acta Phys. Slov.* **1981**, *31*, 229.
41. Ito, M.; Serizawa, H.; Tanaka, K.; Leung, W. P.; Choy, C. L. *J. Polym. Sci., Polym. Phys. Ed.* **1983**, *21*, 2299.
42. Von Meerwall, E.; Ferguson, R. D. *J. Polym. Sci., Polym. Phys. Ed.* **1981**, *19*, 77.
43. Slichter, W. P. *J. Polym. Sci.* **1957**, *24*, 173.
44. Hyndman, D.; Origlio, G. F. *J. Appl. Phys.* **1960**, *31*, 1849.
45. McBrierty, V. J.; McCall, D. W.; Douglass, D. C.; Falcone, D. R. *J. Chem. Phys.* **1970**, *52*, 512.
46. O'Brien, J.; McBrierty, V. J. *Proc. Royal Irish Acad.* **1975**, *25*, 331.
47. Davidson, T.; Gounder, R. N. In "Adhesion and Adsorption of Polymers"; Plenum Press: New York, 1980.
48. Douglass, D. C.; McBrierty, V. J.; Wang, T. T. *J. Chem. Phys.* **1982**, *77*, 5826.
49. Nieman, M. B.; Slonim, I. Y.; Urman, Y. G. *Nature* (London) **1964**, *202*, 693.
50. Olf, H. G.; Peterlin, A. *J. Appl. Phys.* **1964**, *35*, 3108.
51. McBrierty, V. J.; McDonald, I. R. *J. Phys. D., Appl. Phys.* **1973**, *6*, 131.
52. Slonim, I. Y.; Urman, Y. G. *Zh. Strukt. Khim.* **1963**, *4*, 216.
53. Roe, R. J. *J. Polym. Sci.* **1970**, *A2, 8*, 1187.
54. Ito, S. E.; Okajima, S.; Kase, T. *Kolloid-Z.* **1971**, *248*, 899.
55. Kashiwagi, M.; Cunningham, A.; Manuel, A. J.; Ward, I. M. *Polymer* **1973**, *14*, 11.
56. Cunningham, A.; Manuel, A. J.; Ward, I. M. *Polymer* **1976**, *17*, 125.
57. Itoyama, K. *Kobun. Ronbun.* **1976**, *33*, 741.
58. Yamagata, K. *J. Appl. Phys. Japan* **1961**, *30*, 940.
59. Peterlin, A.; Olf, H. G. *J. Polym. Sci.* **1962**, *B2*, 769.
60. Tsutsumi, A.; Hikichi, K.; Kaneko, M. *Polym. J.* **1976**, *8*, 443.
61. Olf, H. G.; Peterlin, A. *J. Polym. Sci.* **1971**, *A2, 9*, 1449.
62. Itoyama, K.; Kashiwagi, M. *Kobun. Ronbun.* **1971**, *9*, 1449.
63. Dimov, K.; Denev, E. *Acta Polym.* **1979**, *30*, 519.
64. McBrierty, V. J.; Douglass, D. C. *J. Magn. Reson.* **1970**, *2*, 352.
65. Gronski, W.; Stadler, R.; Jacobi, M. M. *Macromolecules* **1984**, *17*, 741.
66. Deloche, B.; Samulski, E. T. *Macromolecules* **1981**, *14*, 575.
67. Brandolini, A. J.; Alvey, M. D.; Dybowski, C. *J. Polym. Sci., Polym. Phys. Ed.* **1983**, *21*, 2511.
68. Brandolini, A. J.; Apple, T. M.; Dybowski, C.; Pembleton, R. G. *Polymer* **1982**, *23*, 39.
69. Hempel, G.; Schneider, H. *Pure Appl. Chem.* **1982**, *54*, 635.
70. Hempel, G. *Plaste Kautsch.* **1979**, *26*, 361.

71. Willsch, R.; Schnabel, B.; Scheler, G.; Mueller, R. *Plaste Kautsch.* **1976**, *23*, 735.
72. Garroway, A. N.; Stalker, D. C.; Mansfield, P. *Polymer* **1975**, *16*, 161.
73. Vega, A. J.; English, A. D. *Macromolecules* **1980**, *13*, 1635.
74. Brandolini, A. J.; Dybowski, C. *J. Polym. Sci., Polym. Lett. Ed.* **1983**, *21*, 423.
75. Brandolini, A. J.; Rocco, K. J.; Dybowski, C. *Macromolecules* **1984**, *17*, 1455.
76. Kasuboski, L.; Brandolini, A. J.; Dybowski, C. unpublished data, 1985.
77. Speerschneider, C. J.; Li, C. H. *J. Appl. Phys.* **1962**, *33*, 1871.
78. Alfrey, T. "Mechanical Behavior of High Polymers"; Interscience: New York, 1948.
79. Peterlin, A. *Polym. Eng. Sci.* **1978**, *18*, 488.
80. VanderHart, D. L. *Macromolecules* **1979**, *12*, 1232.

9

HIGH-RESOLUTION SOLID-STATE NMR OF PROTONS IN POLYMERS

Bernard C. Gerstein

DEPARTMENT OF CHEMISTRY, IOWA STATE UNIVERSITY,
AND AMES LABORATORY, EMRRI,
AMES, IOWA 55001

Introduction

Polymers are a class of materials with enormous potential for probing by transient techniques in solid-state nuclear magnetic resonance. The reasons for this judgement lie in the peculiar natures of both the material under study and the technique. Some polymers are not soluble in any solvent and so are not amenable to probing by standard high-resolution liquid-state NMR. This fact, as will be seen, is fortunate, because there is much physics and chemistry that is transparent to a liquid-state measurement. Polymers also represent states of matter that are not, in general, in thermodynamic equilibrium with respect to phase changes. This is to say that polymers exhibit time-dependent phenomena that are quite a delicate probe of their properties and that may be correlated with macroscopic behavior, such as resistance to fracture. These time-dependent characteristics can be exhibited by both microscopic and macroscopic regions of the polymer, with characteristic times ranging from picoseconds to years. Because of the lack of thermodynamic equilibrium in polymers, these materials generally exhibit crystalline and noncrystalline

© 1986 VCH Publishers, Inc.
Komoroski (ed): High-Resolution NMR Spectroscopy of Synthetic Polymers in Bulk

regions, and the sizes and shapes of these regions affect the mechanical properties of the materials. Pulse NMR, by its nature, is a probe that operates in real
time, with the time scale ranging from nanoseconds (the period of a radio
frequency oscillation) to seconds (the lifetime of the nuclear spin states under a
resonant excitation; this lifetime determines the limiting linewidth of the NMR
absorption signal). As will be seen in the section on Interactions that Broaden
NMR Spectra of Solids, protons experience interactions in the solid state that
are sensitive to both short- and long-range order and that are sensitive to
motion. In contrast to the result in a liquid, motion in the solid state can lead
to a broadening of the NMR line, under conditions of "high-resolution, solid-
state NMR," and this broadening can be used as a measure of the frequency
and of the amplitude of the motions. In addition, highly ordered regions of a
polymer in general exhibit molecular motions with frequencies differing from
those in more randomly ordered regions. The differences in time scales of
motions in such regions may be used in conjunction with NMR to probe both
the sizes and the shapes (more accurately the dimensionality) of the ordered
and nonordered regions.

The above remarks apply to any nucleus in a polymer, examples being ^{1}H,
^{2}H, ^{13}C, ^{14}N, ^{15}N, ^{17}O, ^{19}F, ^{29}Si, and ^{31}P. It is the task of this chapter,
however to treat the special case of hydrogen, which is spin 1/2, is 100%
abundant, and has one of the highest magnetic moments in the periodic table.
The theoretical discussion of this chapter also applies to ^{19}F, another 100%
abundant, spin 1/2 nucleus. (High resolution ^{19}F NMR is discussed in Chapter
8 in reference to oriented systems.) First the interactions that abundant, high
magnetogyric ratio nuclei experience are examined, how rotations in spin and
in real space can be used to attack and modify these interactions, and finally
how combinations of rotations in spin and in real space can be used to study
static and time-dependent features of polymers, which may be used to relate
and predict macroscopic properties from microscopic behavior.

The section on Principles is rather mathematical in nature. The mathematics is not new and has achieved the status of being included in mono-
graphs on the field.[1-3] However, the formalism is not so familiar to the
average polymer chemist that it can be assumed to be known generally. It is
impossible in a chapter of the present length to develop this formalism with
rigor. What I will do is to present enough of the material that the connection
between the physics and the arithmetic will appear reasonable. In order to
attenuate the effect of some of the formalism, I shall try to interlace the equations with statements and pictures that give some idea of the underlying
physics. The reader must realize, however, that an understanding of the nature
of the experiments that would be sufficient to allow one to perform such work
in one's own laboratory on commercially available equipment implies a rather
detailed understanding of the mathematics involved. More complete discussions of the basic formalism are available[1-5] and the reader is referred to
them for a deeper understanding than is possible from the development in this
chapter.

Principles

Interactions that Broaden NMR Spectra of Solids

An abundant spin-1/2 nucleus in a solid, ie, ^{1}H or ^{19}F, experiences two major anisotropic interactions. These are: (a) the dipolar coupling between like nuclei, reflecting the fact that in addition to the external magnetic field a given nucleus experiences a local, "through-space" field from all the other nuclear dipoles; and (b) the shielding of a nucleus from the external field by the electronic environment about the nucleus in question. At this point the (also anisotropic) scalar coupling interaction is ignored, because it is negligible compared to the dipolar and shielding interactions for protons.

Each of these interactions is in turn affected by motion in the solid. It is useful to understand that anisotropic interactions of nuclear spins and their environment may be cast in a number of different forms. One form reflects the vector–tensor nature of the interactions;

$$\mathscr{H}_{\mathrm{A}} = \mathrm{k}\,\vec{\mathrm{X}} \cdot \mathbf{A} \cdot \vec{\mathrm{Y}} \qquad (9\text{-}1)$$

Here, $\vec{\mathrm{X}}$ and $\vec{\mathrm{Y}}$ are spin angular momentum operators, expressed as row and column vectors, respectively, and $\mathbf{A}$ is a 3×3 matrix containing real-space (as contrasted to "spin-space") quantities. k is a constant characteristic of the interaction in question; eg, for the shielding interaction, k would be ω_0, the Larmor frequency. In the case of the shielding interaction, $\mathbf{A}$ is the shielding matrix. When appropriate inequalities are satisfied, this shielding matrix may be represented by an ellipsoidal three-dimensional figure, as shown in Figure 9-1. The major three axes of the ellipsoid are the diagonal elements of the shielding matrix, σ_{11}, σ_{22}, σ_{33}. These physically represent the NMR fre-

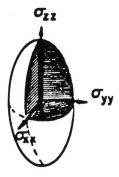

Figure 9-1. Representation of anisotropic shielding as an ellipsoid. Measurement of a static sample with the field, B_0, along an axis σ_{kk} yields an NMR signal at a frequency $\gamma B_0(1 - \sigma_{kk})$. Isotropic rotation of this shielding ellipsoid, with respect to the external magnetic field, in real space with a frequency large compared to half of $(\sigma_{33} - \sigma_{11})$ yields a measured NMR frequency equal to the isotropic value of the shielding matrix, $(\sigma_{11} + \sigma_{22} + \sigma_{33})/3$. In the figure, $\sigma_{xx} = \sigma_{11}, \sigma_{yy} = \sigma_{22}$, and $\sigma_{zz} = \sigma_{33}$.

quencies observed when the external field is parallel to the direction represented by that axis. For example, when the external field, $\vec{B}_o$, is parallel to the direction represented by σ_{11}, with σ_{11} being expressed in parts per million, the observed NMR frequency is $\omega_0(1 - \sigma_{11})$, where ω_0 is the Larmor frequency γB_0. The range of NMR frequencies observed for a given nucleus in a specific environment is therefore the difference between the largest and the smallest diagonal elements of the shielding matrix. In a powdered sample, the direction of the external field with respect to the axes of the shielding matrix ellipsoid will be random. Therefore a given crystallite exhibits sharp lines corresponding to the external field being at some fixed direction with respect to the shielding ellipsoid axes representing nuclei in that crystallite. The spectrum associated with all crystallites, however, is a superposition of all such sharp lines, ie, the powder spectrum is broadened inhomogeneously (see Chapter 2). The possible powder spectra for a nucleus experiencing only inhomogeneous broadening by shielding anisotropies are shown in Figure 2-2 (Chapter 2). An anisotropic shielding environment exhibits a powder spectrum with three discontinuities, representing the three diagonal elements of the shielding matrix. For a shielding environment that is axially symmetrical, ie, with two axes of the shielding ellipsoid the same, the powder spectrum exhibits two discontinuities at the extremes of the spectrum.

When the shielding environment of a nucleus is rotating isotropically with respect to the laboratory-fixed frame of the external magnetic field, at a frequency large compared to the spectral range associated with the shielding matrix, the isotropic value of the shielding matrix is observed in an NMR measurement. The isotropic value is one-third the sum of the diagonal elements of the matrix and is the frequency observed in an NMR measurement of shielding in a liquid. This situation physically corresponds to the shielding ellipsoid rotating isotropically in space. The same statements can be made of dipolar coupling. However, dipolar coupling is an interaction between different nuclei. This means that when a very small number of nuclei is involved (two or three) the powder NMR spectrum is broadened inhomogeneously and exhibits sharp features, as shown for interactions between two or three nuclei at the top of Figure 9-2. However, when many nuclei are involved in dipolar interactions with each other (eg, protons in polyethylene), the powder spectrum is broadened homogeneously, with spectrum as shown in the bottom of Figure 9-2. An interesting fact is that the isotropic value of the dipolar coupling matrix is zero, so isotropic rotation of the internuclear vector between two dipoles, at a frequency large compared to the maximum of the dipolar interaction, results in averaging the dipolar interaction to zero.

Another interesting fact is that when a nucleus experiences both shielding and dipolar interactions, with powder spectra as indicated in Figures 2-2 and 9-2, respectively, the observed NMR spectrum has the features of both spectra of Figures 2-2 and 9-2. The result is to broaden and smooth the details of each figure involved in the process. Thus, when the NMR powder spectrum of protons is observed, eg, in polyethylene, the result is a rather featureless, broad

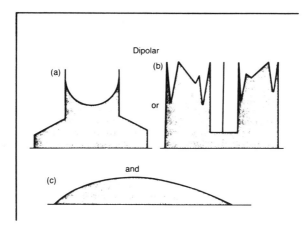

Figure 9-2. Nuclear magnetic resonance spectra from dipolar interactions between: (A) two spin-1/2 systems; (B) three spin-1/2 systems arranged in a triangle; (C) many spin-1/2 systems.

line, wider than either the dipolar coupling, which is roughly 20 KHz, or the shielding anisotropy, with width 4.7 ppm. Because dipolar coupling is independent of static field, its magnitude usually is expressed in hertz. The chemical shift scales linearly with field, so its magnitude is expressed in parts per million.

At this point, the discussion of removing the shielding anisotropy and the dipolar coupling has involved rotation of the real-space portion of the interactions, which may be viewed with the help of picturing this portion as an ellipsoid. Implicit in Eq. (9-1) is the fact that the dipolar and shielding interactions are represented by a "real-space"–"spin-space" product. The ellipsoidal shape representing the anisotropy of the real-space part of the broadening interaction was of help in visualizing how motion in real space could average a broadened powder spectrum to a single line, or to zero. The basic idea in utilizing this shape to understand how motion in real space could average interactions was to picture this motion in terms of rotation of the ellipsoid representing the interaction. Now it is wished to develop a picture for the motion of spin-space portions of internal interactions under experimenter-supplied pulses. This view similarly allows inference of how experimenter-supplied motions in spin space, accomplished by manipulations of carefully timed and phased radiofrequency (rf) pulses, are able to accomplish selective averaging[1] of internal interactions. This means that the dipolar broadening can be removed while the shielding interaction is maintained, and vice versa. The forms of the spin operators are quite simple. That for the shielding anisotropy is simply I_z. That for a dipolar interaction between two spins I_1 and I_2 is $(I_1 \cdot I_2 - 3I_{z1}I_{z2})$. Next, how rotations in spin space and in real space can lead to scaling, or averaging, these interactions in a desired manner is examined in more quantitative detail.

A special consideration in the case of broadening of NMR spectra of polymers is lack of crystallinity in these systems. Even if broadening from dipolar and shielding anisotropy can be exactly removed by rotations in spin and in real space, to leave only isotropic shieldings, noncrystallinity can lead to residual broadening of the remaining spectra. This is because in a glassy polymer, a given chemical species, eg, protons in —CH_2— groups of polystyrene, experiences a dispersion of local geometries associated with the noncrystalline nature of the glass. Thus, each methylene proton in the polymer has a distinct shielding anisotropy, and a distinct isotropic value of the shielding. These isotropic values lie on some distribution curve representing the dispersion of all chemical environments of all methylene protons in the polymer. Sometimes the less crystalline the polymer, the broader this distribution curve, and to the extent that the curve is continuous with a half-width δ the width of all observed isotropic shifts also has this residual broadening, even when dipolar and anisotropic broadening is exactly averaged to zero. Therefore, one measure of the crystallinity of a polymer can be the residual broadening of NMR lines under high-resolution conditions. By "high-resolution conditions" is meant rotations in both spin and real space to remove dipolar broadening, and to average the shielding anisotropy to zero.

Another feature peculiar to polymers is that of differences of motion for nuclei in well-ordered versus poorly ordered regions of the polymer. Provided correlation times for anisotropic motion in the noncrystalline region of the polymer are sufficiently small to average dipolar interactions partially (usually above the glass transition temperature T_g), differences in motion of the crystalline and noncrystalline regions of the polymer lead to a narrowing of the NMR linewidth of the noncrystalline relative to the crystalline region of the polymer. This is to say that nuclei in the noncrystalline region of the polymer experience a smaller dipolar interaction than those in the crystalline region. It will be seen that this difference can be used with multiple-pulse attack of dipolar broadening to measure quantitatively the amount of motionally narrowed relative to rigid portions of the polymer.

Rotations in Spin Space: Scaling

The description of this material involves standard, time-dependent quantum mechanics. For the average polymer chemist, however, the material may be a bit unfamiliar, so the formulation is followed by a physical picture that is hoped to help in visualizing how spin operators can be attacked by rf pulses. The basic equation describing measurable values of observables that vary with time is the time-dependent Schroedinger equation:

$$i\partial\psi/\partial t = \mathscr{H}\psi \quad (\mathscr{H} \text{ is units of s}^{-1}) \tag{9-2}$$

In terms of the density matrix description, this equation becomes:

$$i\partial\rho/\partial t = [\mathscr{H}, \rho] \tag{9-3}$$

where the density matrix, ρ, is obtained from the wave functions, $|\psi\rangle$, by the recipe:

$$\rho = |\psi(t)\rangle\langle\psi(t)| = \psi\psi^* \tag{9-4}$$

The observed quantity in a pulse NMR experiment is the expectation value of a transverse component of angular momentum as a function of time. In terms of the density matrix, this quantity is given by:

$$\langle I_y(t)\rangle = \text{Tr } \rho(t)I_y \tag{9-5}$$

The Fourier transform of this time decay of the transverse magnetization (chosen to be the y component in our example) becomes the NMR spectrum. This decay in general contains dipolar and shielding interactions for protons in polymers, and the spectrum, as indicated before, is a rather featureless combination associated with these two interactions. The accomplishment of selective averaging in solids has been to arrange to remove all or portions of one or more interactions, thus simplifying the interpretation of the resultant spectra. To see how this task is accomplished, examine Eq. (9-5). If ρ, as determined by Eq. (9-3), is not affected by a particular interaction, then the time decay of $\langle I_y(t)\rangle$ also is unaffected, and this interaction does not appear in the spectrum! It is therefore necessary to investigate how $\rho(t)$ behaves under carefully chosen rf pulses, (i.e., rotations in *spin* space) or, under experimenter-controlled motion of the sample in *real* space. In the present section are considered rotations of spin operators by rf pulses. With the Hamiltonian being a sum of rf and internal interactions:

$$\mathcal{H} = \mathcal{H}_{rf} + \mathcal{H}_{int} \tag{9-6}$$

the time development of the density matrix is:

$$\rho(t) = U_{rf}(t)U_{int}(t)\rho(0)U_{int}^{-1}(t)U_{rf}^{-1}(t) \tag{9-7}$$

where the time development operators U_k are determined by the rf and internal Hamiltonians as follows:

$$U_{rf}(t) = T \exp\left[-i\int_0^t \mathcal{H}_{rf}(t)\, dt\right] \tag{9-8}$$

T is a "time ordering" operator that is ignored for purposes of the present discussion. Note that to first order, U_{rf} is the exponential of the integral of $\mathcal{H}_{rf}(t)$:

$$U_{rf}(t) = \exp\left[-i\int_0^t \mathcal{H}_{rf}(t)\, dt\right] \tag{9-9}$$

If the integral in this equation is zero, the exponential reduces to unity, and under such a condition:

$$\rho(t) = U_{int}\rho(0)U_{int}^{-1} \tag{9-10}$$

Now $U_{int}(t)$ is determined by $\mathcal{H}_{int}(t)$ as manipulated by the rf pulses:

$$U_{int}(t) = T \exp\left[-i \int_0^t \tilde{\mathcal{H}}_{int}(t) \, dt\right] \tag{9-11a}$$

where:

$$\tilde{\mathcal{H}}_{int}(t) = U_{rf}^{-1}(t)\mathcal{H}_{int}(t)U_{rf}(t) \tag{9-12}$$

Again, the T in Eq. (9-11a) means time ordering, and for purposes of demonstration all but the first term in the series implied in Eq. (9-11a) is ignored. It is useful, however, to look at another form of Eq. (9-11a), which makes more explicit the implied series. This is the Magnus expansion,[1-3] the form of which is:

$$U_{int}(t) = \exp\left(-i[\overline{\mathcal{H}_{int}^{(0)}} + \overline{\mathcal{H}_{int}^{(1)}} + \cdots]\right) \tag{9-11b}$$

where the terms in the exponential are integrals of the internal Hamiltonian evaluated in the frame of the radio frequency (Eq. 9-12) as before. Again, concentrating on the first term:

$$\overline{\mathcal{H}_{int}^{(0)}} = (1/t) \int_0^t \tilde{\mathcal{H}}_{int}(t) \, dt \tag{9-11c}$$

Now the form of Eq. (9-11b) is quite interesting if just the first term in the exponential is considered and it is realized that formally this "time development" operator for the density matrix, ρ, appears to be time independent. In fact the form of Eq. (9-11b) is just the result for the time development of ρ that would be obtained if the system were developing under a time-independent Hamiltonian.

However, because Eq. (9-2) clearly shows that $\tilde{\mathcal{H}}_{int}$ is time dependent, the physical meaning of Eq. (9-11b) may be thought of as being that the system is affected by an "average" Hamiltonian over the time interval in question, the average Hamiltonian being the time average of $\tilde{\mathcal{H}}_{int}(t)$. Another quite powerful physical picture emerges from this discussion. There are two possibilities for $\overline{\mathcal{H}_{int}^{(0)}}$: (a) it is zero, in which case it is said that to zeroth order, $\mathcal{H}_{int}$ has been averaged to zero—ie, the system behaves at some time as if this particular $\mathcal{H}_{int}$ were not present—and (b) $\overline{\mathcal{H}_{int}^{(0)}}$ is of the form $aI_x + bI_y + cI_z$. Recall the physical fact that under a Hamiltonian $-\gamma\vec{B}_0 \cdot \vec{I} = \omega_0 I_z$, the classical motion is precession of the system, I, about $\vec{B}_0$, with frequency ω_0, the vector form of which is $\vec{\Omega} = \omega\mathbf{k}$, with $\mathbf{k}$ being a unit vector along z. It can be seen that an average Hamiltonian, $\overline{\mathcal{H}_{int}^{(0)}}$, which is linear in components of I, the nuclear spin angular momentum, represents a precession about some effective field. For example, suppose that the average Hamiltonian, under a particular multiple-pulse experiment, for an operator scaling as I_z is $\overline{\mathcal{H}_{int}^{(0)}} = \omega(I_x + I_z)/3$. The effective field corresponding to this operator, being of the form $\gamma\vec{B}_{eff} \cdot \vec{I}$, leads to a frequency of rotation with the vector form $\vec{\Omega}_{eff} = \omega(\mathbf{i} - \mathbf{k})/3$; its scalar value is $\sqrt{2}\,\omega/3$. This means that a magnetization that precesses at a frequency ω under the operator I_z, now precesses at a scaled frequency of $\sqrt{2}\,\omega/3$ under the multiple-pulse experiment in question. In addition, the motion of the spin system, when observed at the times appropriate to the use of $\bar{H}_{int}$, is a precession about the (1, 0, 1) direction. Another way of making this statement is that if the magnetization of the system is initially prepared along the (1, 0, 1) direction by, say a $\pi/4$ pulse along $\bar{y}$ in the rotating frame, then the system is not observed to develop in time because of the operator I_z, under the multiple sequence for which $\bar{I}_z$ becomes $(I_x + I_z)/3$. In other words, the chemical-shift Hamiltonian does not affect the time development of the system in the above-described situation. This fact is used later when discussing studies of relaxation under multiple-pulse sequences. Now, examining Eqs. (9-5) and (9-7), we see that when Eq. (9-9) becomes unity (accomplished by aranging the rf pulse sequence to be "cyclic"), then to zeroth order, $\rho(t)$ is determined by:

$$U_{int}(t) = \exp\left[-i \int_0^t \tilde{\mathscr{H}}_{int}(t)\,dt \right] \qquad (9\text{-}13)$$

It is now necessary to understand what is meant by Eq. (9-12); ie, how is the internal interaction manipulated by the rf pulses through Eq. (9-12)? What is the internal Hamiltonian "in the frame of the rf," $\tilde{\mathscr{H}}_{int}$? This result is quite easy to visualize physically. The internal Hamiltonians in question all have spin parts containing terms I_k (see the end of the last section). The integral in Eq. (9-13) containing $\tilde{\mathscr{H}}_{int}$ implies the integral of a frequency (the broadening implied in $\tilde{\mathscr{H}}_{int}$) multiplied by a time, which is an angle, θ. Therefore the value of $\tilde{\mathscr{H}}_{int}(t)$ inserted into Eq. (9-13) is of the form:

$$\tilde{\mathscr{H}}_{int}(t) = \omega_{int}(r,\ \theta,\ \phi)e^{i\theta I_j}I_k e^{-i\theta I_j} \qquad (9\text{-}14)$$

Here, $\omega_{int}(r,\ \theta,\ \phi)$ is the real-space portion of the interaction in question, including appropriate constants. These terms are explored in discussing the effects of motion on the ability to narrow using pulse sequences.

Equation (9-14) has a very simple physical interpretation. It represents a rotation of I_k by the angle θ about the direction implied by I_j:

$$e^{i\theta I_j}I_k e^{-i\theta I_j} = I_k \cos\theta - \varepsilon_{jkm} I_m \sin\theta \qquad (9\text{-}15)$$

Here, ε_{jkm} is zero when any of j, k, m are the same, $+1$ when they are in cyclic order (x, y, z; z, x, y, etc) and -1 when j, k, m are in anticyclic order (x, z, y, etc). For example when $j = x$, and $k = z$, then Eq. (9-15) yields

$$e^{i\theta I_x}I_z e^{-i\theta I_x} = I_z \cos\theta + I_y \sin\theta$$

This is just I_y when θ is $\pi/2$. A physical picture of the result of an rf pulse in the x direction, designated P_x, at a time t, on the operator I_z therefore may be sketched as shown in Figure 9-3. This picture can be applied to a well-known pulse experiment. This is the spin-echo sequence, with two π pulses, the first along y at time τ and the second along $\bar{y}$ at time 3τ, with observation of the magnetization at a time 4τ. Consider internal interactions to be those which

$$e^{iI_x\theta} I_z e^{-iI_x\theta} = I_z \cos\theta + I_y \sin\theta$$

$$\equiv R_{\theta_x} I_z$$

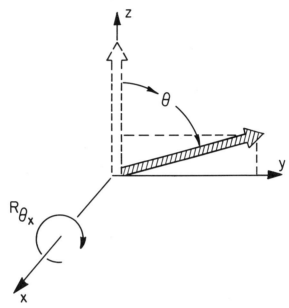

Figure 9-3. The exponential form $e^{i\theta I_x} I_z e^{-i\theta I_x}$ as a rotation of I_z about the x axis by the angle θ.

protons in solids experience: the shielding anisotropy, with spin operator I_z, and dipolar interactions, with spin operator $(I_1 \cdot I_2 - 3I_{z1}I_{z2})$. Equation (9-7) states that the density matrix at 4τ results from the time development operators for the rf and the internal interactions from zero to 4τ. The time development operator for the rf from time zero to time 4τ is an exponential representing a rotation by 2π and therefore is unity: $U_{rf}(4\tau) = U_{rf}^{-1}(4\tau) = 1$. The first approximation to the time development operator for the internal interactions may be inferred from viewing Figure 9-4, and visually integrating the values of $\mathscr{H}_{int}(t)$ thus found. It can be seen that the exponential in $U_{int}(4t)$, when $\mathscr{H}_{int}$ is I_z, becomes:

$$\int_0^{4\tau} \tilde{\mathscr{H}}_{int} \, dt \simeq I_z(\tau - 2\tau + \tau) = 0 \qquad (9\text{-}16)$$

(in this equation constants and real-space multipliers are neglected). If the shielding were the only internal interaction, therefore, then the magnetization, when "stroboscopically" viewed at 4τ (and, it may be shown, at any multiple of 4τ), is unaffected by this operator and at these times, appears to behave as if

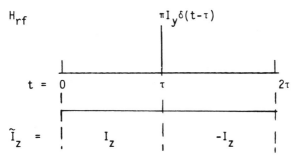

Figure 9-4. Behavior of the operator I_z under an inverting pulse sequence.

this operator were zero. On the other hand, the value of the dipolar interaction as attacked by $\mathcal{H}_{rf}$ is unchanged over the time period in question. That this statement is true may be seen simply by the fact that $I_1 \cdot I_2$ represents a scalar product of two vectors, and is invariant to rotation, Also, whereas I_z changes sign under the π_y rotations, the product $I_{z1}I_{z2}$ does not. Therefore under this spin-echo pulse sequence, the shielding anisotropy is removed while the homonuclear dipolar interactions are maintained.

More complicated pulse sequences can be based on the "dipolar echo" sequence, $(\pi/2_x, \tau, \pi/2_y)$, the simplest of which is the Waugh-Huber-Haeberlen (WAHUHA) sequence (Figure 9-5). The WAHUHA is the above dipolar echo sequence plus its mirror image with the phases inverted to allow $\int H_{rf}(t)\, dt$ to be zero over the cycle time of the sequence. Such a sequence can attenuate dipolar interactions while maintaining shielding. More complicated sequences are capable of removing higher order terms in the exponential expansions implied by the time development equations for the density matrix, and are

A)

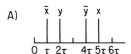

B)

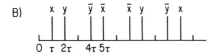

C)

Figure 9-5. Several multipulse sequences for dipolar narrowing: (A) WAHUHA; (B) MREV-8; (C) BR-24. The parameter τ is the cycle time, and the pulses are $\pi/2$ with the phase indicated.

invoked in the work to be described in the section on Applications, Shielding. A feature that must be considered when using multiple-pulse sequences to remove the dipolar interaction is that the dipolar interactions among many spins lead to a line that is broadened homogeneously (see Chapter 2). In this case, there is a time scale over which the multiple-pulse sequences are effectively operative, which is the inverse of the linewidth of the homogeneously broadened line. This is to say that if the homonuclear dipolar broadening from proton–proton interactions in a polymer is of the order of 20 kHz, then the time scale over which the dipolar echo sequence must be applied must be short compared to $1/20,000$ s^{-1}, or 50 μs. Another way of making this statement is that the expansion of the exponentials implied by the operators U_{int} in Eq. (9-10) converges over time scales short compared to the time scale implied by the magnitude of $\tilde{\mathscr{H}}_{int}$, which occurs in the exponential integral in Eq. (9-10). That is, the convergence of the expansion is rapid if $|\tilde{\mathscr{H}}_{int}| t_c \ll 1$, where t_c is the cycle time used in the multiple-pulse attack on the homogeneous broadening interaction. This fact is used later in the discussion of multiple-pulse sequences with varying cycle times used to probe crystallinity in polymers (see Applications, Crystallinity).

It can be seen, therefore, that although the formalism of the mathematics describing the use of pulse sequences appears formidable to the uninitiated, the physics involved is visualized easily. A pulse sequence that involves both the spin-echo, 180° refocusing pulses, inserted into appropriate combinations of dipolar echo sequences, is capable of removing both dipolar and shielding interactions and leaving relaxation under multipulse sequences as the major broadening interaction [the dipolar-narrowed, Carr-Purcell (DNCP) sequence].

Rotations in Real Space

An earlier section (Rotations in Spin Space: Scaling) discussed how the modulation of the spin portion of an internal interaction with time could result in altering the effect of that interaction on the appropriately observed time decay, and thus the observed NMR spectrum of a sample in question. The basic equation was Eq. (9-5), in which the observed value of the transverse magnetization was related to the time development of the density matrix. The "zero-order" removal of a particular interaction involved making its spin portion time dependent, through the use of rf pulses. It was arranged so that the integral of the spin portion of the internal Hamiltonian, in the frame of the rf (which was denoted by $\tilde{\mathscr{H}}_{int}$) and over the cycle time of the rf pulse excitations, would be zero. In the discussion of rotations in real space exactly the same mechanics applies, and it is possible, by observing the magnetization at multiples of the real-space rotational cycle time, to produce selective averaging of some interactions. This aspect, however, is not discussed here. Instead, consider the simpler expedient of applying real-space rotation to the sample in order to examine the result of such rotation on the observed spectrum.

Whereas the last section was concerned with attacking both the shielding and dipolar Hamiltonians, the present section considers only the shielding Hamiltonian. The form of this Hamiltonian, given by Eq. (9-1) is:

$$\mathscr{H}_s = \gamma \vec{I} \cdot \boldsymbol{\sigma} \cdot \vec{B}_0 \tag{9-17}$$

In terms of the angles θ and ϕ, which orient the shielding ellipsoid with respect to the external field, the form of the shielding Hamiltonian is:

$$\mathscr{H}_s = \omega_0 I_z [1 - \bar{\sigma} + (\delta/2)(3\cos^2\theta - 1) - (\eta\delta/4)\sin^2\theta (e^{2i\phi} + e^{-2i\phi})] \tag{9-18}$$

Here, the terms $\bar{\sigma}$, the isotropic value of the chemical shift; δ, the anisotropy; and η, the asymmetry are given in terms of a rearrangement of the diagonal elements of the shielding matrix in order to better illustrate the symmetry of the shielding matrix:

$$\bar{\sigma} = (\sigma_{11} + \sigma_{22} + \sigma_{33})/3$$

$$\delta = \bar{\sigma} - \sigma_{33} \tag{9-19}$$

$$\eta = (\sigma_{22} - \sigma_{11})/\bar{\sigma}$$

For example, an axially symmetrical shielding environment has a value of zero for the asymmetry parameter η.

The form of this Hamiltonian under rotation in real space by a frequency ω_R may be calculated by an equation of exactly the form of Eq. (9-12), except now the rotation operators U are not associated with the rf field, which supplied rotations in spin space, but with operators that supply rotations in real space. These operators are standard 3×3 matrices that are well known; eg, a matrix for a rotation of an angle α about the x axis is given by:

$$R_{x\alpha} = \begin{bmatrix} 1 & 0 & 0 \\ 0 & \cos\alpha & -\sin\alpha \\ 0 & \sin\alpha & \cos\alpha \end{bmatrix} \equiv U_{R_{x\alpha}}$$

The Hamiltonian, found by applying the rotation operators that correspond to physical rotation by an angle $\omega_R t$ about an axis oriented at angle β to the external field, is then used to specify the time development of the density matrix by a generalized form of Eq. (9-10). The time decay of a transverse component of the angular momentum is then calculated from Eq. (9-5). The frequency spectrum of the sample under conditions of sample rotation is then calculated from the Fourier transform of the time decay, as usual. After much algebra, the result is that when a sample is rotated physically at an angle β to the static magnetic field, at a rotation frequency fast compared to half the anisotropy parameter, δ, the observed frequency spectrum is described by:

$$\sigma = \bar{\sigma} + (3\cos^2\beta - 1)(\delta/2)[(3\cos^2\theta - 1) + \delta\eta\sin^2\theta\cos 2\phi] \tag{9-20}$$

This equation contains β, the angle describing the physical rotation about the static field, and the angles θ and ϕ, which describe the orientation of the shielding ellipsoid with respect to the static field. An interesting fact is that the decay in magnetization that leads to this equation for the frequency spectrum has no time dependence associated with the sample rotation and so, in the frequency spectrum, no side bands at the rotation frequency, even though the shielding ellipsoid is being rotated about the static field. This results from the fact that "rapid" sample rotation, ie, a rotation frequency greater than half the shielding anisotropy, has been specified. Under these conditions, the time dependence associated with the sample rotation averages to zero in the observed decay, and the Fourier transform of this decay does not contain frequencies that are multiples of the sample rotation. Equation (9-20) does contain the frequency spectrum of the static, shielding powder pattern (the terms in θ and ϕ) scaled by a term involving the rotation angle, ie, by $3 \cos^2 \beta - 1$. The physical meaning of this result is that a sample spun "rapidly" at the angle β to the external field exhibits a powder spectrum of the form shown in Figure 2-2, scaled by the term $3 \cos^2 \beta - 1$. Note that this term can be positive or negative, depending on whether β is larger or smaller than the magic angle, where $\cos \beta = (\sqrt{3})^{-1}$. At the magic angle, the powder anisotropy disappears from the observed spectrum and just the isotropic value of the shielding, $\bar{\sigma}$, is observed (see Chapter 2). The residual broadening caused by deviations from the magic angle are then calculated easily from Eq. (9-20).

Effect of Molecular Motion

In all that has been said in the two previous sections, it has been assumed that the nuclei in question are not moving in real space. The result of motion on efforts to narrow, in spin or in real space, are now examined when the nuclei and their environments under investigation are moving anisotropically with some correlation time, τ. The effect can be illustrated with Eq. (9-13), in which multiplicative constants and real-space operators are excluded explicitly when the integral of $\mathcal{H}_{int}$ is considered. It is explicitly assumed that the only time dependence exhibited by $\mathcal{H}_{int}$ is supplied to the spin–space portion of $\mathcal{H}_{int}$ by the experimenter through the rf pulses. The real-space part of the shielding Hamiltonian is given in Eq. (9-18). This spatial portion, including constants, can be denoted by $\omega(\theta, \phi, t)$, where now the time dependence of the real-space portion is included explicitly and functionally. If this operator decays with some correlation time τ, then:

$$\omega(\theta, \phi, t) = \omega_0(\theta, \phi)e^{-t/\tau} \tag{9-21}$$

When τ is long compared to t, the exponential may be expanded, and to first order:

$$\omega(\theta, \phi, t) = \omega_0(\theta, \phi)(1 + \cdots) \tag{9-22}$$

which is explicitly time independent. Under this circumstance, the formalism described previously applies, with the only time dependence experienced by $\mathcal{H}_{int}$ being that supplied by the experimenter. However, when this condition is violated, then $\omega(\theta, \phi, t)$ is explicitly time dependent, and this time dependence must be included in the integral in Eq. (9-13). In this circumstance, this integral does not necessarily vanish, and $U_{int}(t)$ is no longer unity at any particular time. The physical result is that the observed value of the transverse component of angular momentum now depends on $\mathcal{H}_{int}$, which is to say that this interaction results in line broadening. Although the effect of motion on averaging in spin space has been discussed here, similar remarks apply to averaging in real space by sample spinning. If in addition to the spinning supplied by the experimenter the nuclei in the sample have a motion with a correlation time shorter than, or on the order of, the sample rotation period, then the effect of sample rotation on averaging of shielding anisotropy is damaged. In fact, the broadening from such motion can be used as a fingerprint of the motion in question.[6]

Applications

Crystallinity

A pulse NMR measurement accomplished by a single excitation of the protons in a polymer leads to a time decay that in some cases (eg, polyethylene, but not polystyrene) has a rapidly decreasing portion near time zero, and a portion with a longer lasting time constant away from time zero, as shown for a sample of polyethylene in Figure 9-6A. That is to say that the decay is characteristic of a sample with at least two different transverse relaxation times, T_2^{short} and T_2^{long}. Because the NMR spectrum is the Fourier transform of this decay, the spectrum consists of a line of approximate full width at half-height of $(3T_2^{long})^{-1}$, as shown in Figure 9-6A. The narrow line has been associated with relatively mobile portions of the sample in which dipolar and

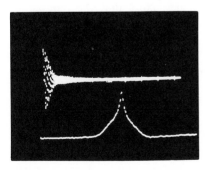
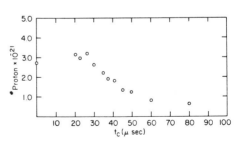

Figure 9-6. (A) Time decay of the magnetization of a sample of polyethylene under a single-pulse excitation and Fourier transform of the decay. (B) Number of protons observed as a function of cycle time. (From reference 7. Reprinted with permission of the American Institute of Physics.)

shielding interactions are attenuated by anisotropic motion. The broad component has been associated with the crystalline fraction of the sample.

One application of high-resolution, solid-state NMR of protons in polymers
has been the use of the cycle-time criterion on narrowing (for multiple-pulse
averaging of a homogeneously broadened line). When the multiple-pulse cycle
time used for attacking the homogeneous proton–proton dipolar interactions
is smaller than the magnitude of these interactions, as measured by the inverse
of the linewidth associated with these interactions, then efficient averaging of
these interactions in spin space takes place. In a polymer, such as polyethylene, a superposition of two lines in the proton NMR spectrum is found frequently, one on the order of 20 kHz wide, the other on the order of 6 kHz
wide. This means that a dipolar-echo cycle time small compared to 6000^{-1} s
$= 166 \ \mu s$, but large compared to $20,000^{-1}$ s $= 50 \ \mu s$ averages dipolar interactions from the noncrystalline but not the crystalline region of the sample.
(One must be aware in making this statement of the limitations of a two-state
model; ie, no intermediate region is recognized. Such intermediate regions are
discussed again later [see Relaxation under Multiple-Pulse Sequences: Hydrogen in Poly(ethylene terephthalate)].) Thus if the initial intensity (the intensity
extrapolated to the zero of time) of the magnetization under the homonuclear
decoupling experiment is plotted as a function of multiple-pulse cycle time,
this value can be expected to represent, for short cycle times, the crystalline
plus noncrystalline region of the polymer and, for longer cycle times, just the
noncrystalline portion of the polymer. At some intermediate time, there is a
dropping off of this intensity as the criterion for spin-space averaging of the
dipolar interactions in the crystalline region of the polymer is violated. These
initial intensities, then, may be used to measure the crystalline fraction of the
polymer quantitatively, and this method of determination may be used as a
check on crystallinity measured by other methods.

This procedure has been used for polyethylene, with the result shown in
Figure 9-6B.[7] The value at $t_c = 0$ is obtained from the known stoichiometry
$(-CH_2-)_n$, density, and weight of the polymer. Note that the number of
protons counted under the multiple-pulse experiment does not change sharply
with variation of the cycle time. One reason for this phenomenon may be that
the violation of the averaging criterion under multiple-pulse dipolar decoupling takes place gradually with increasing cycle time. Another is that the identification of the sample as consisting of two distinct regions, "crystalline" and
"noncrystalline," is too simple a picture and there exists an intermediate, partially ordered region between the two. More is said about this subject in the
section on relaxation under multiple-pulse sequences.

Shielding

If one were to ask a chemist, "Of what use is NMR in studying polymers?"
a probable answer would be "to study sequencing as inferred from chemically
shifted nuclei in solution where sufficiently high resolution can be achieved."

This type of response would characterize that of a large portion of the chemical community vis à vis the use of NMR to probe matter. Another way of stating this response is "to observe the isotropic portion of the shielding and scalar coupling tensors in a situation where there exists sufficient motion to remove the dipolar and electric field gradient tensors." These isotropic values have been used with enormous effectiveness by the chemical community since 1950, the year that Proctor and Yu discovered the chemical shift. This section deals with these parameters, with the caveat that they are a small fraction of the physical and chemical information available from the response of an ensemble of nuclei to resonant radiofrequency excitations. In this section, therefore, are illustrated the basic facts that chemically shifted protons can be detected in polymers, and that anisotropic shieldings can be determined for hydrogen in polymers in which the dipolar broadening is two orders of magnitude larger than the shielding anisotropies. In this discussion of isotropic shieldings, the limits of resolution are probed as a function of the two important experimental parameters: (a) multiple-pulse cycle time, and (b) resonance offset. Comparison of resolution for polymers with nicely crystalline compounds is made to illustrate the effects of dispersion of shielding anisotropies on attainable narrowing.

As stated above, the NMR spectrum of protons in a polymer is a result of both dipolar interactions and shielding anisotropy. In order to select just the isotropic shielding from this broad line, it is necessary to average both the dipolar broadening, and the anisotropy of the shielding to zero. This averaging is accomplished with combinations of rotations in both spin and real space. Multiple-pulse rf sequences that attack the spin portion of the dipolar interaction, as outlined briefly earlier (see Principles, Rotations in Spin Space: Scaling), are used to average the homonuclear dipolar interaction to zero. Combined with this multiple rf pulse attack on the dipolar broadening, the sample is spun at the magic angle at rotation frequencies fast compared to the shielding anisotropy of the protons in question. The latter is not hard to accomplish, because the range of shielding anisotropies for protons in hydrocarbons tends to be not more than 20 ppm,[2] and, eg, at a proton frequency of 300 MHz, half of 20 ppm corresponds to 3 kHz, an easily achievable sample rotation speed, especially if the sample in question can be machined into the rotor, a condition that may be met with many polymers (see Chapter 2).

A question that is asked immediately at this point is the effect of sample spinning on the ability of multiple-pulse sequences to average dipolar interactions. It has been seen above (see Principles, Effect of Molecular Motion) that motion of the nuclei with respect to the external field can hamper the effect of multiple-pulse dipolar decoupling, by making the spatial portion of the dipolar Hamiltonian, $\omega(\theta, \phi)$, time dependent. In this case the integral of both the space and spin portions of $\mathscr{H}_{int}$ must be included in the terms of the Magnus expansion, represented in one form by Eq. (9-11). Then manipulation of the spin portion of $\mathscr{H}_{int}$ alone may not average the total internal dipolar Hamiltonian to zero. It also has been seen, however, that if the correlation time for spatial motion is long compared to that of the multiple-pulse cycle time, the problem reverts to the static case, and only spin terms need be included in the integral of Eq. (9-11b). Such is true for the case at hand. The multiple-pulse attack on the dipolar Hamiltonian must have a cycle time short compared to the inverse of the homogeneous dipolar linewidth. This means cycle times for the dipolar-echo sequence on the order of 5 μs. On the other hand, rotational frequencies are about 2 kHz, which means a rotational cycle time of 500 μs. The factor of 100 is sufficient to insure that the effect of sample rotation on multiple-pulse narrowing is minimal.

Next, therefore, the resolution[8,9] attainable on various polymers is exam-

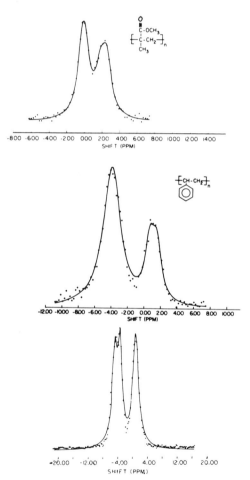

Figure 9-7. Comparison of high-resolution NMR of polymers and crystalline materials. All spectra taken using CRAMPS, with the BR-24 pulse sequence for proton dipolar decoupling. (A) Poly(methyl methacrylate); (B) isotactic polystyrene; (C) 4,4'-dimethylbenzophenone. (From reference 8. Reprinted with permission of the American Institute of Physics.)

ined using multiple-pulse dipolar decoupling of protons, and using magic angle spinning (MAS) of the sample to average shielding anisotropies to isotropic values. The results herein reported were taken at a proton frequency of 56 MHz in the author's laboratory, but the reader should be aware that spectrometers operating at 270 MHz are now being used for combined rotation and multiple-pulse spectroscopy (CRAMPS).[10] Because chemical shifts scale linearly with static field, the resolution at higher fields can be accordingly greater. A problem with the use of very high frequencies in the use of CRAMPS is that the experiments require a relatively uniform B_1 field, which implies a ratio of much greater than one for the length to width of the inductor. As the frequency increases, the inductance used for tuning decreases for reasonable values of tuning capacitors, which means that the length to width ratio of the inductor becomes less favorable for a fixed coil diameter. One means of solving this problem is the use of doubly wound inductances.[11]

Figure 9-7 shows (A) the spectra of hydrogen in poly(methyl methacrylate) (PMMA) and (B) isotactic polystyrene, compared with that of 4,4'-dimethylbenzophenone (DMBP) (C) taken under CRAMPS. The Burum-Rhim-24 (BR-24) pulse sequence (Figure 9-5C) was used for dipolar decoupling of protons. Note that the resolution in the polymers is about 2 ppm, whereas in the crystalline material the two sets of aromatic protons are nicely resolved with linewidths of about 0.5 ppm. Figure 9-8 shows the effect of multiple-pulse cycle time on narrowing of the lines in PMMA, and Figure 9-9 illustrates the effect of resonance offset on the linewidths of protons in DMBP. A summary of the linewidths of protons in four test samples under CRAMPS is given in Table 9-1. Not included in these results are studies of hydrogen in poly(ethylene terephthalate). The latter results are deliberately held until the next section to indicate that isotropic chemical shifts, when combined with

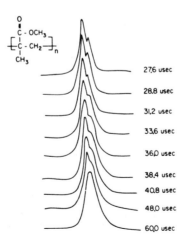

Figure 9-8. Effect of multiple-pulse cycle time on resolution in the proton NMR spectrum of PMMA under CRAMPS. Offset = −1 kHz. (From reference 8. Reprinted with permission of the American Institute of Physics.)

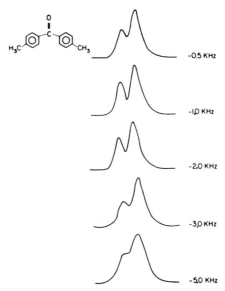

Figure 9-9. Effect of resonance offset on resolution in the proton NMR spectrum of DMBP under CRAMPS. MREV-8 pulse cycle time = 27.6 μs. (From reference 8. Reprinted with permission of the American Institute of Physics.)

other experiments, can be quite powerful informational tools in probing polymer morphologies.

With respect to detection of shielding anisotropies of hydrogen in polymers, the prototype polymer is polyethylene, and there have been two studies of the shielding anisotropy of hydrogen in this compound, prepared in the linear,

TABLE 9-1. LINEWIDTH RESULTS FOR MULTIPLE-PULSE AND CRAMPS MEASUREMENTS[a, b]

	Compound[c]			
	IP	PMMA	DMBP	DMBA
CRAMPS (8-pulse) aliphatic line	157	230	156	145
CRAMPS (8-pulse) aromatic line[d]	160	110	170	131, 74[g]
CRAMPS (24-pulse) aliphatic lines	CH, CH$_2$ 69, 74	CH$_2$, CH$_3$ 79, 65	109	88
CRAMPS (24-pulse) aromatic lines[d]	141	77	62[e] 109[f]	59[e], 65[f] 76[g]
MREV-8	586	431	552	466

[a] Full width at half height in Hz.
[b] From reference 8. Reprinted with permission of the American Institute of Physics.
[c] IP, isotactic polystyrene; PMMA, poly(methyl methacrylate); DMBP, 4,4'-dimethylbenzophenone; DMBA, 2,6-dimethylbenzoic acid.
[d] OCH$_3$ for PMMA.
[e] High-field aromatic line.
[f] Low-field aromatic line.
[g] OH.

high-density form. The first,[9] using the Mansfield, Rhim, Elleman, Vaughan (MREV-8) pulse sequence (Figure 9-5B) at a cycle time of 21 μs, reported the shielding of hydrogen in this compound to be axially symmetrical with an anisotropy of 4.7 ppm. The MREV-8 sequence averages to zero the zero- and first-order terms in the Magnus expansion of the dipolar Hamiltonian under multiple-pulse decoupling, but broadening remains for higher order terms. The second study,[12] on a spherical sample (to minimize the effects of bulk susceptibility) and utilizing the BR-24 pulse sequence, which averages to zero higher order terms in dipolar coupling, reported an axially symmetrical tensor and an anisotropy of 6.9 ppm. The axial character of the anisotropy was defined more clearly than in the first study.

Domain Sizes and Morphologies: Spin Diffusion

This section does not concern exactly high-resolution NMR of protons in solid polymers, but it is basic to material presented later [see Relaxation under Multiple-Pulse Sequences: Hydrogen in Poly(ethylene terephthalate)] in which dipolar decoupling of protons is combined with the type of experiments described in the present section. Because the subject of domain morphologies is of great interest in the study of polymer structures, the present material is included in this section.

The basic spin dynamics used in the present section is that supplied by Goldman and Shen[13] and used on studies of polymers by Assink.[14] This experiment, which consists of three different periods of time evolution for the sample studied, is designed to monitor diffusion of spin information between a mobile (long-T_2) and a rigid (short-T_2) phase of a solid. To accomplish this feat, a 90° pulse places the magnetization of the sample in the transverse plane. A time τ is then allowed to pass such that the magnetization of the rigid phase has lost phase coherence because of homonuclear dipolar interactions. The more mobile phase loses some, but not all, phase coherence. Then an inverse of the first pulse is used to place the remaining magnetization along the static field for a variable time, t. During this period, the rigid phase gains magnetization by spin diffusion from the mobile phase. The recovery of the magnetization of the rigid phase, used with an appropriate solution of the diffusion equation,[15] can then be used to monitor domain sizes and dimensions in such systems.

The spin dynamics of the experiment is shown in Figure 9-10. The pulse sequence is shown in A, and the decay of the magnetization with variable times, t, during which the rigid region is remagnetized by spin diffusion from the mobile region is shown in B. The spin diffusion equation to be solved is:

$$\dot{m}(\mathbf{r}, t) = D\nabla^2 m(\mathbf{r}, t) \qquad (9\text{-}23)$$

Here, D is the diffusion constant (units of cm^2 s^{-1}) and $m(\mathbf{r}, t)$ is the local magnetization density. The solutions for a particular dimensionality of diffusion are expressed in terms of a response function, R(t), which measures the

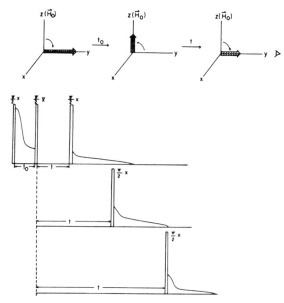

Figure 9-10. (A) Pulse sequence and (B) spin dynamics of the Goldman-Shen experiment. Magnetization in the rigid (crystalline) domain is randomized in a time t_0. Transfer of magnetization from the noncrystalline to the crystalline region is monitored at variable times t. (From reference 15. Reprinted with permission of the American Institute of Physics.)

recovery of magnetization of the crystalline domain, obtained from data such as shown in the bottom of Figure 9-10, as follows:

$$R(t) = 1 - [M_c(t) - M_c(t \to \infty)]/[M_c(t = 0) - M_c(t \to \infty)] \qquad (9\text{-}24)$$

With appropriate boundary conditions, and a Poisson distribution used to describe the spacing between domains, the response function in one, two, and three dimensions may be solved in closed forms as follows [with $\bar{b}$ being an average domain dimension appropriate to the dimensionality of the situation (eg, for one-dimensional diffusion $\bar{b} = b_x \ll b_y, b_z$)]:

(i) one dimensional:

$$R(t) = 1 - \phi(t)$$

$$= (2/\pi^{1/2})(Dt/\bar{b}^2)^{1/2} \qquad \text{for } t \ll \bar{b}^2/D$$

$$= 1 - \pi^{-1/2}(\bar{b}^2/Dt)^{1/2} \qquad t \gg \bar{b}^2/D \qquad (9\text{-}25)$$

(ii) two dimensional:

$$R(t) = 1 - \phi(t)^2$$

$$= (4\pi^{-1/2})(Dt/\bar{b}^2)^{1/2} \qquad \text{for } t \ll \bar{b}^2/D$$

$$= 1 - (\bar{b}^2/\pi Dt) \qquad t \gg \bar{b}^2/D \qquad (9\text{-}26)$$

(iii) three dimensional:

$$R(t) = 1 - \phi(t)^3$$

$$= 6\pi^{-1/2}(Dt/\bar{b}^2)^{1/2} \quad \text{for } t \ll \bar{b}^2/D$$

$$= 1 - (\bar{b}^2/\pi Dt)^{3/2} \qquad t \gg \bar{b}^2/D \qquad (9\text{-}27)$$

The one-dimensional (1D) configurations physically correspond to layer-like domains, with one short dimension for diffusion. The two-dimensional (2D) regions correspond to rodlike domains, and the three-dimensional (3D) regions correspond to cubes or spheres. The plots of the response functions for 1, 2, and 3D are shown in Figure 9-11, where $R(\tau)$ is plotted against the dimensionless quantity $\sqrt{\tau}$, which equals $(Dt)^{1/2}/\bar{b}$. Preliminary results of fits of these equations to data obtained on polyethylene, unconditioned polypropylene film, conditioned polypropylene film, and a vitrain portion of a coal have led to an estimation of sizes and shapes of domains in these systems. Assumptions made to achieve those results were that: (i) D is the same for both domains in all samples, (ii) the distribution on the noncrystalline width obeys a Poisson distribution, and (iii) there is a preferred orientation for the noncrystalline domains in the system. Havens and VanderHart[16] have investigated domains in poly(ethylene terephthalate) and treated the spin-diffusion problem with less restrictive conditions. Their work is reported in the next section.

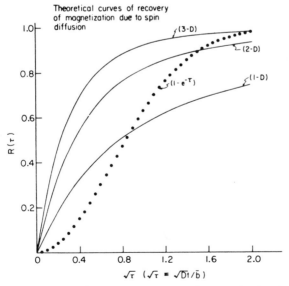

Figure 9-11. Response functions, $R(\tau)$, for recovery of magnetization in the Goldman-Shen experiment in the cases of one-, two-, and three-dimensional diffusion as a function of the diffusion time t. (From reference 15. Reprinted with permission of the American Institute of Physics.)

As a final note to this section, it is necessary to understand that the Goldman-Shen experiment can have consequences not desired when studying domains in polymers. Packer and Pope[17] have pointed out that multiple-quantum coherence can distort NMR spectra under the Goldman-Shen sequence and related sequences.

Relaxation under Multiple-Pulse Sequences: Hydrogen in Poly(ethylene terephthalate)

The present section pursues the idea that achievement of high-resolution NMR of protons in solids places one in a position to use techniques commonly used to study chemical-shift resolved relaxation in liquids. Recall that for a liquid, an inversion–recovery experiment in pulse mode (π_x, τ, $\pi/2_{\bar{x}}$), with Fourier transformation of the decay following the 90° pulse, leads to detection of different longitudinal relaxations for chemically shifted species. A pulse experiment can be performed similarly in which the transverse relaxations of chemically shifted species are obtained. This would be accomplished by a single pulse excitation, followed by a suitable time for decay of transverse magnetization, and then an appropriately phase-shifted π refocusing pulse (for example, the Carr-Purcell-Meiboom-Gill sequence), with Fourier transformation from the peak of the resulting echo.

This type of thinking has led to an elucidation of possible domain structures and morphologies in poly(ethylene terephthalate) (PET) utilizing a number of transient techniques in NMR, including CRAMPS, arranged to probe nuclear relaxation of protons in this polymer.[18,19] One part of this study involved determination of spin–lattice relaxation of protons in the rotating frame characterized by the rate constant $(T_{1\rho})^{-1}$. This parameter is obtained by preparing the system with a 90° pulse, followed by spin locking for a variable time along an axis perpendicular to the preparation pulse, and obtaining the initial magnetization as a function of the spin-lock time (see Chapter 2). Because the effective field in this experiment is the spin-lock field in the rotating frame, the relaxation is along this field, thus the designation $T_{1\rho}$. It was found that there were two different values of $T_{1\rho}$ in this polymer. The variation of this parameter was followed as a function of the locking B_1 field, between 0.5 and 2 mT (5 and 20 G). The field dependence of this parameter indicated that the two $T_{1\rho}$ characterize two mobile phases of the polymer.

In addition, as mentioned earlier (Principles, Rotations in Spin Space: Scaling), a combination of homonuclear dipolar decoupling, and appropriately phased, 180° pulses inserted between the multiple-pulse dipolar-decoupling sequences, the so called DNCP[19] sequence can be used to measure lifetimes under homonuclear dipolar decoupling. As discussed in a previous section (Principles, Rotations in Spin Space: Scaling), however, an alternative to using the refocusing π_y pulses interleaved between homonuclear decoupling sequences to remove dipolar coupling and chemical shift is to prepare the system in a state in which it does not evolve under the chemical shift. For the

MREV-8 pulse sequence, for example, $\bar{I}_z$ becomes $(I_x + I_z)/3$. Therefore, placing the system initially along the (1, 0, 1) direction in the rotating frame (using a $\pi/4$ pulse along $\bar{y}$ before using MREV-8 for homonuclear decoupling) also results in a decay in which lifetime under the multiple-pulse sequence is measured. More will be said about this experiment later in the discussion of proton relaxation in poly(ethylene terephthalate) under homonuclear decoupling.

The first use of combined spin dynamics, which averages both homonuclear dipolar coupling and shielding, was, in fact, in a study of the spin–lattice contribution to the linewidth of polyisoprene.[19] This experiment may be called the ultimate in high-resolution NMR of hydrogen in the solid state, because the residual linewidth is essentially that of the lifetime of the states in question. The residual linewidth for polyisoprene found in that study was roughly 6 Hz at a cycle time for the DNCP experiment of 96 μs. The time constant for the DNCP experiment is denoted T_{1y}.

In the case of hydrogen in PET, it was found that two distinct values of T_{1y} exist. At this point the question arises regarding the relation between portions of the sample exhibiting the long and short values of $T_{1\rho}$ and the long and short values of T_{1y}. This problem is treated later in this section. The combined experiment using MAS and the DNCP sequence, followed by homonuclear dipolar decoupling revealed that both methylene and aromatic protons in this polymer relax with the same two values of T_{1y}. This result implies that spin diffusion takes place even under homonuclear decoupling, a phenomenon traced to three-body correlations in homonuclear dipolar interactions. In addition, the relative intensities of the methylene and aromatic protons detected under CRAMPS, which were 46.6 and 53.4%, respectively, indicated that cross linking between polymer chains in the sample studied was relatively small.

In the presently discussed work on PET, a variant of the Goldman-Shen experiment was used. The pulse sequence is shown in Figure 9-12. In this variation of Goldman-Shen, which combines the features of both $T_{1\rho}$ and T_{1y}, instead of allowing the transverse magnetization of a rigid phase to disappear in a static field, a spin-lock field is turned on for a variable time, t_{SL}, between the initial 90° preparation pulse and its inverse. In this situation, the magne-

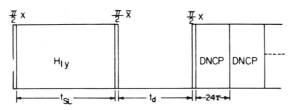

Figure 9-12. Pulse sequence for the spin-lock DNCP experiment, a variation of the Goldman-Shen sequence. (From reference 18. Reprinted with permission of the American Institute of Physics.)

tization of the phase with the shorter value of $T_{1\rho}$ is lost during t_{SL}. Magnetization then is stored along the static field for a time t_d during which diffusion can take place between the phases characterized by different values of $T_{1\rho}$. Then the magnetization is again placed in the transverse plane, and the decay monitored under the DNCP sequence. An example of the result of such an experiment is shown in Figure 9-13. Using such results, the relation between the values of $T_{1\rho}$ and T_{1y} was established, because at t_d = zero, the magnetization observed corresponded to the portion of the sample exhibiting the short value of T_{1y}. Thus, it was established that the portion of the sample giving rise to the long value of $T_{1\rho}$ corresponded to the portion of the sample associated with the short value of T_{1y}. In addition, the fact that spin diffusion occurred over the time periods in question established that discrete regions of the sample existed, separated by some distance over which the magnetization had to diffuse. A rough consideration of the recovery of the magnetization in the above experiment indicated that the region involved in spin diffusion and associated with the longer value of T_{1y} (shorter $T_{1\rho}$), was about 100 Å in extent.

In the second reported study[16] of relaxation of protons in poly(ethylene terephthalate) under multiple-pulse excitation, MAS was not used to separate chemically shifted species. However the measurements were made on a number of different samples, including oriented and nonoriented films, samples

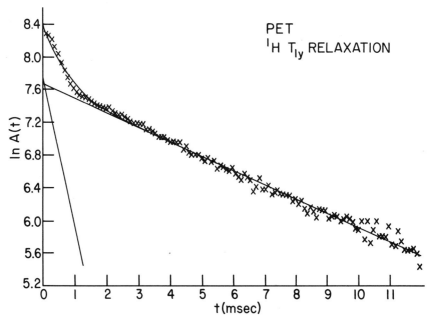

Figure 9-13. The decay of ^{1}H magnetization in PET under the DNCP multiple-pulse sequence. (From reference 18. Reprinted with permission of the American Institute of Physics.)

that had been allowed to shrink "freely," samples that had experienced no shrinkage, annealed samples, and quenched fibers. Shrinkage was controlled by application of a slight tension during annealing. Although the same basic ideas as those reported in the first study were invoked, ie, the use of spin diffusion to monitor domain sizes and morphologies, the details of the spin dynamics were different from those in the first study. A more complete picture of the domain structure and morphologies than that previously drawn became available. The results are summarized as follows:

a. Annealing PET above T_g increases crystallinity, with temperatures above 200°C being most effective. Annealing also increases the minimum crystallite dimension and the long period. It appears that annealing increases the rigidity of chains in the crystallites, which indicates a correlation between crystallite size and chain mobility for semicrystalline PET.

b. A three-region model (the word "phase" is used, but the present author prefers "region" to describe a system obviously not in thermodynamic equilibrium with respect to formation of crystal from glass) is proposed to describe the observed spin diffusion. These regions are: (i) crystal, (ii) constrained noncrystalline, and (iii) nonconstrained, noncrystalline. Annealing causes the system to approach equilibrium by conversion of both noncrystalline regions into crystalline material.

c. Surface areas of crystallites are obtained reliably from initial rates of observed spin diffusion.

Summary

This chapter has presented a brief introduction to the theory and technique of high-resolution, solid-state NMR of abundant spin species. Applications dealing with the acquisition of high-resolution ^1H spectra of polymers, with crystallinity measurement, with spin diffusion as a probe of domain sizes, and with spin relaxation phenomena were reviewed. Although the techniques are difficult experimentally, they can provide a wealth of information. The more widespread availability of commercial instrumentation capable of performing these experiments shoud lead to a wide range of applications.

Acknowledgment

This work was operated for the U.S. Department of Energy by Iowa State University, No. W-7405-Eng-82. The author's research reported in this work was supported by the Office of Basic Energy Sciences, Chemical Sciences Division.

References

1. Haeberlen, U. "High Resolution NMR in Solids: Selective Averaging"; Adv. Magn. Reson. Ser.; Waugh, J. S.; Ed.; Suppl. 1; Academic Press: New York, 1976.
2. Mehring, M. "High Resolution NMR in Solids", 2nd Ed.; Springer-Verlag: Berlin, 1983.
3. Gerstein, B. C.; Dybowski, C. R. "Transient Techniques in the NMR of Solids: An Introduction to the Theory and Practice"; Academic Press: New York, 1985.
4. Slichter, C. P. "Principles of Magnetic Resonance", 2nd Ed.; Springer-Verlag: Heidelberg, 1978.
5. Abragam, A. "The Principles of Nuclear Magnetism"; Oxford University Press: Oxford, 1961.
6. Maricq, M.; Waugh, J. S. *J. Chem. Phys.* **1979**, *70*, 3300.
7. Pembleton, R. G.; Wilson, R. C.; Gerstein, B. C. *J. Chem. Phys.* **1977**, *66*, 5133.
8. Ryan, L. M.; Taylor, R. E.; Paff, A. J.; Gerstein, B. C. *J. Chem. Phys.* **1980**, *72*, 508.
9. Gerstein, B. C. *Phil. Trans. R. Soc. London* **1981**, *A299*, 521.
10. Gerstein, B. C.; Pembleton, R. G.; Wilson, R. C.; Ryan, L. M. *J. Chem. Phys.* **1977**, *66*, 361.
11. Fry, C. G.; Iwamiya, J. T.; Apple, T. M.; Gerstein, B. C. *J. Magn. Reson.* **1985**, *63*, 214.
12. Havens, J. R.; VanderHart, D. L. *J. Magn. Reson.* **1985**, *61*, 389.
13. Goldman, M.; Shen, L. *Phys. Rev.* **1966**, *144*, 321.
14. Assink, R. A. *Macromolecules* **1978**, *11*, 1233.
15. Cheung, T. T. P.; Gerstein, B. C. *J. Appl. Phys.* **1981**, *52*, 5517.
16. Havens, J. R.; VanderHart, D. L. *Macromolecules* **1985**, *18*, 1663.
17. Packer, K. J.; Pope, J. M. *J. Magn. Reson.* **1983**, *55*, 378.
18. Cheung, T. T. P.; Gerstein, B. C.; Ryan, L. M.; Taylor, R. E.; Dybowski, C. R. *J. Chem. Phys.* **1980**, *73*, 6059.
19. Dybowski, C. R.; Pembleton, R. G. *J. Chem. Phys.* **1979**, *70*, 1962.

10

DEUTERIUM NMR OF SOLID POLYMERS

Lynn W. Jelinski

AT&T BELL LABORATORIES
MURRAY HILL, NJ 07974

Scope

This chapter is intended for polymer chemists and polymer physicists. Its purpose is to illustrate the types of detailed, molecular-level information provided by deuterium NMR spectroscopy of bulk polymers. The chapter is divided into two parts. The deuterium NMR experiment is described in the first section, concentrating on the information content of the lineshape and on the information to be obtained from relaxation experiments. In the second part, deuterium NMR is used to address key questions in polymer science. This chapter is restricted to synthetic, rather than biologic, polymers, as there are excellent recent reviews of the latter area.[1-3] In addition, most examples are drawn from work performed at AT&T Bell Laboratories, because the work of Spiess and collaborators has been reviewed extensively in several comprehensive papers.[4-6]

The Deuterium NMR Experiment

Historical Perspective

The contemporary type of solid-state deuterium NMR spectroscopy, namely Fourier-transform quadrupole echo spectroscopy, has been possible only in the past eight years or so. The quadrupole echo NMR pulse sequence[7-9]

© 1986 VCH Publishers, Inc.
Komoroski (ed): High-Resolution NMR Spectroscopy of Synthetic Polymers in Bulk

essentially has revolutionized this field. It is now possible to perform routine experiments and to analyze relaxation times and lineshapes in a detailed and quantitative fashion. Although these new techniques do not obviate the careful results from previous broad-line deuterium NMR work,[10, 11] the pulsed methods are now employed almost universally for obtaining deuterium NMR spectra of solids.

The deuterium NMR lineshapes are very broad (more on this later in this section), and the broad-component information of the free-induction decay signal therefore dies away very rapidly. This decay is so rapid, in fact, that most of the signal is lost during the receiver dead time. The quadrupolar echo pulse sequence refocuses the magnetization, essentially "buying time" until the receiver has had time to recover from the high-powered transmitter pulse and the dead time is over. This pulse sequence is shown in Figure 10-1. The cross-hatched areas schematically represent the time required for the receiver to recover. This pulse sequence and related ones are discussed below (see Experimental Aspects of Deuterium NMR Spectroscopy) when the hardware and software requirements for the solid-state deuterium NMR experiment are described.

QUADRUPOLAR ECHO PULSE SEQUENCE

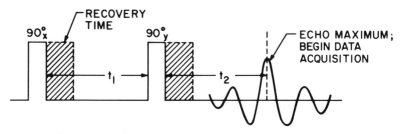

Figure 10-1. The quadrupole echo pulse sequence, used to refocus inhomogeneously broadened lines, such as the solid-state deuterium NMR powder pattern.

During its relatively short existence, quadrupole echo deuterium NMR spectroscopy has been applied to numerous systems—to proteins, lipids, and amino acids, as well as to synthetic polymers, liquid crystals, and plasticizers in polymers. In all of these applications, deuterium NMR has provided new and exceptionally detailed information concerning morphology, molecular motion, and molecular orientation.

The Deuterium NMR Lineshape

The Lineshape in the Absence of Motion. The deuterium nucleus has a spin of 1, which means that in the presence of a magnetic field there are three quantized energy levels, $+1$, 0, and -1. The NMR experiment consists of causing transitions between these levels by the application of energy in the radiofrequency (rf) range. In addition, the deuterium nucleus is quadrupolar and there is a nonspherical charge distribution at the nucleus. The interaction of the quadrupole moment with the electric field gradient tensor at the nucleus causes a substantial perturbation of the Zeeman splitting. This perturbation is so large that the other NMR nuclear spin interactions, such as the scalar J coupling, the dipole–dipole interaction, and the chemical-shift anisotropy (CSA), are negligible (see Chapter 2). Therefore, deuterium solid-state NMR spectra are dominated completely by the quadrupole coupling:

$$H_{total} = H_{Zeeman} + H_{quadrupolar} + H_{dipolar} + H_{CSA}$$

$$55 \text{ MHz} \qquad 250 \text{ kHz} \qquad 10 \text{ kHz} \qquad 0.5 \text{ kHz}$$

The NMR frequencies of the two transitions are given by:[12, 13]

$$\omega = \omega_0 \pm \delta(3 \cos^2 \theta - 1 - \eta \sin^2 \theta \cos 2\phi) \qquad (10\text{-}1)$$

Here, $\delta = 3e^2qQ/8\hbar$, where $e^2qQ/\hbar$ is the quadrupole coupling constant, and ω_0 is the Zeeman frequency. The quantity η is the asymmetry parameter and is usually zero for C—D bonds, meaning that the electric field gradient tensor is axially symmetric. In addition, an axis of the electric field gradient tensor is along the C—D bond direction. The polar angles θ and ϕ specify the orientation of the magnetic field with respect to the principal axis system of the electric field gradient tensor.

If η is taken as zero, the NMR frequencies of the two transitions are given by:

$$\omega = \omega_0 \pm \delta(3 \cos^2 \theta - 1) \qquad (10\text{-}2)$$

This means that the frequencies of the NMR lines depend upon the angle θ that the C—D bond makes with the external magnetic field. If a single crystal or an oriented system is rotated in the magnetic field, different resonance lines are predicted to be obtained for each orientation. This orientation-dependent information could be used to determine the orientation of a specific group (ie, a C—D bond) in a polymer with respect to a fiber draw direction, for example. Alternatively, if the orientation of the C—D bond were changed because of molecular motion, an averaged lineshape would be expected. It will be seen in the following section that this is indeed the case.

Although polymer fibers often are oriented or partially ordered, rarely are polymers obtained as large single crystals. Most cases therefore produce a powder lineshape. This lineshape can be calculated by considering the probabilities of the various C—D bond orientations with respect to the magnetic field.[14] This process is shown schematically in Figure 10-2A–C for a single transition. The two transitions are combined in Figure 10-2D to produce the Pake powder pattern.[15]

How the Lineshape Provides Information about Molecular Motion. First, consider the case where molecular motion is rapid on the deuterium NMR time scale. This situation arises when the motion has a correlation time of less

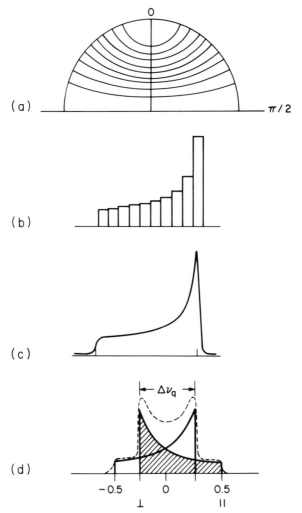

Figure 10-2. The deuterium NMR powder lineshape and how it arises. (A) Consider a sphere divided into latitudes of equal frequency. 0 is the orientation when the C—D bond is parallel to the external magnetic field, B_0, and $\pi/2$ is the perpendicular orientation. (B) A histogram of the areas in the equal-frequency latitudes. (C) A smoothed version of (B). (D) Both transitions.

than about 10^{-7} s. When this is true, the averaged NMR frequency is given by:

$$\omega = \omega_0 \pm \bar{\delta}(3 \cos^2 v - 1 - \bar{\eta} \sin^2 v \cos 2\chi) \qquad (10\text{-}3)$$

where $\bar{\delta}$ is the averaged quadrupole coupling constant, $\bar{\eta}$ is the averaged asymmetry parameter, and the angles v and χ specify the orientation of the magnetic field with respect to the principal axis system of the averaged electric field gradient tensor.

Figure 10-3 shows some explicit lineshapes for various types of motions that often are found in synthetic polymers. Static C—D bonds often occur in crystalline regions of polymers. The static Pake powder pattern is shown in Figure 10-3A. The splitting between the singularities, d, is generally 128 kHz, or three-fourths the quadrupole coupling constant. The following examples illustrate how this Pake pattern is averaged by different types of molecular motions.

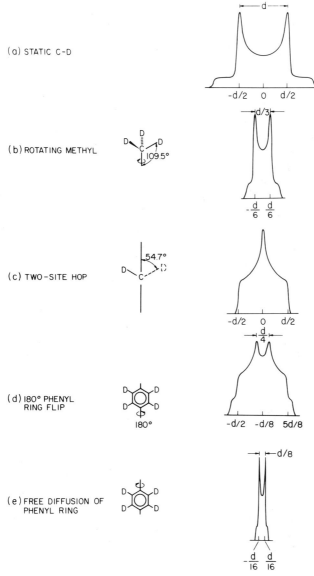

Figure 10-3. Theoretical deuterium NMR lineshapes for various types of rapid ($\tau_c < 10^{-7}$ s) anisotropic motions. See text for details.

Figure 10-3B shows the lineshape predicted for a rapidly rotating methyl group. The C—D bonds in a methyl group form an angle of 109.5° with respect to the rotation axis. Therefore, the $\bar{\delta}$ term in Eq. (10-3) is averaged by the factor $(3 \cos^2 v - 1)$, where v is 109.5°.

The spectrum shown in Figure 10-3C is obtained if the C—D bond is involved in a two-site hop between two positions where the bisector of the hop angle is 54.7°, the magic angle. The component perpendicular to the hop axis is unaffected by the motion, and thus remains at $-d/2$. Another component of the motionally averaged electric field gradient tensor occurs at zero frequency, as $(3 \cos^2 v - 1)$ is zero for the magic angle. The tensor must remain traceless, so the third component occurs at $+d/2$. Note that this type of motion has an asymmetry parameter, η, of 1; ie, the motion produces an axially asymmetric powder pattern.

A slightly more complicated lineshape is shown in Figure 10-3D. This lineshape arises when a deuterated phenyl ring undergoes a 180° flip about the 1,4-phenylene axis. The component of the electric field gradient tensor perpendicular to the flip axis is unaffected by the motion, and remains at $-d/2$. Another component of the tensor is averaged by 120° (ie, the deuteron is seen changing positions by 120° in the flip process), and thus occurs at $-d/8$. Because the tensor must remain traceless, the third component occurs at $+5d/8$. Note that this motionally averaged field gradient tensor is not axially symmetric. In addition, it should be noted that this lineshape pertains only to the *ortho* and *meta* deuterons. A deuteron present in the *para*, or 4 position (as in polystyrene, for example) would be unaffected by motion about the 1,4-phenylene axis and would retain the static Pake pattern shown in Figure 10-3A.

Another possible mode of motion for an aromatic ring is shown in Figure 10-3E. In this case the phenyl ring undergoes free rotation about the 1,4-phenylene axis, in essence sweeping out a cylinder. In this situation the entire pattern would be averaged by a factor of one eighth, as the C—D bond makes an angle of 60° with the rotation axis.

The spectra in Figure 10-3 show theoretical examples of simple motions. Rarely in synthetic polymers are the motions as simple as the ones illustrated here. Generally, 180° ring flips have superimposed on them additional librational modes. Even the lineshapes for methyl rotations can be averaged further if the polymer backbone has motions that cause the methyl group to precess about a cone. Examples of these complications are seen in the following section.

The foregoing examples describe motions that are fast on the deuterium NMR time scale. A particularly interesting situation arises when the rate of molecular motion is intermediate on the deuterium NMR time scale (ie, when the correlation time is less than 10^{-4} s but greater than 10^{-7} s). In this case, the frequency of a single NMR transition fluctuates between two values, in a manner analogous to chemical exchange in high-resolution NMR spectra. This situation is shown schematically (for a single transition) in Figure 10-4.

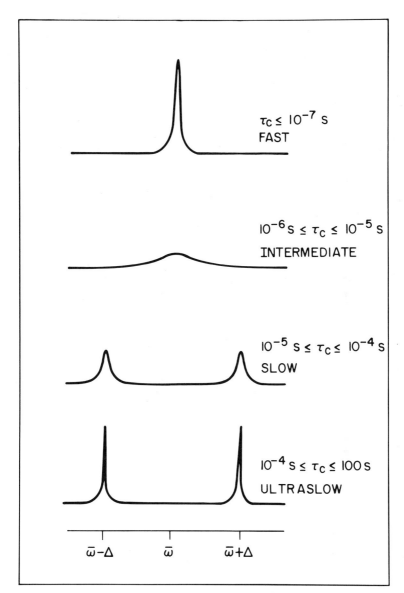

Figure 10-4. Theoretical lineshapes resulting from an exchange between two frequencies, $\bar{\omega} \pm \Delta$. (Adapted from reference 4.)

Deuterium NMR lineshapes can be calculated when motions are intermediate on the deuterium NMR time scale.[14, 16, 17] Such calculated spectra yield rate constants, that, when plotted against inverse temperature, afford apparent Arrhenius activation energies.

Spiess has shown that severe lineshape distortions arise when motions occur in this intermediate frequency range.[17] The quadrupole echo pulse sequence refocusses the inhomogeneously broadened deuterium lineshape—that is, the broadening that occurs because of the frequency dependence of the C—D bond orientation with respect to the magnetic field. However, intermediate rate molecular motion leads to homogeneous broadening. This causes intensity losses because the NMR signal is not refocussed properly by the quadrupole echo pulse sequence. Spiess has shown that intensity losses from homogeneous broadening can be accounted for in the calculated spectra.[17]

A final frequency range is that of ultraslow molecular motions. The technique of spin alignment can be used to probe motions in this regime.[18] This method uses a Jeener-Broekaert pulse sequence[19] to create a state of long-lived order in the spin system. The decay of this order can occur through spin–lattice processes. Ultraslow motions therefore result in a loss of intensity (and a change in lineshape) when the alignment echo is used to " read out " the final state of the system. This method is discussed in greater detail in the section on Software Requirement and Pulse Sequences.

Figure 10-5 shows the approximate characteristic correlation times that can be measured by the spin alignment technique and compares them to those that can be probed through lineshapes. When the motion is rapid on the deuterium NMR time scale (as is the case in Figure 10-3), it is necessary to use spin–lattice relaxation experiments to obtain the rates of motion.

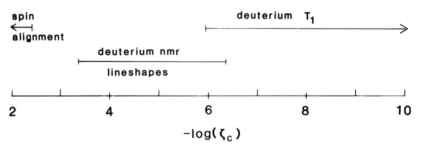

Figure 10-5. A schematic representation of the approximate characteristic correlation times (τ_c) measured by various solid-state deuterium NMR experiments.

Deuterium NMR Relaxation

Many interesting dynamical processes that occur in bulk polymers have correlation times that are less than 10^{-7} s. These include methyl rotations, some aromatic ring flips, diffusion of plasticizers, and some side-chain motions. Although the deuterium NMR lineshape provides information about the types (ie, angular ranges or amplitudes) of these motions, it contains no further information about the rates of these processes when motion is fast on the deuterium NMR time scale. However, Torchia and Szabo have developed and evaluated correlation functions to obtain explicit expressions for solid-state deuterium T_1 values.[20] They show that although the deuterium NMR lineshape is axially symmetric (ie, there is a one to one correspondence between a particular C—D bond orientation and the frequency at which it resonates), there is not necessarily a unique T_1 for each frequency. Therefore, deuterium relaxation is nonexponential in general. (Deuterium T_1 values can be measured using an inversion–recovery pulse sequence, followed by quadrupole echo detection; see the section on Experimental Aspects of Deuterium NMR Spectroscopy.)

Figure 10-6 illustrates the anisotropy observed in the inversion–recovery spectra for the methyl group of labeled alanine. The perpendicular and parallel edges of the Pake pattern do not relax at the same rates, as expected from the frequency dependence of the T_1.

Evaluation of Torchia and Szabo's expressions[20] for two extremes of methyl motion—a discrete three-site hop and free diffusion—affords the T_1 versus correlation time plots shown in Figure 10-7. These plots show that on the fast

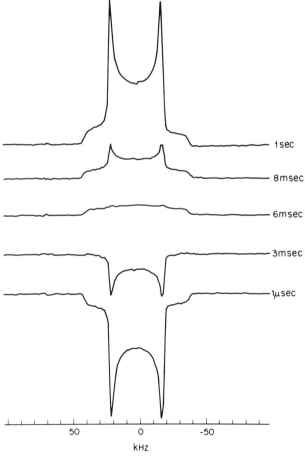

1 sec

8 msec

6 msec

3 msec

1 μsec

50   0   -50

kHz

Figure 10-6. Inversion–recovery solid-state deuterium NMR spectra of polycrystalline L-alanine-d$_3$ obtained at 55.26 MHz and 72°C.

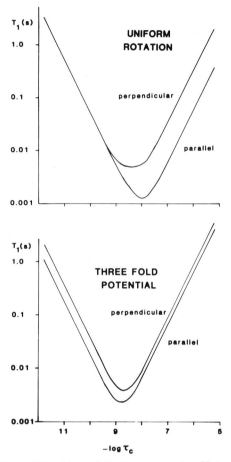

Figure 10-7. Evaluation of Torchia and Szabo's expressions[20] for two extreme models of methyl reorientation. The C—D bond forms a 70.5° angle with the rotation axis. The deuterium frequency is 55.26 MHz.

correlation time side of the T_1 minimum, the parallel and perpendicular edges of the powder pattern have T_1 values that depend on the model for motion. The parallel and perpendicular edges of the powder pattern have T_1 values that are different by a factor of two for the three-site hop model, whereas they are identical for the free-diffusion (uniform rotation) model. Thus, in a single experiment, it theoretically would be possible to determine the model for methyl reorientation. However, most polymer systems have methyl rotations that fall in between these two extremes. It also should be noted that low-amplitude backbone motions of polymers that occur in the megahertz frequency regime can also contribute to the relaxation process, making it difficult to assess the model for methyl reorientation cleanly, except at low temperatures.

Experimental Aspects of Deuterium NMR Spectroscopy

Hardware Requirements. Because the deuterium NMR powder lineshape is over 250 kHz in breadth, the signal dies away rapidly and it is therefore necessary to digitize the free-induction decay very rapidly. Digitization at the Nyquist frequency is not sufficiently rapid for two reasons. First, lineshapes that result when molecular motions have intermediate frequencies on the deuterium NMR time scale often have extremely flat features that are better defined by more rapid digitization.[21] Second, it is essential that digitization begin at the exact top of the echo maximum in order to obtain distortion-free spectra. In most cases, Nyquist frequency digitization does not afford an exact enough definition of the top of the echo maximum. General digitization rates are 200 ns per point (5 MHz) for methyl groups and 100 ns per point (10 MHz) for broader patterns.

Another experimental consideration related to the breadth of the solid-state deuterium NMR powder pattern is the power required to perform an adequate 90° pulse across the entire spectrum. A reasonable 90° pulse width is approximately 3 μs. Even with a 3-μs 90° pulse width, the parallel edges of the pattern suffer intensity losses because of pulse power falloff as a function of frequency. Calculated spectra can be corrected for this experimental problem.[22] Henrichs and co-workers have summarized and quantified some of the sources of distortions in solid-state deuterium NMR spectra, and the interested reader is referred to that paper for additional information.[23]

In addition, the solid-state deuterium NMR probe must be able to withstand the high-power transmitter pulses necessary to achieve the short pulse widths. This usually is accomplished by using nonmagnetic capacitors that have high breakdown voltages at the frequencies of interest.

The deuterium NMR lineshape is symmetrical about zero frequency, so in theory it should be possible to use single-channel detection. (If everything is perfectly balanced, all of a quadrature-detected signal should be in one channel only.) Quadrature detection is generally found to be preferable, particularly in defining the central parts of spectra obtained in the intermediate frequency range.

Software Requirements and Pulse Sequences. Some solid-state deuterium NMR pulse sequences are illustrated in Figure 10-8. The standard quadrupole echo pulse sequence[7-9] (Figure 10-8A) has been described in detail in the previous section on Historical Perspective. Typical times for t_1 and t_2 are 30 and 33 μs, respectively. The data points of the free-induction decay are usually left shifted prior to Fourier transformation, so that the part of the free-induction decay that is transformed begins at the exact top of the echo maximum.

The pulse sequence for the spin alignment, or Jeener-Broekaert pulse sequence, is shown in Figure 10-8B.[18, 24] The 90° pulse in this sequence, or the preparation pulse, generates transverse magnetization. The spin system evolves during τ_1, where the spin isochromats with different C—D bond orientations dephase at different rates. The 45° pulse stores this phase information by cre-

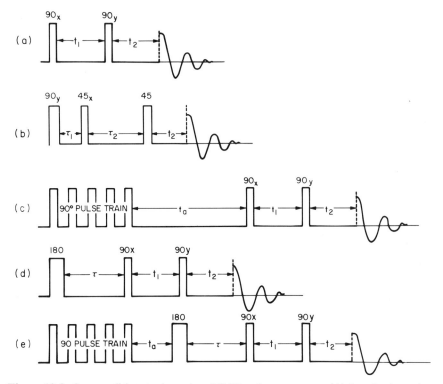

Figure 10-8. Some solid-state deuterium NMR pulse sequences. (A) Standard quadrupole echo; (B) spin alignment; (C) amorphous selection; (D) inversion–recovery quadrupole echo; (E) inversion–recovery of selected amorphous region.

ating a state of spin alignment. The time τ_2 is the mixing time, during which the spin alignment is subject to spin–lattice relaxation with a time constant T_{1Q}. (T_{1Q} is comparable to, but not identical with the T_1).[24] The last pulse reads out the alignment echo. If molecular motion has taken place during τ_2, the echo formation is not complete. Spiess and co-workers have shown that the spin alignment method can be used to distinguish between different types of motion.[24]

Figure 10-8C shows the pulse sequence used to detect selectively the deuterons corresponding to the amorphous regions in polymers.[25] This pulse sequence takes advantage of the fact that the deuterons that belong to the amorphous regions generally have short T_1 values because of molecular motions that are enhanced over those found in the crystalline regions. The 90° pulse train, consisting of five 90° pulses spaced 2 ms apart, is used to saturate the entire spin system. The deuterons from the amorphous region are allowed to relax during t_a, which is set optimally at $3 \times T_{1\,\text{amorphous}}$. Deuterons from the crystalline regions generally undergo negligible relaxation during this time. The standard quadrupole echo pulse sequence is then used to read out the signal from the amorphous deuterons.

The pulse sequence shown in Figure 10-8D is used to measure solid-state deuterium NMR T_1 values. The first pulse, a $180°$ pulse, inverts the magnetization, as in a standard inversion–recovery pulse sequence.[26] The time, τ, is a variable delay, during which time the spin system strives to reestablish its equilibrium. The rest of the sequence is the standard quadrupole echo sequence, which reads out the relaxation information.

The T_1 of the amorphous component (Figure 10-8E) can be measured selectively by using a combination of the previous two pulse sequences (Figure 10-8C and D).[25] As before, the entire spin system is saturated with a train of $90°$ pulses. The amorphous deuterons are allowed to relax during t_a, and they are then subjected to the $180°$ inverting pulse. After a variable delay time, τ, the quadrupole echo sequence is used to read out the relaxation information.

The pulse sequences shown in Figure 10-8 illustrate the most useful of the solid-state deuterium NMR pulse sequences. Other methods, such as progressive saturation,[26] also are used for measuring deuterium NMR relaxation parameters.

Summary of the Experimental Advantages of Deuterium NMR

The primary drawback of deuterium NMR spectroscopy of solids—namely, that specifically labeled material must be used—is offset by the numerous advantages that this technique has to offer. Some of them are summarized here in list form:

1. The quadrupolar mechanism dominates relaxation. The NMR parameters are determined by the quadrupole interaction with the electric field gradient tensor at the deuterium nucleus. Thus, the relaxation mechanism is known with certainty. Furthermore, deuterium NMR spectroscopy is sensitive only to local motions and is therefore extremely selective. In addition, spin diffusion is very inefficient for deuterons, and the relaxation data do not have the sorts of complications often encountered in interpreting relaxation data from other nuclei, such as ^1H, ^{19}F, and ^{13}C.

2. Deuterium has a low natural abundance. Although it is necessary to label a polymer specifically in order to observe its solid-state deuterium NMR spectrum, the labeled site affords an extremely high degree of selectivity. In addition, resonances from nonlabeled deuterons, which occur at the 0.016% level in natural abundance, are not a complicating factor, as is often the case in work involving ^{13}C-labeled material.

3. The electric field gradient tensor is axially symmetric. Because the electric field gradient tensor is axially symmetric to a good approximation, there is a one to one correspondence between C—D bond orientation and frequency. It is therefore straightforward to calculate the deuterium NMR lineshapes in the presence of motion.[27] This is generally not the case for ^{13}C, for example.

4. The tensor orientation is known. The electric field gradient tensor is axially symmetric about the C—D bond direction. Again, the tensor orienta-

tion is generally not symmetrical about bond directions for other nuclei, such as ^{13}C. This obviates the necessity for single-crystal tensor model studies for deuterium NMR tensor assignments.

5. The deuterated samples also can be used for neutron diffraction studies. Neutron diffraction affords information that is complimentary to the deuterium NMR experiment. This provides the opportunity to perform widely different experiments on the same polymer sample, thereby producing data of an informative and uncomplicated nature.

In addition to the above features, which are unique to the deuterium NMR experiment, this method affords an exceptionally high dynamic range over which motions can be measured. Different types of motions can be identified readily and unambiguously, and motional heterogeneity can be assessed and quantified.[4]

Applications to Polymer Science

In this section are described examples of the types of information that can be obtained from the deuterium NMR experiment. This discussion draws primarily upon work performed at AT&T Bell Laboratories, as much of the other work has been reviewed extensively.[4-6] However, the Appendix contains a comprehensive list of all the quadrupole echo deuterium NMR experiments on synthetic polymers that have been reported as of this writing.

Motions about Three Bonds in Polymers

Using solid-state deuterium NMR spectroscopy of selectively labeled poly(butylene terephthalate) (**I**) the mechanism of motion about three bonds has been established.[21] Poly(butylene terephthalate) was used as a model because early carbon NMR relaxation data[28, 29] and chemical-shift anisotropy considerations[30] showed that the terephthalate residue carboxyl groups could be considered static compared to the motions occurring in the alkyl region. In addition, the alkyl portion of this polymer is the shortest sequence that can undergo motions about three bonds,[31] with the terephthalate groups acting as molecular pinning points to prevent longer range motional modes.

I

The deuterium NMR results pertaining to this study have been reviewed,[32, 33] and the interested reader is directed to these references. The main conclusions are stated here.

The central carbons in poly(butylene terephthalate) are found to undergo trans–gauche conformational transitions with a correlation time of 7×10^{-6} s at 20°C. The dihedral jump angle is not as large as the theoretically predicted 120° angle, but is instead 103°. The activation energy for the trans–gauche conformational transition is 5.8 kcal/mol, somewhat greater than a single carbon–carbon bond rotational barrier, but not as large as would be required for a concerted two-bond process. The deuterium NMR data are consistent with the models proposed by Helfand,[34] in which counterrotation occurs about second-neighbor parallel bonds. The mechanisms proposed by Helfand allow trans–gauche conformational transitions to occur in polymers without concomitant large-scale reorientation of the ends of the polymer chain.

Phase Separation in Segmented Copolymers

Segmented Copolyesters. With the rates and types of motion that occur in the poly(butylene terephthalate) homopolymer being established, it is of interest to extend these studies to copolymers that contain poly(butylene terephthalate) as the hard segment. In particular, it is desirable to determine whether the motions of the hard segment are enhanced when it is incorporated in a segmented copolymer, and whether solid-state deuterium NMR spectroscopy can provide evidence about the degree of phase segregation.

Such solid-state deuterium NMR experiments have been performed on the Hytrel copolyesters,[35] **II**. As these results have been reviewed,[32, 33, 36] only the main conclusions from these studies are stated here.

II

The deuterium NMR lineshape from the segmented copolymer in which only the hard segments are labeled can be deconvoluted into two components. One component is identical to that observed for the homopolymer. This is attributed to hard segments that are incorporated properly into phase-separated lamellae. The other component is very sharp and displays more nearly isotropic motion. This component is attributed to those hard segments that are "dissolved" in the soft-segment matrix, or to those segments that form the domain interfaces. The amounts of the two components can be quantified. The results show that 9% of the hard segments are involved in essentially isotropic motions in the segmented copolyester containing 0.87 mol fraction of hard segments. These results provide the first clear evidence that there are two motional environments for the hard segments in this segmented copolyester.

Polyurethanes. Although numerous models have been set forth concerning the morphologic microdomain structure for polyurethanes,[37] there is at present no consensus in this matter. Bonart and Müller,[38, 39] on the basis of small-angle X-ray scattering (SAXS) experiments, have proposed hard-segment packing models in which the hard segments assume fully extended configurations within lamellar or sheetlike domains. Subsequent wide-angle X-ray scattering (WAXS) studies by Blackwell et al supported this model.[40] However, recent results by Van Bogart et al[41] and Koberstein and Stein[42] are inconsistent with these extended sequence models. Using SAXS, both groups of investigators find that the hard-segment chains must be present in either coiled or perhaps folded configurations. Koberstein and Stein developed a new model based on these results.[42] In this model, the hard-segment domain thickness is governed predominantly by the shortest hard-segment sequence length that is insoluble in the soft-segment phase. Sequences longer than this critical length adopt coiled configurations in order to reenter the hard-segment domain and fill space efficiently. Further detailed SAXS experiments on a series of polyurethanes of varying hard-segment content support the Koberstein-Stein model.[37] The extended model and the folded configuration model are illustrated schematically in Figure 10-9.

In order to further address these questions, solid-state deuterium NMR experiments were performed on a series of specifically labeled polyurethanes[43] (structures **III** and **IV**). The deuterium NMR spectra for these polymers are shown in Figure 10-10, where they are compared with the all-hard-segment material. Several points can be made with reference to the spectra shown in Figure 10-10. First, the spectrum of the all-hard polymer (Figure 10-10D–F) is not that of a Pake pattern but instead shows considerable motional narrowing. The lineshape of Figure 10-10D is reminiscent of the lineshapes observed for the Hytrel poly(butylene terephthalate) hard segments,[32] in which trans–gauche conformational transitions occur on the deuterium NMR time scale at room temperature. This result suggests that the hard segments assume not an extended, all-trans conformation, but one in which kink motions can occur.

$$\left[\begin{array}{c}\underset{\text{C-N}}{\overset{\overset{O}{\|} \overset{H}{\underset{|}{}}}{}}-\bigcirc-CH_2-\bigcirc-\underset{\text{N-C-O}}{\overset{\overset{H}{\underset{|}{}} \overset{O}{\|}}{}}+CH_2CD_2CD_2CH_2O+\underset{\text{C-N}}{\overset{\overset{O}{\|} \overset{H}{\underset{|}{}}}{}}-\bigcirc-CH_2-\bigcirc-\underset{\text{N-C-O-}}{\overset{\overset{H}{\underset{|}{}} \overset{O}{\|}}{}}\right]_y$$

III, Hard segment

$$+CH_2-CH_2-O+_{Z_2/2}\left[CH(CH_3)CH_2-O\right]_{Z_1}+CH_2-CH_2-O+_{Z_2/2}$$

IV, Soft segment

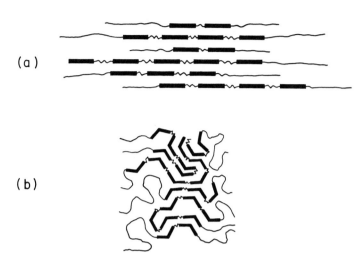

(a)

(b)

Figure 10-9. The fully extended (A) and folded configuration (B) model for the hard segments of polyurethanes. (Reprinted with permission from Dumais, J. J.; Gancarz, I.; Galambos, A.; Koberstein, J. T.; Jelinski, L. W. *Macromolecules* **1985**, *18*, 116. Copyright 1985, American Chemical Society.)

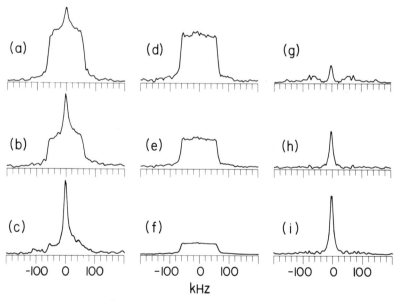

Figure 10-10. Solid-state ^{2}H NMR spectra of hard-segment labeled polyurethanes (see text for structural formula) obtained at 55.26 MHz and 22°C. (A) 70 wt% hard; (B) 60 wt% hard; (C) 50 wt% hard; (D)–(F) all-hard at different gains; (G)–(I) subtraction of spectrum in middle column from that in left column. (Reprinted with permission from Dumais, J. J.; Gancarz, I.; Galambos, A.; Koberstein, J. T.; Jelinski, L. W. *Macromolecules* **1985**, *18*, 116. Copyright 1985, American Chemical Society.)

The spectra shown in Figure 10-10A–C represent polyurethane samples with decreasing amounts of hard segment. Partially relaxed spectra (not shown) indicate that these spectra are composed of two components. The broad component is similar to that observed for the all-hard-segment material. The sharp, nearly isotropic line is attributed to those hard segments too short to enter the hard-segment phase and therefore "dissolved" in the soft-segment phase, as well as to interfacial material.

Subtraction of the spectra in Figure 10-10D–F (ie, the all-hard material) from the spectra in Figures 10-10A–C give clean difference spectra (Figure 10-10G–I, from which the fraction of sharp component can be estimated. These fractions are summarized in Table 10-1.

TABLE 10-1. AMOUNT OF SHARP COMPONENT
IN POLYURETHANE SAMPLES

Hard segment (wt%)	Sharp component (%)
100	0
70	14
60	20
50	50

These results show that solid-state deuterium NMR spectroscopy provides information that is otherwise unobtainable and complimentary to that obtained by other techniques, such as SAXS.

Aromatic Ring Flips in Polymers

Semicrystalline Polymers. Although phenyl ring flips have been observed in a number of biopolymers[2] and synthetic polymers,[25] several fundamental questions concerning aromatic ring flips remain unanswered, particularly in the area of semicrystalline polymers. Among these are: (1) is a pure 180° flip model adequate to explain the motion of phenyl rings in polymers? (2) Do phenyl ring flips occur only in the amorphous regions of semicrystalline polymers? (3) What is the activation energy for the 180° ring-flip process? (4) Is there a broad distribution of ring-flip rates in polymers? Data bearing on the answers to these questions have been obtained from deuterium NMR studies on the specifically labeled poly(butylene terephthalate),[25, 44] V. Solid-state deuterium NMR spectra for this polymer are shown in Figure 10-11. Pulse sequences and recycle delay times were adjusted so that the spectrum in (A) represents aromatic rings from the crystalline, amorphous, and interphase regions of the polymer. The spectrum shown in Figure 10-11B comes primarily from the interphase and amorphous regions, and the spectrum in Figure 10-11C is from the amorphous region alone. Appropriate difference spectra (see figure caption) represent (D) the crystalline region, (E) the interphase region, and (F) the amorphous region. The calculated spectra (Figure 10-11G–I) correspond to the difference spectra (D)–(F). This figure illustrates two

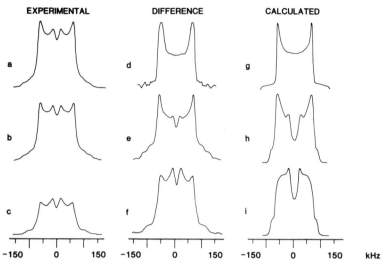

EXPERIMENTAL   DIFFERENCE   CALCULATED

a   d   g

b   e   h

c   f   i

-150   0   150   -150   0   150   -150   0   150   kHz

Figure 10-11. (A)–(C) Solid-state ^{2}H NMR spectra of [aromatic-d$_4$] poly(butylene terephthalate) at 70°C. (A) and (B) are obtained with the quadrupole echo pulse sequence with recycle times of 60 s and 1 s, respectively. Spectrum (C) is obtained with the amorphous quadrupole echo pulse sequence. Difference spectrum D = A − B represents primarily the crystalline region of the semicrystalline polymer. Difference spectrum E = B − C represents the interphase region of the polymer. Spectrum (F) is the same as spectrum (C) but is shown at a different gain. (G)–(I) are calculated spectra. The calculated spectra assume a 156-kHz quadrupole coupling constant and a 180° phenyl ring flip. The rate constants used in calculated spectra are (G) 4.74×10^4 s^{-1}, (H) 6.15×10^4 s^{-1}, (I) 2.84×10^5 s^{-1}. (Reprinted with permission from Cholli, A. L.; Dumais, J. J.; Engel, A. K.; Jelinski, L. W. *Macromolecules* **1984**, *17*, 2399. Copyright 1984, American Chemical Society.)

points. First, there is a considerable amount of motional heterogeneity in this semicrystalline polymer system. Second, the experimental and calculated spectra are in only fair agreement, indicating that there are motions in addition to the pure 180° flip process that are responsible for the observed lineshapes.

$$\left[-C \overset{\overset{\displaystyle O}{\parallel}}{\underset{}{}} \underset{\overset{\displaystyle D \quad\quad D}{}}{\overset{\overset{\displaystyle D \quad\quad D}{}}{\bigcirc}} C \overset{\overset{\displaystyle O}{\parallel}}{\underset{}{}} -O-CH_2\,CH_2\,CH_2\,CH_2-O- \right]_X$$

V

The amount of each component can be estimated from the areas under the curves. These results are summarized in Table 10-2, where they are compared to density measurements. The density measurements and the NMR results are in good agreement.

LYNN W. JELINSKI

TABLE 10-2. POLY(BUTYLENE TEREPHTHALATE)
COMPOSITION

	By NMR[a]	By density[b]
Crystalline	69	75
Interphase	9	
Amorphous	22	

[a] Determination method described in text.
[b] Determined by flotation.[45]

Figure 10-12 shows representative spectra from a complete series of variable-temperature solid-state deutrium NMR experiments on the amorphous component only. The calculated spectra shown in Figure 10-12 were obtained by assuming that the aromatic rings in the amorphous regions undergo pure 180° flips, and that the rate of these flips changes with temperature. The calculated and experimental spectra are in fair agreement, particularly for the low-temperature spectra. However, the centers of the

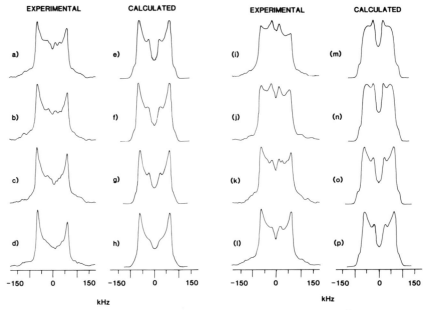

Figure 10-12. Experimental (A)–(D) and calculated (E)–(H) solid-state ^{2}H NMR spectra representing the amorphous deuterons of [aromatic-d$_4$] poly(butylene terephthalate). Spectra (A)–(D) were obtained with the amorphous quadrupole echo pulse sequence at 10°C, 0°C, −10°C, and −20°C, respectively. The rate constants for the calculated spectra are (E) 4.74×10^4, (F) 2.84×10^4, (G) 2.60×10^4, (H) 1.42×10^4 s^{-1}. Experimental (I)–(L) and calculated (M)–(P) solid-state ^{2}H NMR spectra as described above: (I) 70°C, (J) 50°C, (K) 30°C, and (L) 22°C, rate constants: (M) 2.84×10^5, (N) 1.90×10^5, (O) 0.90×10^5, (P) 0.76×10^5 s^{-1}. (Reprinted with permission from Cholli, A. L.; Dumais, J. J.; Engel, A. K.; Jelinski, L. W. *Macromolecules* **1984**, *17*, 2399. Copyright 1984, American Chemical Society.)

experimental spectra are more filled in than predicted, and the expected defini-tion at the edges of the powder patterns is lost. These observations are consis-tent with both a distribution in flip times and the presence of additional high-frequency librational motions that are superimposed on the 180° phenyl-ring flips. These spectra illustrate the heterogeneous nature of the phenyl-ring motions in poly(butylene terephthalate).

When plotted against reciprocal temperature in an Arrhenius fashion, the rate constants obtained by computer simulation of the lineshapes afford an apparent activation energy of 5.9 kcal/mol. The activation energy is for the pure 180° flip process and is in good agreement with the theoretical calcu-lations of Hummel and Flory[46] and Tonelli.[47]

Taken together, these results show that the motions of the phenyl rings in poly(butylene terephthalate) can be attributed to those arising from the crys-talline, interphase, and amorphous regions. Although the phenyl ring motions that occur in poly(butylene terephthalate) are best described by a 180° flipping process, the frequency of these flips is described by a large heterogeneity. (Relaxation data, reviewed elsewhere,[33] support this finding.)

Whether a phenyl ring does or does not undergo a 180° flip is attributed to the conformational space available to that ring. Phenyl-ring flips therefore appear to be a sensitive reporter of local morphology in the material. This finding is used in the next section, where the local structure of a completely amorphous polymer is investigated.

Amorphous Polymers. Using molecular motion as an experimental probe, data are presented that address the question of nonrandom order at the molecular level in a completely amorphous polymer.[48] The selectively deuter-ated poly(arylene ether sulfone) **VI** was used because the material was com-pletely amorphous, eliminating possible complications or ambiguities that could arise from the presence of crystalline domains.

VI
X = 95% ^{1}H
5% ^{2}H

Figure 10-13A shows the solid-state deuterium NMR lineshape obtained with the quadrupole echo pulse sequence and a recycle delay time of 10 s. This spectrum contains a full contribution from the components with short relax-ation times and a partial contribution from components with long relaxation times. The lineshape that corresponds to rapidly flipping phenyl rings can be obtained selectively by using the amorphous quadrupole echo pulse sequence (spectrum not shown).

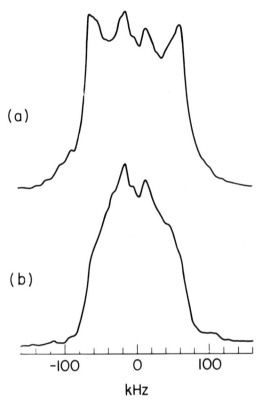

Figure 10-13. Solid-state quadrupole echo deuterium NMR spectra of poly(arylene ether sulfone) obtained at (A) 20°C, and (B) 95°C.

Through a series of progressive saturation experiments, it can be shown that the lineshape in Figure 10-13A arises from a broad distribution of 180° phenyl ring-flip rates. Approximately 50% of the rings are flipping very rapidly and have $\tau_c < 10^{-7}$ s. About 40% of the rings have intermediate flip rates, with $\tau_c \sim 10^{-5}$ to 10^{-6} s. Finally, approximately 10% of the rings are static on the deuterium NMR time scale. This large distribution in phenyl ring-flip rates is attributed to chain packing constraints and thus reflects microscopic fluctuations in packing density.

Spectra obtained at 95°C (Figure 10-13B) show that the constraints that produce the intermediate and static-like deuterium NMR patterns are relaxed easily. This fully relaxed spectrum has essentially no static or intermediate component. The corresponding theoretical lineshape (not shown) can be calculated by assuming that all of the phenyl rings are involved in rapid 180° ring flips. It should be noted that these samples contained approximately 1–2 wt% water, and that they were obtained by precipitation. The spectrum of Figure 10-13A is not fully reversible with temperature cycling; ie, heating to 95°C and returning the sample to room temperature does not immediately produce the same ratio of static to flipping rings as obtained in the freshly precipitated sample.

These results show that (1) there are molecular-level heterogeneities in amorphous polymers, (2) deuterium NMR spectroscopy can be used to probe these environments, (3) the fraction of chains in entangled or in otherwise constrained environments can be quantified, and (4) these constraints are relaxed at temperatures substantially below the glass transition temperature (T_g).

It is appropriate to compare these results with those of Spiess and co-workers, who have observed spectra similar to these for specifically deuterated polycarbonate.[4-6] For polycarbonate, however, all of the phenyl rings undergo rapid 180° ring flips above room temperature. It is only below room temperature that rings with intermediate flip rates are observed.

Interaction of Water with Epoxy Resins

Absorption of small amounts (1–3 wt%) of water by epoxy resins effects a substantial plasticizing action and consequent degradation of their mechanical properties. Solid-state deuterium NMR studies of D_2O-exchanged epoxy resins of known water content provide information concerning the molecular details of the water–epoxy interaction.[44, 49, 50] Figure 10-14A shows a typical quadrupole echo solid–state deuterium NMR spectrum of the exchanged epoxy resin **VII** when it contains 2 wt% D_2O and when this sample is dried (B). The sharp central line is attributed to the sorbed D_2O, because this peak disappears when the sample is heated or dried. This sharp peak cannot be from free water, as it does not freeze at temperatures down to $-20°C$. The outer, broad part of the NMR spectrum is attributed to —OH residues that have undergone exchange with D_2O and have become —OD groups on the polymer backbone.

VII

Spin–lattice relaxation data (Figure 10-15) also show that there is no free water in this system. The spin–lattice relaxation time of pure D_2O under these conditions is 362 ms, whereas the corresponding relaxation time of the sharp signal in the epoxy resin is 12 ms. Although the water is impeded in its motion, it hops from site to site along the polymer backbone with an approximate correlation time of 10^{-10} s. In this respect, the water cannot be considered to be tightly bound.

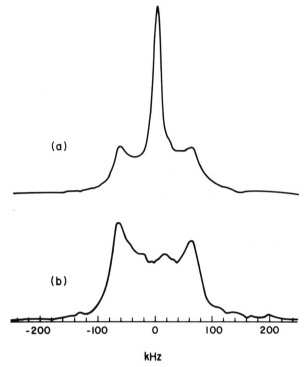

(a)

(b)

| -200 | -100 | 0 | 100 | 200 |

kHz

Figure 10-14. Quadrupole echo deuterium NMR spectra of epoxy resin (A) containing 2 wt% D_2O; and (B) the sample in (A) after it has been dried. The spectra were obtained at 20°C and 55.26 MHz. (Reprinted with permission from Jelinski, L. W.; Dumais, J. J.; Cholli, A. L.; Ellis, T. S.; Karasz, F. E. *Macromolecules* **1985**, *18*, 1091. Copyright 1985, American Chemical Society.)

It is also unlikely that the water disrupts the hydrogen bonding network in the epoxy resin system, as has been proposed on the basis of other studies.[51, 52] This conclusion is based on the deuterium NMR data,[50] which set an upper limit for the —OD/D_2O exchange rate of 10^3 s^{-1} and further show that the water molecules hop from site to site at least six orders of magnitude faster than this.

TABLE 10-3. DEUTERIUM NMR RELAXATION DATA[a] FOR EPOXY–D_2O AND RELATED SYSTEMS

Sample	T_1 (ms)
Neat D_2O	362
Epoxy + 2 wt% D_2O	12
Poly(N-vinylpyrrolidone) + 21 wt% D_2O	12
Neat DMSO-d_6	306
Epoxy + 5 wt% DMSO-d_6	73

[a] All data were obtained at 55.26 MHz for ^2H and at 20°C.

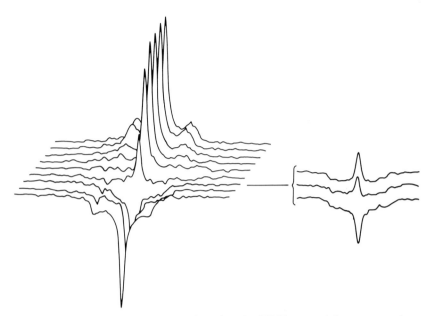

Figure 10-15. Inversion–recovery quadrupole echo NMR spectra for epoxy resin containing 2 wt% D_2O. From top to bottom, the inversion–recovery delay times are 5000, 500, 200, 100, 50, 10, 1, 0.5, and 0.1 ms. The enlargement on the right shows data obtained at inversion–recovery delay times of 9 (top), 7, and 3 ms. All data were obtained at 20°C and 55.26 MHz for deuterium. (Reprinted with permission from Jelinski, L. W.; Dumais, J. J.; Cholli, A. L.; Ellis, T. S.; Karasz, F. E. *Macromolecules* **1985**, *18*, 1091. Copyright 1985, American Chemical Society.)

Table 10-3 shows a comparison of the relaxation data for water in epoxy resins with that for water in poly(N-vinylpyrrolidone), a hydrophilic polymer that contains no exchangeable groups. The results show that the relaxation times for water are identical in both cases. Plasticization of the epoxy resin with dimethyl sulfoxide ($DMSO\text{-}d_6$) also causes the DMSO relaxation time to be depressed (Table 10-3). These results are taken as evidence that the water acts simply as a plasticizer in the epoxy system, just as it does for polymers that contain no exchangeable protons.

The Effect of Tacticity on Methyl Group Motions

Deuterium NMR spectroscopy is an efficient method to study in detail the motions of the side chain groups and the polymer backbone in suitably labeled poly(methyl methacrylate)s[53] (**VIII** and **IX**). The questions of interest are: (1) what effect does tacticity have on the methyl rotation rates of the side-chain α-methyl groups? (2) Can the α-methyl groups be used to study the backbone motions? (3) What differences are there in the rotation rates for the ester methyl and the backbone methyl groups?

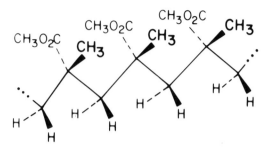

VIII, Isotactic

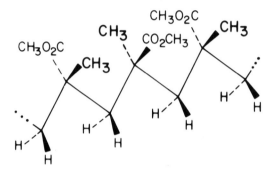

IX, Syndiotactic

Typical inversion–recovery T_1 data for syndiotactic poly(methyl methacrylate), labeled with deuterium at the α-methyl group, are shown in Figure 10-16. These data, obtained at room temperature, show little anisotropy in the lineshapes. This is attributed to low-amplitude megahertz modulation of the methyl groups by backbone motions of the polymer. When the sample is cooled to $-90°C$, the inversion–recovery data show some anisotropy. However, the methyl group lineshapes at this temperature begin to reflect the effect of intermediate exchange.

The data in Table 10-4 show that at $-90°C$, where the backbone motions are essentially frozen out, that the spin–lattice relaxation times for the syn-

TABLE 10-4. RELAXATION DATA[a] FOR POLY(METHYL METHACRYLATE) α-METHYL GROUPS

Temperature (°C)	Syndiotactic T_1 (ms)[b]	Isotactic T_1 (ms)[c]
−90	25	23
22	10	14
90	15	30

[a] Values reported correspond to least-squares analysis of inversion–recovery data for the perpendicular orientation. All spectra were obtained at 55.26 MHz for ^{2}H.
[b] $T_g = 119°C$.
[c] $T_g = 54°C$.

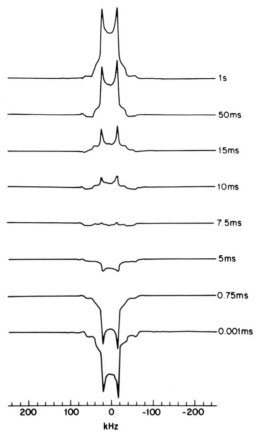

Figure 10-16. Inversion–recovery ^{2}H NMR data for α-methyl labeled syndiotactic poly(methyl methacrylate), obtained at 55.26 MHz and 22°C.

diotactic and isotactic polymer are essentially identical. The methyl lineshapes (not shown) at this temperature are also identical for both tacticities. This indicates that the methyl group rotation rate is not strongly dependent on tacticity. However, as the temperature is raised, the backbone motions begin to influence the spin–lattice relaxation times of the methyl deuterons. At 22°C, the spin–lattice relaxation times for the syndiotactic and isotactic samples are different, with the isotactic material showing a larger relaxation time. Such a difference is anticipated, as the T_g for the isotactic material (54°C) is substantially lower than that for the syndiotactic polymer (119°C). This difference is emphasized further at 90°C, a temperature that is above the T_g for the isotactic polymer but below that for the syndiotactic one.

Note, in passing, that the T_1 for the side-chain ester methyl group is 1000 ms at 22°C, showing that its rotation rate is far faster than that of the α-methyl groups. The plots shown in Figure 10-7 can be used to determine correlation times for the methyl rotations in poly(methyl methacrylate).

Prognosis

It is clear that deuterium NMR spectroscopy is an exceptionally powerful technique for providing detailed information about both the frequency and amplitude of molecular motions in solid polymers. However, solid-state deuterium NMR experiments on polymers are currently at a very early stage of development. Many more systems need to be studied. Much more basic work is required, particularly in the areas of assessing the extent of motional heterogeneities.

Experiments involving stressed, annealed, compressed, extended, and aged samples are anticipated to provide very important information about molecular processes that occur during these mechanical and thermal treatments. With the results from such experiments in hand, it should be possible to clarify the connection between polymer structure and polymer mechanical properties. It is to this end that deuterium NMR studies on bulk polymers are directed.

Appendix

DEUTERIUM NMR STUDIES OF SYNTHETIC POLYMERS

Polymer	Area investigated	References
Epoxy plus D_2O	Water–epoxy interaction	44, 49, 50
Epoxy plus DMSO-d_6	DMSO relaxation time	50
Hytrel	Hard-segment phase separation	32, 35
Liquid crystalline polymers	Structure and dynamics	4–6, 54, 55
Polyacetylene	Relaxtion, morphology	56, 57
Poly(arylene ether sulfone)s	Aromatic ring flips	48
Poly(butylene terephthalate)	Aromatic ring flips	26, 44
	Three-bond motions of alkyl chain	22
Polycarbonate	Methyl motions, ring flips	4–6
Polyethylene	Amorphous regions, oriented material, crystalline regions	4–6, 27, 58
Polystyrene	Aromatic ring motions	4–6
Poly(urethane)s	Hard-segment phase separation	43
Poly(vinyl pyrrolidone) plus D_2O	Polymer–water interaction	50

References

1. Smith, R. L.; Oldfield, E. *Science* **1984**, *225*, 280.
2. Torchia, D. A. *Ann. Rev. Biophys. Bioeng.* **1984**, *13*, 125.
3. Griffin, R. G. *Meth. Enzymol.* **1981**, *72*, 108.

4. Spiess, H. W. *Colloid Polym. Sci.* **1983**, *261*, 193.
5. Spiess, H. W. *J. Mol. Struct.* **1983**, *111*, 119.
6. Spiess, H. W. *Adv. Polym. Sci.* **1985**, *66*, 23.
7. Davis, J. H.; Jeffrey, K. R.; Bloom, M.; Valic, M. I.; Higgs, T. P. *Chem. Phys. Lett.* **1976**, *42*, 390.
8. Blinc, R.; Rutar, V.; Seliger, J.; Slak, J.; Smolej, V. *Chem. Phys. Lett.* **1977**, *48*, 576.
9. Hentschel, R.; Spiess, H. W. *J. Magn. Reson.* **1979**, *35*, 157.
10. Blinc, R., ed. "Magnetic Resonance and Relaxation", *Proc. XIV Colloque Ampère*, Ljubljana, Yugoslavia, 1966; 1967.
11. Mantsch, H. H.; Saitô, H.; Smith, I. C. P. *Prog. NMR Spectrosc.* **1977**, *11*, 211.
12. Abragam, A. "The Principles of Nuclear Magnetism"; Oxford University Press: Oxford, 1961.
13. Spiess, H. W. In "NMR—Basic Principles and Progress", Diehl, P.; Fluck, E.; Kosfeld, R.; Eds.; Vol. 15; Springer-Verlag: New York, 1978.
14. Mehring, M. "High Resolution NMR in Solids", 2nd Ed.; Springer-Verlag: New York, 1983.
15. Pake, G. E. *J. Chem. Phys.* **1948**, *16*, 327.
16. Pschorn, U.; Spiess, H. W. *J. Magn. Reson.* **1980**, *39*, 217.
17. Spiess, H. W.; Sillescu, H. *J. Magn. Reson.* **1981**, *42*, 381.
18. Spiess, H. W. *J. Chem. Phys.* **1980**, *72*, 6755.
19. Jeneer, J.; Broekaert, P. *Phys. Rev.* **1967**, *157*, 232.
20. Torchia, D. A.; Szabo, A. *J. Magn. Reson.* **1982**, *49*, 107.
21. Jelinski, L. W.; Dumais, J. J.; Engel, A. K. *Macromolecules* **1983**, *16*, 492.
22. Bloom, M.; Davis, J. H.; Valic, M. I. *Can. J. Phys.* **1980**, *48*, 1510.
23. Henrichs, P. M.; Hewitt, J. M.; Linder, M. *J. Magn. Reson.* **1984**, *60*, 280.
24. Lausch, M.; Spiess, H. W. *J. Magn. Reson.* **1983**, *54*, 466.
25. Cholli, A. L.; Dumais, J. J.; Engel, A. K.; Jelinski, L. W. *Macromolecules* **1984**, *17*, 2399.
26. Farrar, T. C.; Becker, E. D. "Pulse and Fourier Transform NMR"; Academic Press: New York, 1971.
27. Rosenke, K.; Sillescu, H.; Spiess, H. W. *Polymer* **1980**, *21*, 757.
28. Jelinski, L. W.; Dumais, J. J. *ACS Polym. Prepr.*, **1981**, *22*(2), 273.
29. Jelinski, L. W.; Dumais, J. J.; Watnick, P. I.; Engel, A. K.; Sefcik, M. D. *Macromolecules* **1983**, *16*, 409.
30. Jelinski, L. W. *Macromolecules* **1981**, *14*, 1341.
31. Schatzki, T. F. *ACS Polym. Prepr.*, **1965**, *6*, 646.
32. Jelinski, L. W.; Dumais, J. J.; Engel, A. K. *ACS Symp. Ser.* **1984**, *247*, 55.
33. Bovey, F. A.; Jelinski, L. W. *J. Phys. Chem.* **1985**, *89*, 571.
34. Helfand, E. *J. Chem. Phys.* **1971**, *54*, 4651.
35. Jelinski, L. W.; Dumais, J. J.; Engel, A. K. *Org. Coat. Appl. Polym. Sci. Proc.* **1983**, *248*, 102.
36. Jelinski, L. W. In "Developments in Block Copolymers—2," Goodman, I., Ed.; Applied Sciences Publishers: London, **1986**.
37. For a detailed list of references, see: Leung, L. M.; Koberstein, J. T. *J. Polym. Sci., Polym. Phys. Ed.* **1985**, *1*.
38. Bonart, R.; Müller, E. H. *J. Macromol. Sci. Phys.* **1974**, *B10*, 177, and references cited therein.
39. Bonart, R.; Müller, E. H. *J. Macromol. Sci. Phys.* **1974**, *B10*, 345, and references cited therein.
40. Blackwell, J.; Lee, C. D. *J. Polym. Sci., Polym. Phys. Ed.* **1983**, *21*, 2169.
41. Van Bogart, J. W. C.; Gibson, P. E.; Cooper, S. L. *J. Polym. Sci., Polym. Phys. Ed.* **1983**, *21*, 65.
42. Koberstein, J. T.; Stein, R. S. *J. Polym. Sci., Polym. Phys. Ed.* **1983**, *21*, 1439.
43. Dumais, J. J.; Gancarz, I.; Galambos, A.; Koberstein, J. T.; Jelinski, L. W. *Macromolecules* **1985**, *18*, 116.
44. Jelinski, L. W.; Dumais, J. J.; Cholli, A. L. *ACS Polym. Prepr.* **1984**, *25*(*1*), 348.
45. "Annual Book of ASTM Standards," Part 35, D 1505-68, p. 482; American Society for Testing and Materials: Philadelphia, PA, 1979.
46. Hummel, J. P.; Flory, P. J. *Macromolecules* **1980**, *13*, 479.
47. Tonelli, A. E. *J. Polym. Sci., Polym. Lett. Ed.* **1973**, *11*, 441.
48. Dumais, J. J.; Cholli, A. L.; Jelinski, L. W.; Hedrick, J. L.; McGrath, J. E. *Macromolecules* **1986**, *18*.
49. Jelinski, L. W.; Dumais, J. J.; Stark, R. E.; Ellis, T. S.; Karasz, F. E. *Macromolecules* **1983**, *16*, 1019.

50. Jelinski, L. W.; Dumais, J. J.; Cholli, A. L.; Ellis, T. S.; Karasz, F. E. *Macromolecules* **1985**, *18*, 1091.
51. Kong, E. S. W. *Proc. Org. Coat. Appl. Polym. Sci.* **1983**, *48*, 727.
52. Banks, L.; Ellis, B. *Polym. Bull.* **1979**, *1*, 377.
53. Dumais, J. J.; Jelinski, L. W., unpublished data, 1985.
54. Boeffel, C.; Hisgen, B.; Pschorn, U.; Ringsdorf, H.; Spiess, H. W. *Israel J. Chem.* **1983**, *23*, 388.
55. Samulski, E. T. *Polymer* **1985**, *26*, 177.
56. Greenbaum, S. G.; Mattix, L.; Resing, H. A.; Weber, D. C. *Phys. Lett.* **1983**, *98A*, 299.
57. Ziliox, M.; François, B.; Mathis, C.; Meurer, B.; Spegt, P.; Weill, G. *J. Phys.* **1983**, *44*, C3-361.
58. Hentschel, D.; Sillescu, H.; Spiess, H. W. *Macromolecules* **1981**, *14*, 1605.

Author Index

Subject Index

A

Acrylonitrile-butadiene-styrene copdymer, 59, 145

Activation energy, of motion, 85, 86, 256–258, 260, 349, 352, 355

Adamantane, 154, 165, 192, 193

Adiabatic demagnetization in rotating frame, 43, 44

Adjacent reentry model, 159

ADRF, 43, 44

Affine model, 286–288

n-Alkanes, 234, 235

p-Alkoxybenzoic acids, 227

Aluminum oxide, 60

Angle flipping, 54

Anisotropic interactions, 20–35, 264–270, 287–293, 309–321, 336–340

Anisotropic motion, and coherent averaging, 48

Annealing, effect of, 160, 178, 182–184, 189, 301–303, 333

Antiplasticization, 15

Assignments, peak, 49–52

ATBN, 147, 149

Attached proton test, 51

Autocorrelation function, 105, 248, 254, 255, 259, 260, 272

B

Benzyl radical, 76

Bessel function, 248, 255

Biopolymers, 16, 17

Bisphenol A, 64, 84, 252

Blend, polymer, 4, 14, 103, 114–117

^{11}B NMR, 20

Boltzmann distribution, 40

Bond lengths, in polyacetylene, 79–83

Boron nitride, 60

BPP formalism, 99

BR-24 sequence, 292, 317, 324–327

Branching, in polymers, 160, 210–212

Broadening, of NMR lines, 66, 91–96, 166, 167

Butadiene-acrylonitrile copolymer, 145

Butyl rubber, 140

C

Camphor, 51

Carbonate group, in labeled polycarbonate, 273, 278

Carbon monoxide, 26, 27

Carr-Purcell-Meiboom-Gill sequence, 330

Cellulose, 16, 17, 227

Chain entanglements, 133

Chemical shift, 4, 5, 12, 14

Chemical shift, anisotropic, 8, 9, 25–30, 130, 251, 264–270, 273, 274, 290–293, 309, 310

Chemical shift, anisotropic, and 2D NMR, 53–55

Chemical shift, anisotropic, and motion, 86–91, 264–270

Chemical shift anisotropies, table of, 28

Chemical shift heterogeneity, see Distribution of chemical shifts

Chemical shift, isotropic in solid, 26, 322–327

Chloral polycarbonate, see Polycarbonate chloral

Chlorine, effect on ^{13}C spectrum, 32

Chlorobenzene, 88, 90, 91

Coal, 329

Collagen, 16

Collapse, spectral, 135–138, 219, 220

Conformation, 5, 12, 175ff, 227ff

Conformational exchange, 248, 349

Cooperative motions, 103, 104

Copolyesters, 349, 350

Copolymers, block, 4, 117

Correlated states model, 260

Correlation function, see Autocorrelation function

Correlation time, 105–107, 126, 127, 139, 140, 255, 257, 342

Correlation time, distribution of, 129

Coupling, spin-spin, 4, 35, 53

CPMAS NMR, 20, 44–46, 64ff, 123–125, 153–155, 249–250

CP-nutation NMR, see Nutation, CP

CRAMPS, 324–326, 330, 331

Crosslinks, 3, 12, 74, 122, 125, 140–142

Cross polarization, 2, 6, 39–46

Cross polarization, double, 43, 51–52

Cross polarization rate, 42, 163, 185, 188, 189

Crystallinity, degree of, 204–205, 321, 322